Neuroethics in Practice

Neuroethics in Practice

EDITED BY

ANJAN CHATTERJEE

MARTHA J. FARAH

OXFORD
UNIVERSITY PRESS

OXFORD
UNIVERSITY PRESS

Oxford University Press is a department of the University of Oxford.
It furthers the University's objective of excellence in research, scholarship,
and education by publishing worldwide.

Oxford   New York
Auckland   Cape Town   Dar es Salaam   Hong Kong   Karachi
Kuala Lumpur   Madrid   Melbourne   Mexico City   Nairobi
New Delhi   Shanghai   Taipei   Toronto

With offices in
Argentina   Austria   Brazil   Chile   Czech Republic   France   Greece
Guatemala   Hungary   Italy   Japan   Poland   Portugal   Singapore
South Korea   Switzerland   Thailand   Turkey   Ukraine   Vietnam

Oxford is a registered trademark of Oxford University Press in the UK and certain other
countries.

Published in the United States of America by
Oxford University Press
198 Madison Avenue, New York, NY 10016

Library of Congress Cataloging-in-Publication Data
Neuroethics in practice / edited by Anjan Chatterjee, Martha J. Farah.
p. cm.
Includes bibliographical references.
ISBN 978-0-19-538978-4 (alk. paper)   I. Chatterjee, Anjan.   II. Farah, Martha J.
[DNLM: 1. Neurosciences—ethics.   2. Bioethical Issues.   3. Biomedical   Enhancement—ethics.
4. Brain—physiology.   5. Brain Damage, Chronic—therapy. WL 21]
174.2′9689—dc23
2012029361

9 8 7 6 5 4 3 2 1
Printed in the United States of America
on acid-free paper

CONTENTS

CONTRIBUTORS

Tamara B. Bockow
University of Pennsylvania
Philadelphia, PA

Anjan Chatterjee
Department of Neurology
    Center for Cognitive Neuroscience
    and Center for Neuroscience &
    Society
University of Pennsylvania
Philadelphia, PA

Breehan Chancellor
Department of Neurology
The University of Pennsylvania
Philadelphia, PA

Thomas Cochrane
Department of Neurology
Brigham and Women's Hospital
Boston, MA

Peter Conrad
Department of Sociology
Brandeis University
Walthan, MA

John Detre
The Mahoney Institute of Neurological
    Sciences
University of Pennsylvania
Philadelphia, PA

Martha J. Farah
Center for Neuroscience & Society
University of Pennsylvania
Philadelphia, PA

Joseph J. Fins
Division of Medical Ethics
    Departments of Medicine, Public
    Health and Psychiatry
Weill Medical College of Cornell
    University
New York, NY

Lachlan Forrow
Beth Israel Deaconess Medical
    Center
Harvard Medical School
Boston, MA

Felipe Fregni
Laboratory of Neuromodulation and
    Center of Clinical Research
    Training
Harvard Medical School
Boston, MA

Seth J. Gillihan
Department of Psychiatry
University of Pennsylvania
Philadelphia, PA

Leigh Hochberg
Massachusetts General Hospital
Brigham and Women's Hospital
Boston, MA

Jared Horvath
Berenson-Allen Center for
    Noninvasive Brain Stimulation
Beth Israel Deaconess Medical
    Center
Harvard Medical School
Boston, MA

Allan Horwitz
Department of Sociology
Rutgers University
Newark, NJ

Steven E. Hyman
Stanley Center for Psychiatric
    Research
Broad Institute
Cambridge, MA

Jason Karlawish
Professor of Medicine
Medical Ethics, and Health Policy
University of Pennsylvania
Philadelphia, PA

Kelly Kelleher
Adolescent Medicine Fellowship
Center for Innovation in Pediatric
    Practice
Developmental/Behavioral Pediatrics
    Fellowship
Nationwide Children's Hospital
Columbus, OH

Scott Y. H. Kim
Department of Psychiatry
University of Michigan
Ann Arbor, MI

Jonathan Kimmelman
Biomedical Ethics Unit
Departments of Social Studies of
    Medicine and Human Genetics
McGill University
Montreal, Canada

Steven Laureys
Coma Science Group
Cyclotron Research Center
Department of Neurology
Sart Tilman Liège University Hospital
Liège, Belgium

Alvaro Pascual-Leone
Professor of Neurology
Harvard Medical School
Cambridge, MA

Michael B. Russo
Division of Neuropsychiatry
Walter Reed Army Institute of Research
Silver Spring, MD

Nicholas D. Schiff
Departments of Neurology and
    Neuroscience
Weill Medical College of Cornell
    University
Adjunct Faculty
The Rockefeller University
New York, NY

Ilina Singh
Department of Sociology King's College
London, UK

Melba C. Stetz
International Association of
    CyberPsychology, Training, and
    Rehabilitation
Brussels, Belgium

Thomas A. Stetz
National Geospatial-Intelligence
  Agency
Honolulu, HI

Matthis Synofzik
Department of Neurodegenerative
  Diseases
Hertie Institute for Clinical Brain
  Research
German Centre for Neurodegenerative
  Diseases (DZNE)
University of Tübingen
Tübingen, Germany

Stacey Tovino
William S. Boyd School of Law
University of Nevada, Las Vegas
Las Vegas, NV

PREFACE

Allow us to introduce ourselves and our book. We have been colleagues and friends since the late 1990s, when Anjan joined the faculty at the University of Pennsylvania. By some sort of convergent evolution we both ended up working in neuroethics, although our paths into the field and our perspectives on it differ somewhat.

Martha, a cognitive neuroscientist, had always been fascinated by the profound questions of mind-brain relations that lie at the heart of cognitive neuroscience, but found herself longing for a way to relate this intellectually fascinating subject to real-world social problems and policies. When bioethicists at Penn and elsewhere began talking about neuroscience, she found what she was looking for.

Anjan, a practicing neurologist as well as cognitive neuroscience researcher, was confronting some professional unease of his own. Like many physicians, he noted that expanding bureaucratic burdens on clinicians made it increasingly difficult to be thoughtful about patient care. He noticed that patients and their caregivers, regardless of their illness, wanted to improve their quality of life. Some looked to him to directly enhance their cognitive and emotional lives. His engagement with this problem led him to think and write on the problem of cosmetic neurology.

These different paths and perspectives informed the planning and editing of the book. Our goal was not an exhaustive survey of neuroethics. There are already several books that attempt this, and we felt no drive to produce another. Our goal was a more circumscribed book that would highlight those aspects of neuroethics with tangible relevance to health care. We also wished to address emerging concerns that arise from recent advances in translational neuroscience. By focusing on neuroethical issues with real-world relevance, which are at present largely to be found in medicine, both of us satisfied our interests. We chose the title *Neuroethics in Practice* to signal the relevance of the book's content to health care.

Of course, matters of practice and principle are closely intertwined, and *Neuroethics in Practice* includes at least as much writing about basic science and ethical principles as it does writing about patient care. The book is not intended as an operational guide for clinicians. We would not presume to dictate the correct approach to the problems discussed here, which are interesting precisely because well-informed and well-intentioned individuals can disagree about how best to proceed. Rather, we have tried to equip readers with an understanding of relevant scientific and bioethical precedents and perspectives with which to approach important real-world clinical dilemmas. Examples of these dilemmas, from among the many discussed in this book, are patient requests for enhancement, decisions about competence or legal culpability for patients with dementia or addiction, family inquiries about the mental life of severely brain-damaged patients, and the use (therapeutic or experimental) of new methods in clinical neuroscience, from brain imaging for psychiatric diagnosis to brain stimulation for a wide range of psychiatric indications.

These issues will be of interest to many readers who are not physicians, psychologists, or other health care professionals. One reason is that health care is personally relevant to everyone; we are all patients at some point or other. Another reason is that health care neuroethics encompasses some of the most well-specified and thoroughly analyzed problems of neuroethics. If the core issues of neuroethics arise from the intersection of individual, societal, and commercial interests with our growing ability to understand and manipulate brain function, then neurology, neurosurgery, and psychiatry are the fields in which we face the most immediate and concrete manifestations of these core issues.

The authors of the chapters that follow share our vision for *Neuroethics in Practice* as an exploration of neuroethics through real-world clinical problems. The book's success is entirely due to their deep understanding of their assigned topics, their broad perspective on neuroethics in general, and their crystal-clear writing. We are supremely grateful to them for their contributions. We also thank our editors at Oxford University Press for their expert guidance and encouragement: Catharine Carlin, who helped us get started with OUP, and then Joan Bossert, who shepherded the book through production. We'd work with either of them again in a heartbeat. Finally, we thank our fellow members of the Penn Center for Neuroscience & Society for their unfailing collegiality and support.

Anjan Chatterjee
Martha J. Farah

PART ONE

Brain Enhancement

1

Brain Enhancement in Healthy Adults

ANJAN CHATTERJEE

Anabolic athletes, medicated musicians, and souped-up students seem to abound. To this list we can add pill-popping professors (Chatterjee, 2008; Sahakian & Morein-Zamir, 2007). "Cosmetic neurology" is the term I used to describe the practice of using neurologic interventions to improve movement, mood, and mentation in healthy individuals (Chatterjee, 2004; 2006). The term has historic roots going back to the 1940s when it started to be used for reconstructive and plastic surgery in healthy people (Himmel, 1948). Elsewhere, I have argued that many of the factors driving the rise of cosmetic surgery also apply to cosmetic neurology (Chatterjee, 2007). The term is also related to Kramer's earlier use of the term "cosmetic pharmacology" (Kramer, 1993), although neurology encompasses a wider range of possible interventions (Hamilton, Messing, & Chatterjee, 2011). This chapter does focus on pharmacological interventions in anticipation of the fact that physicians will face increasing pressure from their "patients" to dole out drugs for nontherapeutic uses. In what follows, I describe what is available currently and in the near future and outline the ethical conundrums that ensue.

Physicians might reflexively want to rely on a treatment versus enhancement distinction in their practice. Therapy is treating disease, whereas enhancement is improving normal abilities, and physicians are in the business of treating disease. On scrutiny, the distinction between therapy and enhancement is murky particularly when disease lacks clear boundaries, and conditions are continuous rather than categorical. For example, if short people can be treated with growth hormone (Cuttler et al., 1996), does it matter if they are short because of a growth hormone deficiency or because of other reasons (Daniels, 2000)? Furthermore, the widespread use of cosmetic surgery to enhance normal physical attributes suggests that physicians will embrace such practices (Chatterjee, 2007). At the root of cosmetic neurology is a more difficult question: If one purpose of medicine is to improve the quality of life of individuals who happen to be sick, then should medical knowledge be applied to those who happen to be healthy?

ENHANCEMENTS

Enhancements fall into three general categories: improvement of motor systems, cognition, and mood and affect. Interventions such as alcohol, tobacco, and caffeine have been available for a long time. Many others are on the horizon. For novel medications the effects in clinical populations are often not known, and their efficacy and safety in healthy individuals are relatively unexplored. However, we can anticipate that such interventions will be increasingly available, efficacious, and safe.

Movement

Medicine can make people stronger, swifter, and more enduring. Professional athletes use anabolic steroids to improve their strength and quickness. Beyond steroids, new ways of improving motor performances are around the corner. Insulin-like growth factor (IGF) produced by the liver may improve the quality of life of people without disease. IGF given to men over the age of 60 for six months increased their muscle mass, decreased body fat, and improved skin elasticity (Rudman et al., 1990). In mice, injection of recombinant viruses containing the IFG-1 gene directly into muscle also increased muscle mass and strength and prevented declines observed in untreated old mice (Barton-Davis, Shoturma, Musaro, Rosenthal, & Sweeney, 1998). Recent reports in rodents of drugs that increase muscle mass by altering cellular metabolism offer a couch potato's dream of exercise in a pill (Narkar et al., 2008).

Maximizing blood oxygenation optimizes muscle activity, which enhances athletic performance. In the past, athletes trained at high altitudes and used autologous blood transfusions to increase their oxygen-carrying capacities (Gaudard, Varlet-Marie, Bressolle, & Audran, 2003). Human erythropoietin (EPO), used to treat anemia, has been used as a form of athletic doping (Gaudard et al., 2003; Varlet-Marie, Gaudard, Audran, & Bressolle, 2003). New transfusion methods, motivated by blood supply shortages and contaminants, are likely to have implications for performance when endurance is critical (Gaudard et al., 2003).

Finally, learning motor skills may be improved by medications that enhance neural plasticity. For example, amphetamines in small doses promote plasticity and accelerate motor learning (Grade, Redford, Chrostowski, L, & B, 1998; Walker-Batson, Smith, Curtis, Unwin, & Greenlee, 1995). Their effects are most pronounced when paired with training, as seen in patients with weakness following stroke. Could amphetamines also be used in normal subjects at the time of skilled motor learning, such as learning to swim or ski, or playing the piano?

Cognition

We now have unprecedented research in therapeutic options for degenerative and developmental cognitive disorders. Currently, available treatments target the catecholamine and cholinergic systems.

The effects of amphetamines on plasticity may also apply to cognitive systems (for a recent review, see Repantis, Schlattmann, Laisney, & Heuser, 2010). Amphetamines improve the effects of speech therapy in aphasic patients (Smith, & Farah, 2011; Walker-Batson et al., 2001). Might similar effects occur in normal subjects? Modafinil improves arousal and ameliorates deficits of sustained attention associated with sleep deprivation (Caldwell, Caldwell, Smythe, & Hall, 2000; Lagarde, Batejat, Van Beers, Sarafian, & Pradella, 1995). Methylphenidate is used widely to improve attention, concentration, spatial working memory, and planning (Mintzer & Griffiths, 2007; Pary et al., 2002; Weber & Lutschg, 2002; Zeeuws, Deroost, & Soetens, 2010). Students commonly use amphetamines despite the fact that it may also impair previously established performance (Babcock & Byrne, 2000; Diller, 1996); the actual empirical data in support of its effects are far from clear (Smith & Farah, 2011; Ilieva, Boland, & Farah, 2012). Newer nonaddictive drugs such as atomoxetine are likely to increase off-label use of such medications.

Cholinesterase inhibitors also improve attention and memory (Repantis, Laisney, & Heuser, 2010). These medications are used widely in Alzheimer's disease, and their use in older individuals is on the rise. The effects of cholinesterase inhibitors on normal subjects are not well studied. However, one intriguing report suggests an effect in the setting of highly skilled performance. Yesavage and colleagues (Yesavage et al., 2001) reported that commercial pilots taking 5 mg of donepezil for one month performed better than pilots on placebo on demanding Cessna 172 flight simulation tasks, particularly when responding to emergencies. These drugs may have beneficial effects on semantic processing (FitzGerald et al., 2008), memory (Grön, Kirstein, Thielscher, Riepe, & Spitzer, 2005), and in counteracting the effects of sleep deprivation (Chuah & Chee, 2008).

Two new classes of drugs for memory, ampakines and cyclic AMP response element binding protein (CREB) modulators, are on the horizon (Hall, 2003). These drugs capitalize on recent advances in understanding of the intracellular events that contribute to structural neural changes associated with the acquisition of long-term memory.

Facilitation of glutamatergic transmission promotes long-term potentiation, presumed to foster synaptic plasticity and memory formation. Ampakines augment AMPA type glutamate receptors by depolarizing postsynaptic membranes in response to glutamate. Since NMDA receptors crucial to induction of long-term potentiation (Kemp & McKernan, 2002) are sensitive to this depolarization, ampakines are thought to facilitate the acquisition and consolidation of new memories (see Lynch, 2003 for a review). Early studies show that ampakines improve memory in rats (Granger et al., 1993; Staubli, Perez, Xu, Rogers, & Ingvar, 1994) and normal humans (Ingvar et al., 1997). The NMDA receptors themselves may ultimately be a target of genetic modification. Mice genetically altered to overexpress NMDA receptors have superior learning and memory abilities (Tang et al., 1999).

Neurogenetic studies suggest that CREB is a critical molecular "switch" in forming long-term memories (Tully, Bourtchouladze, Scott, & Tallman, 2003). Gene expression is promoted by activation of CREB, which itself is dependent on NMDA receptor activation. Specific protein kinases activate CREB. CREB then sets off a transcription cascade that produces specific structural changes at the synapse. Drosophila genetically altered to overexpress CREB demonstrate long-term conditioning to odor-shock pairings after only one exposure, a conditioning that normally takes 10 trials (Yin, Del Vecchio, Zhou, & Tully, 1995). Similar effects are seen in mammals (Scott, Bourtchouladze, Gossweiler, Dubnau, & Tully, 2002). Mice given rolipram, a phosphodiesterase inhibitor that enhances CREB, form long-term memories in fewer than half the trials needed by untreated mice (Tully et al., 2003).

Mood and Affect

The aisles of most drugstores testify to the public's appetite for mood regulators such as St. John's Wort, kava kava, and valerian. Antidepressants, most notably selective seratonin reuptake inhibitors (SSRIs), are used widely for depression but also for anxiety, obsessive compulsive disorders, and oppositional behaviors. Some estimate between 9.5 and 20% of Americans are depressed (Health, 2003). SSRIs may selectively dampen negative and not positive affect (Knutson et al., 1998) and increase affiliative behavior in social settings (Tse & Bond, 2002). If SSRIs improve a general sense of well-being, regardless of illness or health, might more than 20% of Americans wish to take them?

New approaches to treating affective illnesses will probably expand our therapeutic options (Holmes, Heilig, Rupniak, Steckler, & Griebel, 2003; Salzano, 2003). Blocking gluco-corticoids may be of benefit in a subset of depressed patients. Corticotropin releasing factor (CRF) seems to mediate long-term stress effects through the stria terminalis, a structure related to the amygdala (Davis, 1998; Walker, Toufexis, & Davis, 2003). Around the corner are several new ways of potentially controlling affective states by modulating neuropeptides. Neuropeptides are small proteins in the brain that influence how information is processed and can be linked to quite specific behaviors. Corticotropin release factor (CRF) seems to mediate the long-term effects of stress, and blocking CRF may blunt these effects. In addition to CRF, other neuropeptides seem to play a role in depression and anxiety. These include substance P, vasopressin, neuropeptide Y, and galanin. Clinical trials of neuropeptide agonists and antagonists that cross the blood–brain barrier are just beginning (Holmes et al., 2003). We may even be able to modulate our emotional states in more subtle ways. For example, oxytocin might be used to induce trust (Kosfeld, Heinrichs, Zak, Fischbacher, & Fehr, 2005).

Pharamacologic agents can also modulate the way emotional events are remembered (Cahill, 2003). In animals, consolidation of emotional memories is strengthened by epinephrine and dampened by beta blockers injected within the amygdala. Similar effects occur in normal people. Subjects given propanolol recall emotionally arousing stories as if they were emotionally neutral (Cahill, Prins, Weber, & McGaugh, 1994). Propanalol also enhances the memory of events surrounding emotionally charged events that are otherwise suppressed (Strange, Hurlemann, & Dolan, 2003). In one pilot study, patients in an emergency room given propanalol after a traumatic event suffered fewer posttraumatic stress disorder symptoms when assessed one month after later (Pitman et al., 2002). Most would agree with treating posttraumatic stress disorder to help individuals that are paralyzed by their disturbing memories. However, these studies suggest that less disturbing memories might also be muted, if we so desired.

ETHICAL DILEMMAS

Pharmacological enhancements raise deep ethical dilemmas. These dilemmas coalesce around four concerns: safety, distributive justice, coercion, and the erosion of character.

Safety

Virtually all medications have potential side effects that range from minor inconveniences to severe disability or death. For example, amphetamines, often used for cognitive enhancements, have FDA black box warnings, particularly in regard to the risk of addiction and serious cardiac side effects including sudden death (Chatterjee, 2009). In disease states, one weighs risks against potential benefits. Thus a patient with glioblastoma multiforme might be willing to endure toxic chemotherapies, because the alternative is so grim. In healthy states any risk seems harder to accept, because the alternative is normal health. For some interventions the risks are known or suspected. EPO improves endurance but increases the risk of stroke. Modafinil enhances alertness on some tasks but may compromise performance on others (Caldwell et al., 2000). Genetically modified mice may have terrific memories (Tang et al., 1999) but are more sensitive to pain (Tang, Shimizu, & Tsien, 2001).

A subtler version of the safety concern is that of trade-offs rather than side effects. Would some cognitive enhancements be accompanied by detriments to other cognitive processes? For example, medications that enhanced attention and concentration might conceivably deter imagination and creativity. There is very little research in the kinds of trade-offs (but see Farah, Haimm, Sankoorikal, & Chatterjee, 2009) that might follow from long-term use of enhancements.

Character and Individuality

This concern takes two general forms, one about eroding character and the other about altering the individual. The erosion of character concern is wrapped around a "no pain, no gain" belief (Chatterjee, 2008a; 2008b). Struggling with pain builds character, and eliminating that pain undermines good character. Similarly, getting a boost without doing the work is cheating, and such cheating cheapens us (Kass, 2003). To some extent the question of cheating depends on whether we emphasize the process or the outcome. Goodman argues that in situations where outcomes are important and the activity is not a zero-sum game, enhancements do not cheapen us (Goodman, 2010).

While the concerns about character run deep, they are mitigated by several factors. Which pains are worth the hypothetical gains they might bring? We live in homes with central heat and air, eat food prepared by others, travel vast distances in short times, take Tylenol for headaches and H2 blockers for heartburn. Perhaps these conveniences have eroded our collective character and cheapened us. But few choose to turn back.

A fundamental concern is that chemically changing the brain threatens our notion of personhood. The central issue may be that such interventions threaten essential characteristics of what it means to be human (*President's Council on Bioethics*, 2003). For example, would selectively dampening the impact of our painful memories change who we are, if we are to some degree the sum of our experiences? This is a difficult issue to grapple with, and consensus on the essence of human nature may be elusive (Elliott, 2003b; Fukayama, 2002; Wolpe, 2002). Invasive surgical procedures such as sex-change operations are used to express one's individuality. Elliott (Elliott, 2003a), in reviewing such practices, suggests that "in America, technology has become a way for some people to build or reinforce their identity (and their sense of dignity) while standing in front of the social mirror."

Distributive Justice

If we can enhance ourselves, who gets to? New drugs are expensive, and there is no reason to expect insurance companies or the state to pay for them for nontherapeutic purposes. Only those that can afford to pay privately would get enhancements. A familiar counter to the worry of widening inequities is that this is not a zero-sum game. With widening disparities, even those at the bottom of the hierarchy receive some benefit and improve from their previous state in some absolute sense (*President's Council on Bioethics*, 2003). This argument assumes that people's sense of well-being is determined by an absolute level of quality rather than by a recognition of one's relative place. However, beyond worries about basic subsistence, well-being seems mostly affected by expectations and relative positions in society (reviewed by Frank, 1987).

One might argue that the critical issue is access and not availability (Caplan, 2003). If access to such enhancements were open to all, then differences might even be minimized. This argument may have logical merit, but in practice (in the United States) it skirts the issue. We tacitly accept wide disparities in modifiers of cognition, as demonstrated by the acceptance of inequities in education, nutrition, and shelter.

Coercion

The concern here is that matters of choice often evolve into forces of coercion. Coercion takes two forms. One is the implicit coercion to maintain or better one's position in some perceived social order. Such pressure increases in a winner-take-all environment in which more people compete for fewer and bigger prizes (Frank & Cook, 1995). Many professionals work sixty, eighty, or more than a hundred hours a week without regard to their health and hearth. Emergency department residents use zolpidem especially, but also modafinil, to regulate sleep and effectiveness (McBeth et al., 2009). Athletes take steroids to compete at the highest levels, and children at high-end preparatory schools take methylphenidate in epidemic proportions (Hall, 2003). To not take advantage of enhancements might mean being left behind. Students frequently refer to academic assignments or grades as reasons to take amphetamines (Arria, O'Grady, Calderia, Vincent, & Wish, 2008; DeSantis, Webb, & Noar, 2008).

A second form of coercion, which has received less attention, is the explicit demand of superior performance by others. Such coercion could take regulatory forms. For years, those in the armed forces have been encouraged to take enhancements for the greater good. Might this logic extend to civilian domains? Yesavage and colleagues' (Yesavage et al., 2001) findings that pilots taking donepezil performed better in emergencies than those on placebo could have wide implications. If these results are reliable and significant, should pilots be expected to take such medications? Can airline executives require this of pilots? Would they offer financial incentives to pilots willing to take these medications? Would those fearful of flying pay more for cholinergic copilots? Closer to home, should medical students and postcall residents take stimulants to attenuate deficits in sustained attention brought on by sleep deprivation (Webb, Thomas, & Valasek, 2010)? Will hospital administrators require this practice? Insurance companies? Patients?

FUTURE CONSIDERATIONS AND POLICY IMPLICATIONS

The armamentarium of drugs that could be used to enhance healthy individuals is growing. We can anticipate that this growth will continue for the indefinite future. While the ethical concerns run deep, some form of this practice seems inevitable. Countervailing social pressures are overwhelming. The print media and bioethicists generally discuss enhancements positively (Forlini & Racine, 2009). Individuals overestimate the efficacy of enhancers like methyphenidate or modafanil (Repantis, Schlattmann et al., 2010). Pharmaceutical companies have significant economic incentives in expanding their markets to healthy individuals. Since 1997, the FDA has allowed direct advertising to consumers. Television advertisements now give permission to indulge in a pepperoni pizza without the fear of heartburn, because one can take an H2 blocker prophylactically. One would be surprised if similar advertisements did not recommend getting an edge with cognitive enhancers or a boost with mood manipulators.

Treatments to enhance normal abilities are likely to be paid for privately. If social pressures encourage wide use of medications to improve quality of life, then pharmaceutical companies stand to make substantial profits and they are likely to encourage such pressures. According to Elliott (Elliott, 2003a), in 2001 GlaxoSmithKline spent $91 million dollars in direct advertising to consumers for its medication Paxil—more than Nike spends on its top shoes. Gingko biloba, despite its minimal affects on cognition (Solomon, Adams, Silver, Zimmer, & DeVeaux, 2002), is a billion-dollar industry. Pharmaceutical companies, undoubtedly encouraged by sales of Viagra, are not oblivious to the marketing possibilities of new interventions that could apply to the entire population (Hall, 2003; Langreth, 2002).

Physicians who hope that medicine will serve as meaningful break on dispensing enhancement mediation to healthy individuals are likely to be disappointed. The historical precedent of the widespread use of cosmetic surgery demonstrates a willingness of physicians to engage in nontherapeutic practices. With appropriate incentives and cultural frameworks in place, cosmetic surgery went from being considered frivolous in the early part of the twentieth century to logging over nine million procedures in 2004 by licensed physicians (Chatterjee, 2007) and over 12 million by 2009. With easy access to medications, especially over the Internet, it is inconceivable that enhancements will not be used widely.

Strict prohibition of the use of enhancements is unlikely to be an effective policy. This approach would simply move the market for such medications underground, and inhibit guiding the actual practice of cosmetic neurology in an informed way. Establishing professional norms molded by cultural values and communal discussions—and with proper safeguards—will be needed. The American Academy of Neurology has begun this process. Their guidance does not prohibit neurologists from prescribing enhancements (Larriviere, Williams, Rizzo, Bonnie, & on behalf of the AAN Ethics, 2009).

Clearly, we need adequate research in the use of enhancement medication in nondiseased individuals. Whether results from diseased populations will generalize to normal individuals is not clear. For example, would the benefits of stimulants for individuals with attention deficit disorders generalize to those without attention deficits? What cognitive trade-offs might occur? There are significant impediments to acquiring these data necessary for individuals to make a well-informed choice. Institutional review boards may be reluctant to endorse such research. After all, why should the institution accept any risk of severe side effects, however small that risk might be, when participants are healthy? Additionally, who would fund such research? In the United States, the major source of biomedical research funding, the National Institutes of Health have been reluctant to fund research into nontherapeutic interventions.

Several policies to maximize benefits and minimize harm would be helpful to mitigate the ethical concerns raised by cosmetic neurology (Appel, 2008; Greely et al., 2008). Enforceable policies concerning the use of cognitive-enhancing drugs to support fairness, protect individuals from coercion, and minimize enhancement related socioeconomic disparities should be implemented. Physicians, educators, regulators, and others professional groups will need to establish their own positions as cultural norms are debated and made explicit.

REFERENCES

Appel, J. M. (2008). When the boss turns pusher: A proposal for employee protections in the age of cosmetic neurology. *Journal of Medical Ethics, 34*(8), 616–618.

Arria, A., O'Grady, K., Calderia, K., Vincent, K., & Wish, E. (2008). Nonmedical use of prescription stimulants and analgesics: Associations with social and academic behaviors among college students. *Pharmacotherapy, 38*(4), 1045–1060.

Babcock, Q., & Byrne, T. (2000). Student perceptions of methylphenidate abuse at a public liberal arts college. *J Am College Health, 49,* 143–145.

Barton-Davis, E., Shoturma, D., Musaro, A., Rosenthal, N., & Sweeney, H. (1998). Viral mediated expression of insulin-like growth factor I blocks the aging-related loss of skeletal muscle function. *Proceedings of the National Academy of Sciences America, 95,* 15603–15607.

Cahill, L. (2003). Similar neural mechanisms for emotion-induced memory impairment and enhancement. *Proceedings of the National Academy of Sciences America, 100,* 13123–13124.

Cahill, L., Prins, B., Weber, M., & McGaugh, J. (1994). Beta-adrenergic activation and memory for emotional events. *Nature, 371,* 702–704.

Caldwell, J. J., Caldwell, J., Smythe, N. r., & Hall, K. (2000). A double-blind, placebo-controlled investigation of the efficacy of modafinil for sustaining the alertness and performance of aviators: a helicopter simulator study. *Psychopharmacology, 150,* 272–282.

Caplan, A. (2003). Is better best? *Scientific American, 289,* 104–105.

Chatterjee, A. (2004). Cosmetic Neurology: The controversy over enhancing movement, mentation and mood. *Neurology, 63,* 968–974.

Chatterjee, A. (2006). The promise and predicament of cosmetic neurology. *The Journal of Medical Ethics, 32,* 110–113.

Chatterjee, A. (2007). Cosmetic neurology and cosmetic surgery: Parallels, predictions and challenges. *Cambridge Quarterly of Healthcare Ethics, 16,* 129–137.

Chatterjee, A. (2008). "Cosmetic neurology" and the problem of pain. In C. Read (Ed.), *Cerebrum 2008. Emerging ideas in brain science* (pp. 81–93). New York: Dana Press.

Chatterjee, A. (2008). Framing pains, pills, and professors. *Expositions, 2.2,* 139–146.

Chatterjee, A. (2009). A medical view of potential adverse effects. *Nature, 457*(7229), 532–533.

Chuah, L. Y. M., & Chee, M. W. L. (2008). Cholinergic augmentation modulates visual task performance in sleep-deprived young adults. *Journal of Neuroscience, 28*(44), 11369–11377.

Cuttler, L., Silvers, J., Singh, J., Marrero, U., Finkelstein, B., Tannin, G., et al. (1996). Short stature and growth hormone therapy: A national study of physician recommendation patterns. *JAMA, 276,* 531–537.

Daniels, N. (2000). Normal functioning and the treatment-enhancement distinction. *Cambridge Quarterly, 9,* 309–322.

Davis, M. (1998). Are different parts of the extended amygdala involved in fear versus anxiety? *Biological Psychiatry, 44,* 1239–1247.

DeSantis, A., Webb, E., & Noar, S. (2008). Illicit use of prescription ADHD medications on a college campus: a multimethodologiical approach. *Journal of American College Health, 57*(3), 315–324.

Diller, L. (1996). The run on Ritalin: Attention deficit disorder and stimulant treatment in the 1990s. *Hastings Center Report, 26,* 12–14.

Elliott, C. (2003a). American bioscience meets the American dream. *The American prospect, 14,* 38–42.

Elliott, C. (2003b). *Better than well: American medicine meets the American dream.* New York: Norton.

Farah, M. J., Haimm, C., Sankoorikal, G., & Chatterjee, A. (2009). When we enhance cognition with Adderall, do we sacrifice creativity? A preliminary study. *Psychopharmacology, 202,* 541–547.

FitzGerald, D. B., Crucian, G. P., Mielke, J. B., Shenal, B. V., Burks, D., Womack, K. B., et al. (2008). Effects of donepezil on verbal memory after semantic processing in healthy older adults. *Cognitive and Behavioral Neurology, 21*(2).

Forlini, C., & Racine, E. (2009). Disagreements with implications: diverging discourses on the ethics of non-medical use of methylphenidate for performance enhancement. *BMC Medical Ethics, 10*(1), 1–13.

Frank, R. (1987). *Choosing the right pond.* New York: Oxford Press.

Frank, R., & Cook, P. (1995). *The winner-take-all strategy.* New York: The Free Press.

Fukayama, F. (2002). *Our posthuman future.* New York: Farrar, Straus & Giroux.

Gaudard, A., Varlet-Marie, E., Bressolle, F., & Audran, M. (2003). Drugs for increasing oxygen transport and their potential use in doping. *Sports Medicine, 33,* 187–212.

Goodman, R. (2010). Cognitive enhancement, cheating, and accomplishment. *Kennedy Institute of Ethics Journal, 20*(2), 145–160.

Grade, C., Redford, B., Chrostowski, J., L, T., & B, B. (1998). Methylphenidate in early poststroke recovery: a double-blind, placebo-controlled stud. *Archives of Physical Medicine & Rehabilitation, 79,* 1047–1050.

Granger, R., Deadwyler, S., Davis, M., Perez, Y., Nilsson, L., Rogers, G., et al. (1993). A drug that facilitates glur=tamergic transmission reduces exploratory activity and improves performance in a learning dependent task. *Synapse, 15,* 326–329.

Greely, H., Sahakian, B., Harris, J., Kessler, R. C., Gazzaniga, M., Campbell, P., et al. (2008). Towards responsible use of cognitive-enhancing drugs by the healthy. *Nature, 456*(7223), 702–705.

Grön, G., Kirstein, M., Thielscher, A., Riepe, M., & Spitzer, M. (2005). Cholinergic enhancement of episodic memory in healthy young adults. *Psychopharmacology, 182*(1), 170–179.

Hall, S. (2003). The quest for a smart pill. *Scientific American, 289,* 54–65.

Hamilton, R., Messing, S., & Chatterjee, A. (2011). Rethinking the thinking cap: Ethics of neural enhancement using noninvasive brain stimulation. *Neurology, 76*(2), 187–193.

Health, T. N. I. o. M. (2003). *The numbers count: Mental disorders in America*. Washington, DC. (Document Number)

Himmel, J. (1948). Cosmetic Plastic Surgery: Its relationship to personality. *Ohio Med, 44*(7), 711–713.

Holmes, A., Heilig, M., Rupniak, N., Steckler, T., & Griebel, G. (2003). Neuropeptide systems as novel therapeutic targets for depression and anxiety disorders. *Trends in Pharmacological Sciences, 24*, 580–588.

Ilieva, I., Boland, J. & Farah, M.J. (2012). Objective and subjective cognitive enhancing effects of mixed amphetamine salts in healthy people. Neuropharmacology.

Ingvar, M., Ambros-Ingerson, J., Davis, M., Granger, R., Kessler, M., Rogers, G., et al. (1997). Enhancement by an ampakine of memory encoding in humans. *Experimental Neurology, 146*, 553–559.

Kass, L. (2003, October 16). *The pursuit of biohappiness*. Washington Post, p. A25.

Kemp, J., & McKernan, R. (2002). NMDA receptor pathway as drug targets. *Nature Neuroscience, 5* Suppl, 1039–1042.

Knutson, B., Wolkowitz, O., Cole, S., Chan, T., Moore, E., Johnson, R., et al. (1998). Selective alteration of personality and social behavior by serotonergic intervention. *American Journal of Psychiatry, 155*, 373–379.

Kosfeld, M., Heinrichs, M., Zak, P. J., Fischbacher, U., & Fehr, E. (2005). Oxytocin increases trust in humans. *Nature, 435*(7042), 673–676.

Kramer, P. (1993). *Listening to prozac*. New York: Penguin.

Lagarde, D., Batejat, D., Van Beers, P., Sarafian, D., & Pradella, S. (1995). Interest of modafinil, a new psychostimulant, during a sixty-hour sleep deprivation experiment. *Funf Cin Pharmacol, 9*, 1–9.

Langreth, R. (2002). *Viagra for the brain*. Forbes, *February*.

Larriviere, D., Williams, M. A., Rizzo, M., Bonnie, R. J., & on behalf of the AAN Ethics, L. a. H. C. (2009). Responding to requests from adult patients for neuroenhancements. Guidance of the Ethics, Law and Humanities Committee. *Neurology, 73*(17), 1406–1412.

Lynch, G. (2003). Memory enhancement: The search for mechanism-based drugs. *Nature Neuroscience Supplement, 5*, 1035–1038.

McBeth, B. D., McNamara, R. M., Ankel, F. K., Mason, E. J., Ling, L. J., Flottemesch, T. J., et al. (2009). Modafinil and zolpidem use by emergency medicine residents. *Academic Emergency Medicine, 16*(12), 1311–1317.

Mintzer, M., & Griffiths, R. (2007). A triazolam/amphetamine dose–effect interaction study: dissociation of effects on memory versus arousal. *Psychopharmacology, 192*(3), 425–440.

Narkar, V. A., Downes, M., Yu, R. T., Embler, E., Wang, Y.-X., Banayo, E., et al. (2008). AMPK and PPAR¥ Agonists Are Exercise Mimetics. *134*(3), 405–415.

Pary, R., Lewis, S., Matuschka, P., Rudzinskiy, P., Safi, M., & Lippman, S. (2002). Attention deficit disorder in adults. *Annals of Clinical Psychiatry, 14*, 105–111.

Pitman, R., Sanders, K., Zusman, R., Healy, A., Cheema, F., Lasko, N., et al. (2002). Pilot study of secondary prevention of posttraumatic stress disorder with propanolol. *Biological Psychiatry, 51*, 189–192.

President's Council on Bioethics. (2003). Beyond therapy: Biotechnology and the pursuit of happiness.

Repantis, D., Laisney, O., & Heuser, I. (2010). Acetylcholinesterase inhibitors and memantine for neuroenhancement in healthy individuals: A systematic review. *Pharmacological Research, 61*(6), 473–481.

Repantis, D., Schlattmann, P., Laisney, O., & Heuser, I. (2010). Modafinil and methylphenidate for neuroenhancement in healthy individuals: A systematic review. *Pharmacological Research, 62*(3), 187–206.

Rudman, D., Feller, A., Nagraj, H., Gergans, G., Lalitha, P., Goldberg, A., et al. (1990). Effects of human growth hormone in men over 60 years old. *New England Journal of Medicine, 323,* 1–6.

Sahakian, B., & Morein-Zamir, S. (2007). Professor's little helper. *Nature, 450,* 1157–1159.

Salzano, J. (2003). Taming stress. *Scientific American., 289,* 87–95.

Scott, R., Bourtchouladze, R., Gossweiler, S., Dubnau, J., & Tully, T. (2002). CREB and the discovery of cognitive enhancers. *Journal of Molecular Neuroscience, 19,* 171–177.

Smith, M.E. & Farah, M.J. (2011). Are prescription stimulants "smart pills"? The epidemiology and cognitive neuroscience of prescription stimulant use by normal healthy individuals. Psychological Bulletin, 137, 717-741.

Solomon, P., Adams, F., Silver, A., Zimmer, J., & DeVeaux, R. (2002). Ginkgo for memory enhancement: arandomized controlled trial. *JAMA, 288,* 835–840.

Staubli, U., Perez, F., Xu, G., Rogers, M., & Ingvar, S. (1994). Facilitation of glutamate receptors enhance memory. *Proceedings of the National Academy of Sciences America, 91,* 771–781.

Strange, B., Hurlemann, R., & Dolan, R. (2003). An emotion-induced retrograde amnesia in humans is amyugdala- and B-adrenergic-dependent. *Proceedings of the National Academy of Sciences America, 100,* 13626–13631.

Tang, Y.-P., Shimizu, E., & Tsien, J. (2001). Do "smart" mice feel more pain, or are they just better learners. *Nature Neuroscience, 4,* 453–454.

Tang, Y.-P., Shimizy, E., Dube, G., Rampon, C., Kerchner, G., Zhuo, M., et al. (1999). Genetic enhancement of learning and memory in mice. *Nature, 401,* 63–69.

Tse, W., & Bond, A. (2002). Serotonergic intervention affects both social dominance and afiliative behavior. *Psychopharmacology, 161,* 373–379.

Tully, T., Bourtchouladze, R., Scott, R., & Tallman, J. (2003). Targeting the CREB pathway for memory enhancers. *Nature Reviews Drug Discovery, 2,* 267–277.

Varlet-Marie, E., Gaudard, A., Audran, M., & Bressolle, F. (2003). Pharmacokinetics/Pharmacodynamics of recombinant human erythropoietins in doping control. *Sports Medicine, 33,* 301–315.

Walker, D., Toufexis, D., & Davis, M. (2003). Role of the bed nucleus of the stria terminalis versus amygdala in fear, stress, and anxiety. *European Journal of Pharmacology, 463,* 199–216.

Walker-Batson, D., Curtis, S., Natarajan, R., Ford, J., Dronkers, N., Salmeron, E., et al. (2001). A double-blind, placebo-controlled study of the use of amphetamine in the treatment of aphasia. *Stroke, 32,* 2093–2098.

Walker-Batson, D., Smith, P., Curtis, S., Unwin, H., & Greenlee, R. (1995). Amphetamine paired with physical therapy accelerates motor recovery after stroke: Further evidence. *Stroke, 26,* 2254–2259.

Webb, J. R., Thomas, J. W., & Valasek, M. A. (2010). Contemplating cognitive enhancement in medical students and residents. *Perspectives in Biology and Medicine, 53*(2), 200–2014.

Weber, P., & Lutschg, J. (2002). Methylphenidate treatment. *Pediatric Neurology, 26*, 261–266.

Wolpe, P. (2002). Treatment, enhancement, and the ethics of neurotherapeutics. *Brain and Cognition, 50*, 387–395.

Yesavage, J., Mumenthaler, M., Taylor, J., Friedman, L., O'Hara, R., Sheikh, J., et al. (2001). Donezepil and flight simulator performance: effects on retention of complex skills. *Neurology, 59*, 123–125.

Yin, J., Del Vecchio, M., Zhou, H., & Tully, T. (1995). CREB as memory modulator: Induved expression of a dCREB2 activator isoform enhances long-term memory in Drosophila. *Cell, 81*, 105–115.

Zeeuws, I., Deroost, N., & Soetens, E. (2010). Effect of an acute d-amphetamine administration on context information memory in healthy volunteers: Evidence from a source memory task. *Human Psychopharmacology: Clinical and Experimental, 25*(4), 326–334.

2

The Case for Clinical Management of Neuroenhancement in Young People

ILINA SINGH AND KELLY KELLEHER

Should doctors prescribe neuroenhancers to healthy patients? The American Academy of Neurology says "yes"—at least for healthy adults. Cognitive under-performance due to fatigue, age, or simply having too much to do can cause stress and anxiety. Doctors have an obligation to relieve suffering in all its forms. "'Live with it!' is never an acceptable response," says Gordan Brown, a member of the American Academy of Neurology's Ethics, Law, and Humanities Committee that wrote the guidelines on neuroenhancement (Larriviere et al., 2009; www.aan. com/elibrary/neurologytoday/?event=home.showArticle&id=ovid.com:/bib/ovft db/00132985-200911190-00005).

But what about neuroenhancement in children and adolescents? Does their suffering—the need to juggle demanding social schedules, sports commit-ments, family obligations and other extracurriculars with homework, exams, and papers—also oblige the physician to prescribe neuroenhancers, if requested? The slower and simpler childhoods of previous decades are not likely to come back. Why, therefore, should young people have to "live with it" when adults do not? If left to their own devices, young people in the United States clearly do not choose to suffer; their use of prescription medications to enhance attention, focus, alert-ness, and relieve performance anxiety has risen sharply and rapidly over the past several years (Johnston et al., 2006). Should we leave them to it, or should we take seriously the demand for neuroenhancement among US young people and their parents?

In this chapter, we take neuroenhancement in US young people seriously. We believe that over time, stimulants and other neuroenhancers will increasingly be used to enhance young people's cognitive and behavioral functioning, along-side growing general public acceptability of neuroenhancers as tools to improve academic, social, and workplace performance (Schermer, 2009; Greeley et al., 2009).

We focus on the most common current neuroenhancers used by US young people: stimulant drugs. We outline the key social and ethical concerns raised by the use of stimulant drugs for enhancement in young people, and we make specific research, practice, and policy recommendations. We also suggest a rationale for clinical management of psychotropic neuroenhancers in US young people, attending closely to the necessary boundaries on such practice asserted by structural and clinical factors, as well as by potential ethical conflicts. This outline and the subsequent rationale for management focuses on stimulants, but it can serve as a template for novel neuroenhancers that reach the US child market.

DEFINITION OF NEUROENHANCEMENT

We define neuroenhancement as the use of neurotechnologies (e.g. stimulant drugs) to improve cognitive and/or behavioral functioning and performance where cognitive and/or behavioral functioning is not judged to be impaired. Enhancement is often differentiated from treatment; in reality, distinctions between treatment and enhancement of human cognitive and behavioral functioning are often continuous rather than categorical (Daniels, 2000; see also Conrad & Horwitz, chapter 4 in this volume). A better distinction between treatment and enhancement takes into account the level of contextual impairment caused by behaviors and/or cognitive functioning. Where functioning is not impaired at clinically significant levels in a particular social context, the motivation for use of neurotechnologies can be considered enhancement.

An emphasis on context suggests that impairment is a fluid concept. For example, it is possible that increasing use of neuroenhancers by young people will over time shift the threshold of impairment such that young people whose functioning in a particular context was previously in the normal range would now be considered impaired. This kind of shift may already be happening. Anecdotal reports suggest that in some highly competitive US secondary schools parents feel that young people who are not taking stimulants are at a disadvantage due to the high percentage of young people who are taking stimulants.

We expect that initially neuroenhancement of young people will be relatively localized to well-resourced families and communities, whose young people attend fairly competitive secondary schools. Indeed, given the reports that a high proportion of US young people at such schools use stimulants and other psychotropic drugs to improve academic performance, it is likely that a diagnosis of mild attention deficit hyperactivity disorder (ADHD) already functions as a pathway to neuroenhancement in these communities.

PSYCHOTROPIC DRUG USE AND ENHANCEMENT IN YOUNG PEOPLE

Diversion of Stimulant Drugs Among Young People

The proliferation of the number and types of psychotropic drugs over the past two decades has provided several potential products for neuroenhancement among children and adolescents (to whom we will refer, collectively, as "young people"[1]). These products address anxiety, memory, attention, focus, and alertness; they have applications in academic settings, in sports, and as recreational drugs. However, the only pharmaceutical agents in widespread use as neuroenhancers are the stimulant medications, including methylphenidate (e.g., Ritalin) and dex-amphetamine compounds (e.g., Adderall). Since the introduction of psychotropic drug treatments for ADHD more than fifty years ago, pediatric stimulant medications have been used as appetite suppressants, sleep suppressants, and study aides throughout the United States. Still, estimates of non-prescription use of stimulants were below 0.5% until 1995 across the age range from high school to adults. Since the mid-1990s, 2.5% of high school students, college students, and young adults consistently report misuse of stimulants. In grades 10–12, a larger proportion of students (4.1%) acknowledge non-prescription use, with boys reporting greater misuse than girls (Teter et al., 2005). College students have the highest rates, with estimates ranging from 5 to 50% (Smith & Farah, 2011). Among students engaging in non-prescription use at US colleges, estimates of their reported goals in misuse vary significantly, although cognitive or academic enhancement generally ranks highest. In a review by Smith and Farah (2011) other reasons included to get high, to party longer, to enhance athletic performance, and to suppress appetite.

From which sources do students get their stimulants? School and social networks have long served as primary sources of non-prescription drugs, but prescription drugs are becoming as desirable as nonprescription drugs. Some observers suggest that prescription drugs, including tranquilizers, stimulants, and painkillers, are now preferred by US young people over street drugs (Johnston et al., 2005). Prescription drugs are associated with a proven track record of safety and efficacy, and young people often have easy and safe access to them. Indeed, researchers and journalists report that university chat sites and listserves are filled with offers of prescription drugs for nonmedical use. (Friedman, 2006; Talbot, 2009; S. Vrecko, personal communication). Approximately one-fifth of all children, adolescents, and young adults prescribed ADHD medications report giving, selling, or being forced to hand over their medications to other students (Poulin 2001). Outside these networks, the Internet serves as another readily available means of obtaining prescription drugs.

Some young people get stimulants for purposes of neuroenhancement via an entirely legal means: physicians' prescriptions. Surveys of physician prescribing practices suggest that prescribing of psychotropic drugs to young people has risen sharply since 1999, and that increased prescriptions for stimulants and antidepressants are especially notable (Thomas et al., 2006; Delate et al., 2004). At the same time, physicians are frequently criticized for failing to perform adequate assessments before making a diagnosis, particularly of ADHD (Friedman, 2006). Moreover, prescription of psychotropic medications to young people is not necessarily associated with a mental illness diagnosis; between 1994 and 2001, approximately one-fifth of office visits that resulted in a prescription of a psychotropic drug excluded a diagnosis of mental illness (Thomas et al., 2006). Pressure on physicians from parents, pressure on physicians from insurance providers and the structure of national health care systems, greater parent awareness about drugs through pharmaceutical company advertising, and the lack of interaction among professionals involved in the care of young people are all factors that have been identified as potential drivers in the increase in prescribing of psychotropic medications to US adolescents. Campus health clinics in the United States have come under particular scrutiny for contributing to the flow of prescription drugs among students, since physicians in such clinics often receive requests to refill prescriptions for psychotropics without opportunity or support to confirm the diagnosis (Rimsza & Moses, 2005).

Physicians are not only prescribing more psychotropic drugs to young people; they are also prescribing more such drugs to adults (Glied & Frank, 2009). In the United States polypharmacy is common, and both adults and young people are likely to have more than one prescription for psychotropic drugs (Zonfrillo et al., 2005). Adults frequently do not take all the medication prescribed to them, and their medicine cabinets may be full of unused drugs. Young people report that these cabinets are another easy source of prescription drugs (Friedman, 2006).

ETHICAL ISSUES AND PRECEDENTS

In recent years, attitudes toward performance enhancement among adults have changed substantially, such that an increasing proportion of the public now supports and uses neuroenhancers (see Chatterjee, chapter 1 in this volume). We believe that the current level of use of stimulants for purposes of enhancement among young people (which is almost certainly under-reported) will also increase, not least because the use of psychotropic cognitive enhancing agents will become more normal in future generations. Use of these agents by young people for purposes of enhancement has social and ethical implications that require scrutiny and analysis. It is particularly important to note that the ethical and practical issues surrounding pediatric neuroenhancement are not the same as for neuroenhancement in adults.

Research on neuroenhancement practices among young people is still in its earliest stages; it does not provide sufficient ground for a prospective analysis of social and ethical harms and benefits of neuroenhancement in young people. We therefore need a proxy ground on which to conduct this analysis. In the following section we use existing analyses of stimulant drug treatment of ADHD in young people to identify and evaluate the relevant social and ethical harms and benefits of stimulant drug use in young people for purposes of enhancement. At the end of the chapter, we will make recommendations for policy, practice, and further research.

SAFETY AND PHYSICAL EFFECTS OF STIMULANT DRUGS

Stimulant drugs have an extensive history of use in pediatrics and child psychiatry, and there is good anecdotal information about safety and side effects. However, there is little systematic longitudinal safety data. Long-term impact of stimulant drug use on the human brain is unknown (Vitiello, 1998), and there are likely to be different developmental implications of treatment use and enhancement use of methylphenidate. This is because dosing practices for these two different uses of the drug will likely differ, with enhancement doses typically focused on short-term events and treatment emphasizing ongoing use over long periods. In addition, pre-existing structural and functional differences in diagnosed and undiagnosed persons may affect the impact of stimulants on brain development. There is some data from animal models on the effects of stimulants. Chronic administration of stimulants does appear to have lasting biobehavioral effects in adult animal brains (Martins et al., 2006), but the impact of these effects is unclear. Stimulant use in adolescence is associated with biochemical modifications in brain reward-related circuits in adult rats, which may account for the positive effect of increased capacity for self-control in adulthood (Grund et al., 2006; Adriani et al., 2007). Yet another study found that the structural and biochemical changes in reward-related areas associated with methylphenidate use in mice overlapped with changes associated with cocaine use (Kim et al., 2009). The director of the US National Institute for Drug Abuse (NIDA) subsequently voiced concern that non-medical use of methylphenidate could lead to addiction (www.medicalnewstoday. com/articles/137454.php). However, animal models may bear little relation to the complex developmental processes of human brain development.

Common side-effects of stimulant drugs in young people taking stimulants for ADHD include appetite suppression and insomnia. Extended use of stimulants may have a small effect on young people's growth and exacerbate pre-existing heart conditions. Young people taking stimulant drugs for ADHD may be at greater risk for substance abuse and delinquent behaviors, although it is not possible to know whether stimulants are causal factors in these outcomes (Jensen et al., 2007). Despite widespread reports of safety associated with proper dosing and a long history of use, there has been political, medical, and public pressure to curb the increasing use of stimulants in young people. Since February 2007, all drug treatments for ADHD have carried the FDA's most severe warning—a black box label. The affected drugs include methylphenidate (e.g., Ritalin and Concerta), dexamphetamine (e.g., Adderall), and atomoxetine (Strattera). The warnings emphasize the risks of serious cardiovascular side effects (including death), growth suppression, and the development of psychosis or other psychiatric conditions.

Abuse Potential

It is currently impossible to know the extent to which diversion of stimulant drugs occurs for enhancement purposes. Abuse potential of stimulants is a serious safety concern; however, it can be minimized if a young person is prescribed sustained release formulations, which have a lower potential for abuse (Volkow & Swanson, 2003).

Use of Multiple Psychotropic Drugs

Data from the Monitoring the Future Survey (www.monitoringthefuture.org) suggest that young people currently use many different psychotropic prescription drugs for non-medical purposes, including tranquillizers, pain killers, stimulants, and hypnotics (Johnston et al., 2005). These are used both for a "high" and for practical purposes such as performance enhancement (Friedman, 2006). It is unknown whether young people use multiple drugs simultaneously, but multiple drugs are available to them in their communities. There is no available safety data for simultaneous use of current psychotropic drugs by young people, for treatment purposes or for enhancement purposes. Moreover, as the neuroenhancement market expands, for example, into memory enhancement, novel neuroenhancers will become available, and it is possible that these will be used in combination with other neuroenhancers. Dosing practices involving multiple current and novel psychotropic drugs may lead to physical and developmental harms as these drugs interact with each other in unknown ways.

AUTONOMY AND COERCION RELATED TO STIMULANT DRUG USE

Prescriptions for stimulant drugs to treat ADHD have raised concerns that fall into two broad categories: autonomy and coercion. We will discuss autonomy first.

Legally and practically, young people are not autonomous agents. However, young people do have rights to protection of aspects of their developing autonomy, in line with their rights as persons. Potential threats to autonomy include the impact of stimulant drug treatments on a young person's developing sense of personal identity, particularly the experience of an "authentic" (unmedicated) self, and her sense of personal responsibility. These autonomy concerns are associated with the physical and psychological impacts of alterations in behavior, motivation, attention, interaction with others, and performance, which are associated with stimulant drug treatments and may or may not be perceived as self-altering by young people. There is, as yet, little empirical evidence to support or deny these concerns, although there is at least one study underway that investigates the social and ethical impacts of stimulant drug treatments among young people with ADHD (www.adhdvoices.com).

Personal Identity

The available research on the ethical impacts of stimulant drug treatments suggests in that there are, in fact, benefits to such treatment. To the extent that notions of personal authenticity are affected by stimulant drug medication, the impact is more positive than negative, at least until adolescence. Young people with ADHD who take stimulant medication tend to feel that they have increased agency when on medication in determining the outcome of their immediate actions, and also in forging their life trajectories (Singh et al., 2010; Singh, 2012).

Studies among young people taking stimulants for ADHD suggest that stigma is a primary threat to self-perceptions and self esteem. Young people with ADHD are more likely to be bullied and teased by peers for their behaviors than for taking stimulants, and this kind of teasing appears to be more likely among young people in under-resourced settings. In middle-class US settings, there is certainly awareness of young people who are "different," but an ethos of valuing differences minimizes overt teasing at least within schools (although cyberbullying is an increasing problem outside school). There is a strong desire among younger children in the United States to keep their ADHD diagnosis and treatment secret from their peers, even though the majority of these children agree that stimulants are beneficial to them. Among teenagers, however, stimulant dosing often becomes part of peer group negotiations; friends will tell friends whether and when they should take their medication (Singh, unpublished data).

It seems likely, then, that in settings where neuroenhancement is accepted or even desirable, there will be minimal stigma attached to neuroenhancement practices for older children, and that enhancement practices will be communally decided and negotiated. As is the case with other prescription and non-prescription drugs, adolescent peer groups are likely to put pressure on individuals to use neuroenhancers. Among younger children, there may be shame and secrecy associated with neuroenhancement, especially if neuroenhancement is primarily driven by parental desire for improved performance.

Personal Responsibility

Stimulant drug treatment of young people with ADHD can complicate a younger child's ability to develop an appropriate conception of personal responsibility for behavior. Studies suggest that younger children taking stimulants attribute successes to their own actions, but attribute failures to the drug not working properly (Pelham et al., 2002). It is possible that this attribution could be in part strategic: Interviews with young people suggest that they do use stimulants to explain their performance or actions, but they tend to be aware of times when they are using these explanations as an excuse for poor performance or behavior (Singh, 2011).

When using stimulants for neuroenhancement, young people are more likely to use stimulants for short-term, event-specific purposes rather than as a daily regimen. Such dosing practices could increase attributions of successes and failures to medication, and may consequently create a greater threat to a young person's sense of personal responsibility for academic performance, athletic prowess, or social interactions. There may also be a greater potential for a young person to become psychologically dependent on stimulants, if he or she comes to view successes as brought about largely by stimulants.

Coercion

Another set of concerns involves young people's capacity to participate in decision making about stimulant drug treatments, and the potential for coercion of young people in relation to neuroenhancement.

Young People's Capacity and Rights to Assent and Dissent to Neuroenhancement

In treatment cases, parents or a legal guardian are empowered to make medical decisions on behalf of the child until the child is the age of majority. In most settings, young people are asked for their *assent* to treatment or participation in research, signaling that they are not considered competent to give fully informed *consent*. Despite this widespread practice of distinguishing between assent and consent, the literature on young people's participation in treatment decisions suggests that young people have the capacity to understand their medical conditions, and are competent to judge treatment decisions from quite a young age (Alderson, 1993; Kuther, 2003; Miller et al., 2004). However, parents have caretaking liability for young people, and decisions around enhancement involving long-term risks may be particularly challenging for younger children because they involve calculation of future risk-benefit ratios.

Should parents and caregivers have the right to override a child's dissent to using psychotropic drugs for purposes of enhancement? Equally important: Should a child's request for neuroenhancers ever be granted without the parents/caregivers' consent? Such questions are ethically appropriate, and they reflect an awareness of children's rights and dignity as persons, as set out in, for example, The Geneva Convention on Rights of the Child (www.hrweb.org/legal/child.html). However, these rights are interpreted and applied in different ways in different medical settings, without much oversight, and the rights of parents to determine what is in the best interests of the young person may take precedence in cases where intervention or non-intervention is not obviously harmful to the young person.

Coercion and Parents

Parents have received a good deal of social and ethical scrutiny in relation to stimulant drug treatment for young people. Parents can be driven by performance pressures or personal parenting goals to produce highly successful children at the expense of the child's physical or mental health. Individual parenting goals can be shaped or exacerbated by social ideologies; for example, cultural notions of good mothering place a burden on mothers in consumer-driven societies to access and exploit all available resources to ensure a child's success. These factors can have a coercive influence and can drive uptake of stimulant drug treatments for young people. The extent to which parents and young people are vulnerable to these social influences varies with demographic and environmental factors. Studies investigating stimulant use for enhancement purposes among young people suggest that demographic influences affect outcomes: In the enhancement case, stimulant use is related to environmental factors such as region of the country, competitiveness of the school, and number of students at a school being treated for ADHD.

Parenting factors can create a coercive context for young people around performance-enhancing drugs. However, if neuroenhancers come to be considered a valuable resource for young people's learning and success, then it is possible that demographic factors associated with lower stimulant drug use (such as race and ethnicity) could come to be interpreted as restrictive rather than protective.

Coercion and Schools

School practices have also been scrutinized in relation to stimulant drug treatment of children. US schools are alleged to exert so much pressure on families to pursue stimulant drug treatment that some US states have passed legislation which makes it illegal for schools to accuse parents of educational neglect for refusing to give their children stimulants (www.cchr.org/press_room/press_releases/Over_500_Parents_Say_Schools_Coerced_Them_to_Administer_Psychiatric_Drugs_to_Children.html; www.edwatch.org/). Some of these legal moves against schools have been spearheaded by entities with active biases against psychotropic drugs, such as the Church of Scientology. Still, the great majority of research evidence indicates that long-term use of stimulants does not improve overall academic achievement (Jensen et al., 2008). Skeptics therefore believe that teachers may be using stimulants, which are known to improve classroom behavior, as a means of classroom management, rather than as a means of improving an individual child's chances of academic success. Teachers are often the first professionals to suggest to parents that a child would benefit from stimulant treatment (Sax & Kautz 2003). While teachers increasingly receive training on how to recognize symptoms of psychiatric disorders that will benefit from psychotropic drug treatment, it is not clear that these trainings involve a discussion of ethical issues such as coercion. Indeed, in several US states teachers are now legally prohibited from discussing stimulant drug treatment with families (www.ablechild.org/slegislation.htm).

It is unclear whether the enhancement case of stimulant drug use raises unique or new ethical concerns for schools. The concerns outlined above about use of stimulants by young people are relevant to both the treatment and the enhancement case. In both cases, subtle coercive factors may influence teachers' encouragement of stimulant use, such as the school's need to perform well on national standardized tests and league tables, the results of which have resource and incentive implications for teachers and for schools.

If neuroenhancement becomes a more common practice, schools can assist in protecting young people from harms and can help to increase the benefits of neuroenhancement for young people. For example, schools are often the site where abuse and diversion of stimulant drugs occurs, and some schools monitor these practices as part of drug abuse prevention programs.

Coercion and the Pharmaceutical Industry

The pharmaceutical industry drives research and development in psychotropic drugs, enabling people around the world to live active and productive lives despite mental illness. But to justify the investment into research and development of many drugs that never make it to market, the industry puts billions of dollars into mar-keting the drugs that do make it. These marketing activities have been the target of ethical scrutiny, particularly in the US context (Conrad & Horwitz, this volume).

Many observers of the rise in psychotropic drug use in the American context (as compared to Europe and other parts of the world) have identified the US phar-maceutical industry marketing programs, its political lobbying power, and its close interactions with physicians as key factors in rising use and increasing pub-lic acceptance of psychotropic drug interventions. Psychotropic drugs are among the most profitable agents for young people, at least in the United States (http://health.howstuffworks.com/health-illness/treatment/medicine/medications/10-most-profitable-drugs3.htm). The industry stands to benefit from a neuroenhance-ment drug market for young people, as neuroenhancement practices in young people increase the consumer base for psychotropic drugs as well as increasing demand for a particular drug because of its dual-use (treatment and enhance-ment) applications.

Direct-to-consumer (DTC) advertising is prohibited in all countries except the United States and New Zealand. However, the worldwide web means that con-sumers anywhere can access drug company sites, where so-called educational information about the drug is frequently conflated with advertising for the drug. Advertising is a primary social means of increasing desire for various kinds of individual enhancement, and stimulant drug advertising is no exception.

Advertising for stimulant drugs in the United States encourages consumers to view drugs as a means of lifestyle enhancement. Print advertisements for ADHD drugs tend to be directed at mothers and promise women a highly idealized rela-tionship with their newly successful little boys and girls. This connection sustains an oppressive mothering ideology that creates a moral obligation on the part of the good mother to pursue the depicted ideals for her child (Singh, 2007). In this sense, advertising of stimulant drugs operates coercively to create both the desire for and the obligation to take up performance-enhancing resources.

Young people are particularly susceptible to advertising for controlled sub-stances (Saffer, 2002); advertisements for tobacco (the Joe Camel ads) and for alcohol have come under considerable scrutiny for their appeal to young people. If psychotropic neuroenhancement for young people becomes a common social practice, then the pharmaceutical industry could initiate advertising campaigns that target young people as potential consumers of psychotropic enhancers. Such advertisements are likely to sell a highly appealing lifestyle and personal identity to young people as part of the effort to sell drugs.

THE "MEDICAL HOME" FOR NEUROENHANCEMENT OF YOUNG PEOPLE

As neuroenhancement in US young people becomes an increasingly common prac-tice, we need to consider how best to protect young people from ethical and physical risks of stimulant drugs, and how to maximize the benefits of neuroenhancement. In our view, it is imperative to locate a relevant site where US young people's use of neuroenhancement technologies can be safely, objectively, and ethically evaluated and managed. We propose that this site should be the primary care clinic.

High-quality health care is increasingly understood to involve a partnership among families, patients, and the health care system. The best outcomes are achieved when primary care clinicians provide longitudinal coordination for medical services, especially those that involve medication (Rittenhouse & Shortell, 2009). Adapted from a term coined in the middle of the last century, the "patient-centered medical home" has been identified as a promising approach to delivering more effective chronic illness care, preventive care, and pharmaceutical management for patients of every age (Fisher, 2008). The movement toward medical homes has benefited primary care clinicians as well as patients, since there are growing incentives to support these coordination activities and primary care clincians report greater awareness and knowledge of their patients' concerns and treatments.

The inclusion of neuroenhancement activities in the medical home is a natural extension of current activities and will likely be required by any providers of such activities going forward. Primary care clinicians prescribe the majority of stimulants for ADHD treatment, and as trusted professionals with access to these controlled substances, it is likely that families interested in neuroenhancement will approach them (Preen, 2008; Goldman, 1998). Because of strong relationships, primary care clinicians are well situated to assist families and young people in considering neuroenhancing agents (Heneghan, 2004). Primary care clinicians are highly trusted by parents and young people, especially for advice on sensitive topics. They often have longstanding relationships with their patients and families that engender trust and insight into family and child dynamics (Freeman & Richards, 1990; Baker & Streatfield, 1995). Because they are embedded in the community, primary care physicians are often aware of other resources for the family and the child as well as of local expectations that may be encouraging neuroenhancement (Etz et al., 2008).

Primary care practices are also an essential setting for decision making and information necessary for initiation and monitoring of neuroenhancing drugs. Child and adolescent prior and current drug treatments, drug allergies, and history of side effects are all necessary pieces of information for decision making around enhancement with stimulants. This information is often available exclusively in primary care records. In addition, primary care clinicians can provide less biased opinions because they are independent agents who are unlikely to benefit financially from most neuroenhancing drugs or other neuroenhancing interventions. This is frequently not the case for other mental health professionals (Bush, 2006). For these reasons, primary care clinicians are the cornerstone to effective assessment, initiation, and monitoring of neuroenhancing drugs in children and adolescents.

ETHICAL CHALLENGES FOR PRIMARY CARE OVERSIGHT OF STIMULANTS FOR PURPOSES OF ENHANCEMENT IN YOUNG PEOPLE

Although primary care clinics are the logical sites for assisting parents and young people with issues related to neuroenhancement, they are not without bias.

Primary care clinicians are frequently influenced disproportionately by pharmaceutical detailing, even for off-label indications for children and adolescents. Such detailing induces changes in prescribing behaviors for clinicians that affect patient receipt of psychotropic drugs (Fugh-Berman & Ahiri, 2007; Huskamp et al., 2008; Donohue et al., 2004). Second, primary care clinicians are influenced by parent and child demand for drugs even when the evidence for effectiveness of drugs is low or nonexistent. For example, antibiotics are still often prescribed for viral upper respiratory infections, even though they are not effective for these conditions, because of parent and patient demand (Butler et al., 1998). In short, there are important factors that may encourage some primary care clinicians to prescribe neuroenhancing drugs for children and adolescents in situations where it may not be in the best interests of those patients.

PRACTICAL CHALLENGES TO PRIMARY CARE OVERSIGHT OF STIMULANTS FOR PURPOSES OF ENHANCEMENT IN YOUNG PEOPLE

The bigger challenge for primary care management of neuroenhancement will be the implementation of a sound program in primary care settings for initiating such medications, monitoring outcomes and side effects, and tracking abuses. Several specific obstacles loom large. First, no readily available assessment tools for measuring which young people would most benefit from neuroenhancement (and how often) are available. Second, if patients choose to pay cash for these medications, no system-wide records will be available to make sure such medications are not diverted for resale or abuse. Current systems for detecting excessive numbers of prescriptions for diversion rely on insurance and administrative claims records (Wysowski, 2007; Pradel, 2009).

One of the advantages of bringing neuroenhancement into primary care is the possibility of follow-up care. However, following young people prescribed neuroenhancing drugs will require specific outcome measures, and these have not yet been developed. Side-effect monitoring will also be challenging, and will have to allow for different dosing practices that may be related to neuroenhancement, as compared with treatment for conditions like ADHD. It is also possible that parents who support neuroenhancement will be less likely to observe side-effects in their children, or they may be more reluctant to report side-effects from medications they sought for their children.

CONCLUSION AND OUTLOOK

Growing use of neuroenhancers by young people is likely inevitable. Although models of ethical use of neuroenhancers by adults are being created by entities like the American Academy of Neurology, there are at present no such models for neuroenhancement in young people. The lack of attention to the particular issues and vulnerabilities related to young people and neuroenhancement is indicative of a general absence of research on young people and psychotropic drugs. Most such drugs are still prescribed to young people off-label, due to the absence of clinical trials with children and adolescents. This is changing with the FDA pediatric exclusivity program (Devaugh-Geiss et al., 2004), but still, as we have shown in this chapter, in terms of safety of drugs for young people, as well as relevant ethical issues, there is, in many cases, inadequate research upon which to base specific recommendations. Because drugs are being developed faster than long-term studies can be conducted and because long-term developmental outcomes may be decades off, it is imperative that parents, caregivers, clinicians, schools, and pharmaceutical firms operate with the utmost caution and careful monitoring in the case of pediatric neuroenhancement. In order to bring neuroenhancement of young people into primary care, professional societies that provide primary care medical services for children must develop position statements and policies on primary care use of neuroenhancement drugs for and adolescents. This will require careful attention to the potential for individual and ethical harms and benefits to neuroenhancement in young people outlined in this chapter, as well as support for and coordination of research and monitoring programs.

Neuroenhancement in young people raises significant concerns among many observers; we hope that the research, policy, and practice recommendations outlined here realistically contribute to minimizing the risks and maximizing the benefits of current and future neuroenhancing practices in this population.

RECOMMENDATIONS

Because young people are more vulnerable to many effects of neuroenhancers than adults and rely on proxies for their care, the barriers to use of neuroenhancers should be higher than those for adults. Harms and benefits specific to young people—of particular neuroenhancers, as well as of performance enhancement itself—must be made explicit in order to inform relevant policy and practice. Barriers should be age-appropriate; that is, they should take into account the developmental and cognitive status of the child, as well as the decision-making capacities of young people at different ages. Where not enough is known to establish valid, evidence-based barriers to use of neuroenhancers for young people, further research will need to be undertaken, although sufficient research to ensure safety of the developing child will require a long time.

Recommendations Related to Physical Safety of Stimulants for Neuroenhancement in Young People

At the clinical level, stimulants should be prescribed in such a way as to minimize the abuse potential of these drugs, including a preference for low-abuse potential long-acting preparations where these are available and suitable for the young person. Young people should not be given simultaneous prescriptions for multiple neuroenhancers. At the health care organization level, health care systems and pharmacies should institute systems to monitor the quantity and location of requests for neuroenhancers to avoid abuse and monitor total dosage. At the regulatory level, government agencies charged with drug safety should mandate full disclosure of all data in trials for neuroenhancement. This data should be analysed by independent scientists.

More research is needed on clinical trials involving psychotropic drugs for purposes of neuroenhancement in young people. We also need to know more about patterns of diversion and abuse of stimulants among young people, including more valid tests for differentiating diversion for academic and other purposes. Animal models are suitable for basic research on neuroenhancement because of the long gestation for human research in such studies.

Recommendations Related to Consent in Young People's Use of Stimulants for Neuroenhancement

If demands for neuroenhancement through the clinic become a reality, child and parent assent and consent to neuroenhancement with stimulants should be mandatory. Before physicians can become gatekeepers to neuroenhancement, they should receive training in the assessment of a young person's capacity to participate in decision making and consent processes.

Further research needs to be undertaken to better understand age thresholds and competency tests for young people's consent to neuroenhancement. Much can be learned in this regard from research on consent and assent procedures in the context of children's involvement in clinical trials and other paediatric research. Research on shared decision-making processes is particularly notable here (e.g., Coyne & Harder, 2011). Decision aids should also be developed to help young people better understand the risk and benefits, and to assist clinicians in understanding external pressures experienced by young people in the enhancement process.

Recommendations Related to Identity and Moral Self-Understandings in the Use of Stimulants for Neuroenhancement in Young People

Prescribers of stimulant drugs for purposes of enhancement should work as part of a multidisciplinary team that enables evaluation of a young person's self-perceptions and self-understandings, particularly around issues of identity, autonomy, and independent decision making. Existing standardized measures can be used initially, such as the Harter's Self-Perception Profiles (Harter, 1988). For those receiving neuroenhancers, this evaluation should be followed by focused counseling on these issues.

Motivations and goals of parents and young people in relation to neuroenhancement are still poorly understood and require investigation. It is also important to better understand and measure threats to identity development, autonomy, and independent decision making from stimulants for young people at varying ages. Demographic factors are likely to be a critical component of robust models of motivation and goals among parents and young people, and of threats to identity among young people.

Recommendations Related to Parents and Coercion in Relation to Stimulant Use for Neuroenhancement in Young People

We believe that if neuroenhancement becomes clinically available, all young people and families ought to have equal access to stimulant drugs to enhance performance and to the social, ethical, and clinical supports we have outlined in this chapter. Parents' vulnerability to social coercion and the impact of parent coercion on the young person should be discussed as part of the clinical evaluation, and found to be appropriate before neuroenhancers for a young person are prescribed.

Further research is needed to identify the demographic contexts in which young people, parents, and caregivers are most vulnerable to socially based coercive influences. It is also important to ascertain the qualitative impacts on young people of a society in which use of psychotropic agents to enhance abilities becomes normative.

Recommendations Related to Schools and Coercion, in Relation to the Use of Stimulants for Purposes of Neuroenhancement in Young People

Schools should be prevented from coercion of young people or families into adopting stimulants for enhancement purposes. It is also important for schools to manage the administration of stimulants for neuroenhancement with the same policies and procedures used for the administration of any psychiatric drugs. Therefore, teachers' concerns about the physical or emotional consequences of enhancement practices in the case of a particular young person should be raised through standard mechanisms already available in schools. As is the case with other drugs, teachers and schools should cooperate in efforts to prevent abuse of stimulant drugs by young people.

Incentives to teachers and school administrators to promote achievement on state and national student examinations should not be so disproportionate that they motivate these adults to encourage large-scale use of enhancement among young people.

In relation to academic achievement, further research is necessary to understand how use of stimulants for neuroenhancement impacts short- or long-term academic achievement or motivation at different stages of schooling. It is also important to analyse whether the case of stimulants for neuroenhancement in young people represents a qualitatively different set of ethical concerns as compared to the cases of widely used non-drug neuroenhancement practices adopted by parents and schools, such as tutoring, music lessons, and early language education.

Recommendations Related to Pharmaceutical Industry and Coercion

In those countries where DTC advertising is legal, advertisements for neuroenhancers that target young people as consumers should be prohibited. In addition, DTC advertising for neuroenhancement in young people should not be allowed in any media to which young people are likely to have regular access (e.g., the Internet, television, radio, and young people's literature).

Recommendations Related to Primary Care

Primary care clinicians should act as the primary gatekeepers or medical home for neuroenhancement among young people. If this comes to pass, pharmaceutical detailing and free sample provision around enhancement in primary care facilities should be prohibited. Clinician uncertainty about ethical constraints in neuroenhancement of young people should be dealt with through hospital ethics boards as other similar issues are.

Better understanding is needed of the capacity and willingness of primary care clinicians to conduct and coordinate neuroenhancement for young people. In line with the capacity question, research is needed to develop and validate clinical screening, assessment, and monitoring tools for short-term enhancement in young people. Systems should be in place to track requests for off-label use of drugs or devices developed for neuroenhancement in adults, so as to enable the health care sector, working with research and regulatory bodies, to forecast future demand for neuroenhancement products among young people.

ACKNOWLEDGMENTS

An earlier, extended version of this chapter was published in the *American Journal of Bioethics-Neuroscience* 2010, with the title "Neuroenhancement in young people: Proposal for research, policy and clinical management."

NOTES

1. We note that our use of this term is not entirely satisfactory, given the importance of recognizing the developmental status of young people at different stages of childhood and adolescence.

REFERENCES

Adriani, W., Canese, R., Podo, F., & Laviola, G. (2007). H MRS-detectable metabolic brain changes and reduced impulsive behavior in adult rats exposed to methylphenidate during adolescence. *Neurotoxicology and Teratology, 291*, 116–125.

Alderson, P. (1993). *Children's consent to surgery*. London: Open University Press.

Baker, R., & Streatfield, J. (1995). What type of general practice do patients prefer? Exploration of practice characteristics influencing patient satisfaction. *The British Journal of General Practice, 45401*, 654–659.

Bush, S. (2006). Neurocognitive enhancement: ethical considerations for an emerging subspecialty. *Applied Psychology, 132*, 125–136.

Butler, C.C., Rollnick, S., Pill, R., Maggs-Rapport, F., & Stott, N. (1998). Understanding the culture of prescribing: qualitative study of general practitioners' and patients' perceptions of antibiotics for sore throats. *General Practice, 317*, 637–642.

Coyne, I., & Harder, M. (2011). Children's participation in decision-making: Balancing protection with shared decision-making using a situational perspective, *Journal of Child Health Care, 15* (4), 312–319.

Daniels, N. (2000). Normal functioning and the treatment-enhancement distinction. *Cambridge Quarterly 93*, 309–322.

Delate, T., Gelenberg. A. J., Simmons, V. A., et al. (2004). Trends in the use of antidepressants in a national sample of commercially insured pediatric patients, 1998 to 2002. *Psychiatric Services, 55*, 387–391.

Deveaugh-Geiss, J., March, J., Shapiro, M., et al. (2004). Child and adolescent psychopharmacology in the new millennium: A workshop for academia, industry, and government. *Journal of the American Academy of Child & Adolescent Psychiatry, 45*(3), 261–270.

Donohue, J. M., Berndt, E. R., Rosenthal, M., Epstein, A. M., & Frank, R.G. (2004). Effects of pharmaceutical promotion on adherence to the treatment guidelines for depression. *Medical Care, 42*(12), 1176–1185.

Epstein, J. N., Hoza, B., March, J. S., Molina, B.S.G., Newcorn, J. H., Severe, J. B., Wigal, T., Gibbons, R. D., & Hur, K. (2007). 3-year follow-up of the NIMH MTA study. *J Am Acad Child Adolesc Psychiatry, 468*, 989–1002.

Etz, R. S., Cohen D. J., Woolf, S. W., Holtrop, J. S., Donahue, K. E., Iasscson, N. F., Stange, K. C., Ferrer, R. L., & Olson, A. L. (2008). Bridging primary care practices and 40 communities to promote healthy behaviors. *American Journal of Preventative Medicine, 355* Suppl, S390–S397.

Fisher, E. S. (2008). Building a medical neighborhood for the medical home. *N Engl J Med, 359*(12), 1202–1205.

Freeman, G., & Richards, S. (1990). How much personal care in four group practices. *British Medical Journal, 3016759*, 1028–1030.

Friedman, R. A. (2006). The changing face of teenage drug abuse—The trend toward prescription drugs. *N Engl J Med, 354* (14), 1448–1450.

Fugh-Berman, A., & Ahari, S. (2007). Following the script: How drug reps make friends and influence doctors. *PLoS Med*, April *4*(4): e150. www.ncbi.nlm.nih.gov/pmc/articles/PMC1876413.

Glied, S. A. & Frank, R. G. (2009). Better but not best: Recent trends in the well-being of the mentally ill. *Health Affairs, 28*, 637–648.

Goldman, L. S., Genel, M., Bezman, R. J., & Slanetz, P. J. (1998). Diagnosis and treatment of attention-deficit/hyperactivity disorder in children and adolescents. Council on Scientific Affairs, American Medical Association. *JAMA, 27914*, 1100–1107.

Greely H, Sahakian B, Harris J, Kessler RC, Gazzaniga M, Campbell, P. & Farah, M. (2008). Towards responsible use of cognitive-enhancing drugs by the healthy. *Nature 456*, 702–705.

Grund, T., Lehmann, K., Bock, N., Rothenberger, A., & Teuchert-Nood, G. (2006). Influence of methylphenidate on brain development—an update of recent animal experiments. *Behavioral and Brain Functions, 21*, 2.

Heneghan, A. M., Mercer, M., & DeLeone, N. L. (2004). Will mothers discuss parenting stress and depressive symptoms with their child's pediatrician? *Pediatrics, 1133*, 460–467.

Hickson, G. B., Allemeier, W. A., & O'Connor, S. (1983). Concerns of mothers seeking care in private pediatric offices: Opportunities for expanding services. *Pediatrics, 725*, 619–624.

Huskamp, H. A., Donohue, J. M., Koss, C., Berndt, E. R., & Frank, R. G. (2008). Generic entry, reformulations and promotion of SSRIs in the US. *Pharmacoeconomics, 26*(7), 603–616.

Jensen, P., Arnold, L. E., Swanson, J. M., Vitiello, B., Abikoff, H. B., Greenhill, L. L., et al. (2007). 3-year follow-up of the NIMH MTA study. *Journal of the American Academy of Child and Adolescent Psychiatry, 46*(8), 989–1002.

Johnston, C., Fine, S., Weiss, M., Weiss, J., Weiss, G., & Freeman, W. S. (2000). Effects of stimulant medication treatment on mothers' and children's attributions for the behavior of children with attention deficit hyperactivity disorder. *Journal of Abnormal Child Psychology, 284*, 371–382.

Johnston, L. D., O'Malley, P. M., Bachman, J. G., & Schulenberg, J. E. (2006). *Monitoring the Future national survey results on drug use, 1975–2005: Volume I, Secondary school students (NIH Publication No. 06–5883).* Bethesda, MD: National Institute on Drug Abuse.

Kim, Y., Teylan, M. A., Baron, M., S and s, A., Nairn, A. C., & Greengard, P.(2009). Methylphenidate-induced dendritic spine formation and _FosB expression in nucleus accumbens. *Proceedings of the National Academy of Sciences.* Doi. 10.1073/pnas.0813179106.

Kuther, T. L. (2003). Medical decision-making and minors: issues of consent and assent. *Adolescence 3815*, 343–358

Livingstone, S., van Couvering, E., & N. Thumim. (2005). Adult media literacy: A review of the research literature on behalf of Ofcom. http://legacyreports.spectrumaudit.org.uk/advice/media_literacy/medlitpub/medlitpub rss/aml.pdf

Miller, V. A., Drotar, D., & Kodish, E. (2004). Children's competence for assent and consent: a review of empirical findings. *Ethics & Behavior, 143*, 255–295.

Pelham, W. E., Hoza, B., Pillow, D. R., Gnagy, E. M., Kipp, H. L.,Greiner, A. R.,et al. (2002). Effects of methyphenidate and expectancy on children with ADHD: Behavior, academic performance, and attributions in a summer treatment program and regular classroom settings. *Journal of Consulting and Clinical Psychology, 70*(2), 320–335.

Poulin, C. (2001). Medical and nonmedical stimulant use among adolescents: from sanctioned to unsanctioned use. *Canadian Medical Association Journal, 1658*, 1039–1044.

Pradel, V., Frauger, E., Thirion, X., Ronfle, E., Lapierre, V., Masut, A., et al. (2009). Impact of a prescription monitoring program on doctor-shopping for high dosage buprenorphine. *Pharmacoepidemiol Drug Safety, 181*, 36–43.

Preen, D. B., Calver, J., Sanfilippo, F. M., Bulsara, M., & Holman, C. D. (2008). Prescribing of psychostimulant medications for attention deficit hyperactivity disorder in children: differences between clinical specialties. *Medical Journal Australia, 1886*, 337–339.

Rimsza, M. E., & Moses, K. S. (2005). Substance abuse on the college campus. *Pediatr Clin North Am, 52*(1):307–319.

Rittenhouse, D. R., & Shortell, S.M. (2009). The Patient-Centered Medical Home: Will It Stand theTest of Health Reform? *JAMA, 301*(19), 2038–2040.

Saffer, H. (2002). Alcohol advertising and youth. *Journal of Studies on Alcohol.* Supplement 14, 173–181. www.collegedrinkingprevention.gov/media/journal/173-saffer.pdf.

Sax, L., & Kautz, K.L. (2003). Who first suggests the diagnosis of attention deficit/hyperactivity disorder? *Annals of Family Medicine, 1*, 171–174.

Schermer, M., Bolt, I., de Jongh, R., & Olivier, B. (2009). The future of psychopharmaco-logical enhancements: Expectations and policies. *Neuroethics, 2,* 75–87.

Singh, I. (2007). Not just naughty: 50 years of stimulant drug advertising. In E. Watkins & A. Toon (Eds.), *Medicating Modern America.* New York, NY: New York University Press.

Singh, I. (2008). Beyond polemics: science and ethics of ADHD. *Nature Rev Neuroscience, 9,* 957–964.

Singh, I. (2011). A disorder of anger and aggression: Children's perspectives on ADHD in the UK. *Social Science and Medicine, 73*(6), 889–896.

Singh, I. & Kelleher, K. J. (2010). Neuroenhancement in young people: Proposal for research, policy and clinical management. *American J of Bioethics- Neuroscience. 1*(1), 3–16.

Singh, I. (2012). Not Robots: Children's perspectives on authenticity, moral agency and stimulant drug treatments. Journal of Medical Ethics. Advance on-line publication 28 August. doi:10.1136/medethics-2011-100224

Singh, I., Kendall, T., Taylor, C., Hollis, C., Batty, M., Mears, A., & Keenan, S. (2010). The experience of children and young people with ADHD and stimulant medication: A qualitative study for the NICE guideline. *Child and Adolescent Mental Health.* doi: 10.1111/j.1475–3588.2010.00565.x

Smith, M. E., & Farah, M. J. (2011). Are prescription stimulants "smart pills"? The epide-miology and cognitive neuroscience of prescription stimulant use by normal healthy individuals. *Psychological Bulletin, 137,* 717–741.

Talbot, M. 2009. Brain gain. *The New Yorke*r. www.newyorker.com/reporting/2009/04/27/090427fa_fact_talbot.

Thomas, C. P., Conrad, P., Casler, R., & Goodman, E. (2006). Trends in the use of psy-chotropic medications among adolescents, 1994 to 2001. *Psychiatric Services, 57*(1), 63–69.

Volkow, N. D., & Swanson, J. H. (2003). Variables that affect the clinical use and abuse of methylphenidate in the treatment of ADHD. *American Journal of Psychiatry, 160,* 1909–1918.

Wysowski, D. (2007). Surveillance of prescription drug-related mortality using death certificate data. *Drug Safety, 306,* 533–540.

Zonfrillo, M. R., Penn, J. V., & Leonard, H. L. www.psychiatrymmc.com/displayArticle.cfm?articleID=article22

3

Ethical Considerations

Cogniceuticals in the Military

MICHAEL B. RUSSO, MELBA C. STETZ, AND
THOMAS A. STETZ

Pharmaceutical agents that affect alertness, thought, and emotional processes are ubiquitous in society. Some agents are used socially, such as caffeine and nicotine. Others are obtained easily by prescription for treatment of common disorders, such as antidepressants and hypnotics. Still others are used under highly regulated conditions, and include amphetamines and modafinil. Each of these types of psychoactive agents has the potential to sustain or potentially enhance aspects of cognition in healthy individuals. This chapter primarily focuses on cogniceuticals that alter alertness, attention, perception, and sleep, which are relevant to military settings and operations. For the purpose of a discussion of ethics, differences among pharmaceutical classes will be removed by referring to them collectively as *cogniceuticals*.

This chapter will discuss the ethical arguments both for and against the use of cogniceuticals, discuss cross-cultural considerations, and suggest guidelines for use of cogniceuticals, all within a military context. The arguments discussed are revised and extended from those published earlier by Russo and colleagues (2007, 2008; Russo, Caldwell & Thomas, 2006).

ETHICAL ISSUES ARISING IN THE MILITARY CONTEXT

What ethical issues arise when considering the use of cogniceuticals by military personnel? Most of the issues encountered in other contexts will arise here. For example, safety is an important concern that should govern the practice of cognitive enhancement in both military and civilian contexts. As Chatterjee (chapter 1, this volume) reviews these concerns, they will not be further discussed here.

INDIVIDUAL CHOICE

Wrye Sententia, director of the Center for Cognitive Liberty and Ethics, argues that a healthy individual's freedom over his or her own mind is an essential component in defining personal identity in a civilized and enlightened society. Sententia defines the term "cognitive liberty" as "every person's fundamental right to think independently, to use the full spectrum of his or her mind, and to have autonomy over his or her own brain chemistry" (Sententia, 2004). She adds that "the individual, not corporate or government interests, should have sole jurisdiction over the control and/or modulation of his or her brain states and mental processes," and she articulately expresses concern that the power of the government, such as the military has over its members, could lead to forced or coerced medication.

Jeffrey Wilson, a professor at the United States Military Academy in West Point, New York, also adroitly argues that an environment wherein a military member is or feels forced to take an enhancement medication may result in alienation of his or her values and could lead to irrevocable psychological harm (Wilson, 2004). Wilson asserts that even if forced medication use was banned, the military is clearly an organization where conformity is admired and peer pressure is strong. In addition, an ethical use of cogniceuticals should mitigate conditions in which military members, often formed into small close-knit units and dependent upon one another for their individual and collective survival, would feel undue pressure to utilize an agent that purports to enhance warfighting performance abilities.

Two general moral assumptions appear relevant here. First, in a moral nation the primacy of an individual's right to his or her body and mind, relative to group or national interests, is respected even in times of war or emergency. This respect for individual rights is consistent with individuals' agreements, in all-volunteer militaries, to relinquish control over some individual rights.

The second assumption is that ethical standards apply equally in civilian and military settings and that military service does not confer an excepted status to the warfighter. Thus, a standard such as informed voluntary consent applies to the warfighter. Yet even something as simple as informed consent can be difficult to achieve in a battlefield setting.

PERSONAL RESPONSIBILITY

Ronald Bailey and Carol Freedman present arguments that neurological enhancements could undermine personal responsibility (Bailey, 2003; Freedman, 2000). They assert that cogniceuticals may result in relinquishment of personal responsibility could be devastating to the ability of the military to maintain discipline. Furthermore, a breakdown could result in service members performing atrocities attributed to pharmacologically altered minds, and potentially could undermine the Uniform Code of Military Justice if defense attorneys successfully argue that enhanced service members cannot be held responsible for their actions.

The potential relinquishment of personal responsibility cannot be applied equally to every cogniceutical agent. In the case of amphetamine, concerns have been raised about its effects on judgment, for example in the trial of American servicemen for killing Canadians in the Tarnack Farms friendly fire incident (Friscolanti, 2005). With regard to caffeine, there is no evidence in the literature that caffeine renders the individual more or less personally responsible. Even when a cogniceutical is capable of affecting capacity for responsible action, it is difficult to attribute behavior occurring in the midst of battle to medications as opposed to fear, stress, and fatigue.

NECESSITY

Even the safest of pharmacologic agents are not necessary for existence by healthy humans. As such, they should not be used routinely as a substitute for normal human biological needs. Consider the example of suppressing appetite to facilitate productivity under conditions of food scarcity. Food is a necessary substrate for normal human function, although clearly an individual can fast for days without experiencing irreversible biochemical effects. To the military organization, providing food is costly and logistically challenging, while to the individual, carrying the meals constitutes a considerable additional weight. If medications were developed that specifically suppressed appetite for extended periods, and considered for use in service members as a substitute for providing and carrying meals, that would constitute an unethical application with regard to necessity. However, if a service member found him or herself without food for unanticipated or unintended reasons, the use of an appetite suppressant to reduce discomfort and suffering may not necessarily be unethical.

Relating necessity to individual choice, a second rationale beyond the cognitive liberty earlier described by Sententia exists for an individual to determine for him- or herself whether a cogniceutical should be employed at a particular point in time. Individuals vary greatly in their resilience to sleep deprivation, mental stress, and environmental extremes. Some may know from experience that they are capable of doing without a cogniceutical in a specific circumstance—and they should have the liberty to exercise this judgment. However, in some circumstances the deployment of a cogniceutical may have to be suggested because the ability by the impaired individual to recognize the need may be compromised.

In the future, neurophysiological monitoring of cognition may alert the command to the need for employment of a cogniceutical, and possibly doctrine will evolve that permits a commander to order a cogniceutical for the benefit of the organization. As an example of command-dictated preventive medicine, commanders protect the individual and the group by ordering flu shots for everyone, despite protests by individuals who sometimes prefer to pass on the opportunity for inoculation. Regarding operational recommendations of the argument over necessity, leaders and individuals should weigh the costs of using the cogniceutical against the risks of not employing the agent. The use of the cogniceutical may be ethically supported when not employing the medication places the individual or those around him at risk of loss of life or deterioration of health.

Cross-Cultural Considerations on a Multinational Battlefield

We assume that actions taken by the military member during a conflict are in accordance with their own national laws, with the concepts of *jus in bello* (justice in war), and with the Geneva Conventions. This assumption may be difficult to apply when national laws differ among countries with regard to the use of psycho-active pharmaceuticals, and modern warfare often occurs within a coalition and many nations join to battle under a single flag.

On contemporary coalition battlefields, arguments can occur easily among commanders of multinational forces over the acceptability of the various national policies. Coalitions can have sociocultural conflicts that result in degradation of both capability and cohesion. For example, agreement to forbid the use of ethanol across all coalition forces because of its universally accepted deleterious effects on cognition may be more easily achieved than forbidding the use of an agent unique to a single nationality. In one example, Fiji Islanders were forbidden by a coalition's commanding general from using any alcohol, but they were free to make and drink kava, containing psychoactive kavapyrones, which they did every single night. Does the primacy of the individual rights of a nation within a coalition parallel the primacy of the individual rights of the members within their militaries? Allied nations may decide through internal rules of engagement to relinquish some national rights in specifically defined circumstances when they agree to participate in coalition warfare. Where allied service members of many countries increasingly fight shoulder-to-shoulder, broad recommendations for coalition doctrine should be developed regarding ethical uses of cogniceuticals.

Canada, the Netherlands, Denmark, and Germany all consider the military use of amphetamines unacceptable. However, all consider caffeine an acceptable fatigue countermeasure, and some give qualified endorsement to caffeine's use as a military cogniceutical (Jaeger, 2007; Meijer, 2007; Nielsen, 2007; Roehers & Roth, 2008). It is important to note that in Denmark, Germany, and the Netherlands it is *illegal* to prescribe a controlled medication for uses other than those for which it is oficially approved (i.e., there is no provision for off-label use of amphetamine or modafinil as a military fatigue countermeasure). In the United States, the Food and Drug Administration (FDA) recognizes that off-label use of select compounds is "often appropriate and may represent the standard of practice" (Woodcock, 1997). Emphasizing the argument for the use of caution for all cogniceuticals is the fact that even caffeine is not necessarily benign, as caffeine has in some individuals unwanted adverse effects, the development of tolerance, and physiological withdrawal after chronic use (2001; Juliano & Griffiths, 2004).

SUGGESTED GUIDELINES

Does the US government have official recommendations on the ethical use of cogniceuticals as applied to military situations? One source of guidance (which stopped well short of regulation) was a report on enhancement prepared by President George W. Bush's President's Council on Bioethics. In "Beyond Therapy:

Biotechnology and the Pursuit of Happiness" (2003), the council provides insight into the military use of performance enhancing pharmaceuticals when they note that "there may indeed be times when we must override certain limits or prohibitions that make sense in other contexts—offering steroids to improve the strength of service members while rejecting them for athletes, offering amphetamines to improve the alertness of fighter-pilots while rejecting them for students, offering antianxiety agents to steady the hands of surgeons while rejecting them for musicians. When we override our own boundaries, we do so or should do so for the sake of the whole, and only when the whole itself is at stake, when everything human and humanly dignified might be lost. And we should do so only uneasily, overriding boundaries rather than abandoning them, and respecting certain ultimate limits to ensure that men remain human even in moments of great crisis. For example, even if they existed, and even in times of great peril, we might resist drugs that eliminate completely the fear or inhibition of our service members, turning them into 'killing machines' (or 'dying machines'), without trembling or remorse. Such biotechnical interventions might improve performance in a just cause, but only at the cost of making men no different from the weapons they employ." In sum, the council's inspirationally written report suggests possible ethical uses of cogniceuticals in the military but also points out the moral risks for individuals and society.

Consideration of more specific criteria may permit the ethical use of cogniceutical agents in today's dynamic and often chaotic operational environments. Although no criteria in themselves can be sufficient to embrace all potential operational conditions, if they are applied within the framework of individual dignity then they may provide pragmatic guidance. Six questions may be considered:

(1) Have all available nonpharmacologic alternatives been fully utilized?
(2) Is the circumstance such that not using a cogniceutical risks life or health?
(3) Is the medication safe for use in this individual, and safe within the context of the operational environment?
(4) Is the use of the medication consistent with its dosage and pharmacological function—that is, is an alertness promoting medication being used at appropriate doses and for a specified period during a period when wake is desirable?
(5) Is use truly voluntary and informed—is an individual service member freely accepting the medication with full understanding of its effects and side effects?
(6) Is appropriate medical supervision and oversight provided?

If the answers are yes to all of these criteria, then the authors suggest that use of a cogniceutical is ethically supportable.

An example to illustrate the context in which a cogniceutical agent may be used ethically is found in a common scenario often seen during military deployments. Transporting service members from one continent to another frequently involves an extended airline flight under cramped conditions. Service members often fly across multiple time zones, and they are frequently asked to perform difficult and extended duty upon arriving into the operations theater. Many dread the flight itself, and many are anxious about what awaits them upon arrival. An intercontinental troop transport flight most likely would have a representative of the medical community on board—possibly a physician, physician's assistant, or an enlisted medic. If a service member asked specifically for a sleeping pill, and a medical representative dispensed a short-acting hypnotic, would this be ethically acceptable? Although the use of a hypnotic in this situation possibly may be justified under a sleep disorder category, for example environmental sleep disorder (Caldwell & Caldwell, 2005), the salient ethical question within the context of this discussion regards the legitimacy of providing the cogniceutical to treat a healthy individual performing military duties under adverse, stressful conditions.

Applying the above-discussed criteria, we may ask if all nonpharmacologic alternatives have been utilized. Nonpharmacological alternatives for those susceptible to the anxiety of cramped and uncomfortable troop transport conditions could be self-hypnosis or deep meditation techniques; however, these techniques would be limited to those individuals expert in such behavioral control mechanisms. Clearly the service member is not going to get a bed or a cot—the crowded seat is all that is available. With regard to health or loss of life, acute sleep deprivation could place the life of the troop or the lives of others at risk at a later time, and chronic sleep deprivation is in itself unhealthy. Especially in sustained operations, adequate sleep should be obtained whenever circumstances permit. The request appears truly voluntary and the conditions for which the service member is requesting the cogniceutical are those in which any individual, military or not, might request a sleep-inducing medication—for example, a crowded, long-distance flight. Taking the medication would not endanger the individual or others. The effective period of the medication would be within the duration of the flight and the medication would not cause drowsiness in the postflight period, and the dose and use of the hypnotic would be consistent with FDA guidelines. Finally, the medication would be dispensed by a knowledgeable medical authority.

If one accepts troop transport as an example of a circumstance in which a healthy individual—or multiple individuals—could take a hypnotic ethically, then one may conclude that there are circumstances under which a cogniceutical may be used ethically in an operational military setting. This example highlights an artificial distinction between the voluntary use of a cogniceutical by an individual versus the voluntary use by a group of individuals. As long as each service member within the group is deciding for him or herself whether to utilize a cogniceutical, and a group is not en masse ordered or coerced to take a cogniceutical, then the ethical consideration of individual choice may be supported.

A common hypothetical military situation may present a circumstance for consid-ering the ethical use of one or the other of these alerting compounds. A small number of service members become cut off from their unit and find themselves temporarily surrounded by enemy. The operational situation requires them to remain alert and situationally aware over night until reinforcements arrive the following day.

For example, let us assume that if the surrounded service members fall asleep, they would be overrun and killed. Let us also assume that they are all equally fatigued, and at least one service member volunteers to take a pharmacological fatigue countermeasure. Applying the suggested guidelines, (1) not employing the medication is almost certainly going to result in death of the service member and his or her colleagues; (2) there is no opportunity to sleep at this point in time; (3) the service member desiring the cogniceutical is voluntarily requesting the medication—recognizing the benefit of remaining awake relative to the cost of falling asleep; (4) the effects of either compound would be safe over this one-night period for both the service member and for his fellow service members; (5) the use of either caffeine or amphetamine at therapeutic doses to stay awake is consistent with their pharmacological functions; and (6) written or oral authority by medical personnel may be provided at the time, or in advance of the scenario in anticipa-tion of its possibility. What is the result of applying the suggested guidelines?

The use of either alertness-enhancing compounds in the surrounded service member example may be considered ethical as it conforms with the proposed criteria. Doctrine to support the use of cogniceuticals in the surrounded service member example could include the issuance of an emergency cogniceutical kit—possibly to the medic—that included alertness enhancing agents. Those cogni-ceuticals with smaller therapeutic windows and lower therapeutic indices would be more closely controlled, while those with large therapeutic windows and high therapeutic indices could be more widely distributed. In summary, the specific choice of cogniceutical is most likely not part of the ethical argument so long as the compound is safe at the intended doses and is used appropriately.

It is important to note an important difference between the troop transport and the surrounded service member scenarios. In the troop transport scenario, the circumstance was planned and the requests for the hypnotic cogniceutical could be anticipated. However, there is no alternative other than air transport for quickly delivering large numbers of service members to a distant location, and the envi-ronment of the troop transport is unavoidable. In the surrounded service member scenario, the use of the alertness enhancing agents is unplanned and occurs under emergency circumstances—the occurrence is unintended, although most likely not entirely unanticipated. The likely consequence of not using the cogniceuticals in the surrounded service member scenario could be mission failure and death of the friendly service members. In sum, considerations based upon a planned ver-sus an unplanned circumstance, or on an anticipated versus unanticipated condi-tion, do not appear to be necessary for the ethical application of cogniceuticals.

Furthermore, planned or likely anticipated circumstances provide more oppor-tunity for the unethical use of cogniceuticals. For example, would it be justifiable to provide the pilots of the troop transport with alertness-enhancing agents? One could argue that the troop transport pilots should not need cogniceuticals. The pilots fly the transport routes routinely, can plan sleep both prior to and following flight duty, and can even alternate responsibilities for aircraft control between the pilot and copilot so as to reduce in-flight fatigue. A flight plan whose success depended partly upon the pilot requesting a pharmacologic alerting agent would be difficult to support ethically if alternative nonpharmacologic aircrew rest scheduling alternatives exist. Thus, planned missions would not conform with the criterion that cogniceuticals be employed only after nonpharmacologic options were determined to be inapplicable.

A significant consideration regarding the surrounded service member scenario pertains to the ethics of not providing the desperate individuals with an emergency supply of alertness-enhancing cogniceuticals. As caffeine is moving from nutritionally based packaging in meals to adjunctive use in chewing gum, let us consider the ethics of providing—or not providing—caffeine. Caffeine is found in rations—such as coffee powder and high-calorie power bars. Once the surrounded service member uses all of the sustenance-based caffeine, the ethical question is whether to provide a nonsustenance form of caffeine, possibly chewing gum, possibly tablets—both of which are available off the shelf at just about every US pharmacy.

Doctrine regarding the field use of caffeine has been suggested by the US National Academies National Academic Press (2001). The Institute of Medicine Committee on Military Nutrition Research wrote, "because caffeine is commonly consumed in the military and most individuals are familiar with its effects, a clearly labeled caffeine product that permits self-dosing to obtain effective dose levels would appear to be appropriate ... Labeling would permit the few individuals who might experience adverse effects from the use of caffeine, or whose religious beliefs precluded its use, to avoid it" (Jaeger, 2007). The committee on Military Nutrition Research specifically addressed the ethics of caffeine use: "In the committee's judgment, it is unethical to coerce any individual to consume caffeine, if for religious or health reasons that individual does not wish to consume stimulants." As caffeine is the first prepackaged cogniceutical to be provided en masse to service members in the US military National Stock Number, the application of doctrine for the field use of caffeine would be a milestone in the process of development of doctrine directed toward fielding new cognitive performance enhancing technologies.

CONCLUSION

Guidelines for ethical usage of cogniceuticals should be developed and employed as a component of their operational application. Among enhancement technologies, use of psychoactive pharmaceuticals to alter cognitive performance in healthy individuals is among the most controversial, and development of guidelines for the ethical use of cogniceuticals is as essential as it is challenging.

AUTHOR NOTE

The views expressed in this chapter are those of the author and do not reflect the official policy or position of the Department of the Army, Department of Defense, or the US government.

ACKNOWLEDGMENTS

The author would like to thank Maria Thomas of Johns Hopkins School of Medicine for her considerable guidance, and John Caldwell for his expertise on the use of pharmacologic fatigue countermeasures in the US military. The author would also like to thank James McGhee; Elmar Schmeisser, director of Neurosciences Research, Army Research Organization, Research Triangle, North Carolina; Louis Banderet, US Army Research Institute of Environmental Medicine; and Jeffrey Wilson, professor of ethics and English, US Military Academy, West Point, New York for providing constructive philosophical and conceptual guidance.

REFERENCES

Bailey, R. (2003). The battle for your brain: science is developing ways to boost intelligence, expand memory, and more. But will you be allowed to.... *Reason, 34*, 9, 25 (7).

Caffeine for the sustainment of mental task performance—Formulations for Military Operations. Institute of Medicine Committee on Military Nutrition Research. National Academies Press (2001)

Caldwell, J. A., & Caldwell, J. L. (2005). Fatigue in military aviation: An overview of military-approved pharmacological countermeasures. *Aviation, Space, and Environmental Medicine. 76* (7 Suppl.), C39–C51

Chatterjee, A. (2004). Cosmetic neurology: The controversy over enhancing movement, mentation, and mood. *Neurology, 29* 63/6, 968–974.

Conrad, P., & Potter, D. (2004). Human growth hormone and the temptations of biomedical enhancement. *Sociology of Health and Illness, 26* (2), 184–215.

Daniels, N. (1994). The genome project, individual differences, and just health care. In T. F. Murphy & M. A. Lappe (Eds.), *Justice and the human genome project* (pp. 110–132). Berkeley: University California Press.

Department of Defense (2000). Directive 6200.2. August 1. Use of investigational new drugs for force health protection. Washington, DC: Department of Defense.

Department of Defense (2000). Directive 3216.2 March 25. Protection of human subjects and adherence to ethical standards in DoD-supported research. Washington, DC: Department of Defense.

Freedman, C. (2000). Aspirin for the mind? Some ethical worries about psychopharmachology. In E. Parens (Ed.), *Enhancing human traits: Ethical and social implications* (p. 135). Washington, DC: Georgetown University Press.

Friscolanti, M. (2005). *Friendly fire: The untold story of the U.S. bombing that killed four Canadian soldiers in Afghanistan.* Mississauga, Ontario: John Wiley and Sons Canada, Ltd.

Guarino, A. M. (1982). Basic principles of toxicology. In C. R. Craig & R. E. Stitzel (Eds.), *Modern Pharmacology* (pp. 81–95). Boston: Little, Brown.

International Classification of Sleep Disorders, Second Edition, American Academy of Sleep Medicine Press, 2005

Jaeger, H. (2007). A glance at the tip of a big iceberg: Commentary on "recommendations for the ethical use of pharmacologic fatigue countermeasures in the U.S. military." *Aviation, Space, and Environmental Medicine, 78* (5, Suppl.), B128–B130.

Juliano, L. M., & Griffiths, R. R. (2004). A critical review of caffeine withdrawal: Empirical validation of symptoms and signs, incidence, severity, and associated features. *Psychopharmacology (Berlin), 176* (1), 1–29.

Lieberman, H. (1999). *Effect of caffeine on cognitive function and alertness.* Presentation at the Institute of Medicine Workshop on Caffeine Formulations for Sustainment of Mental Task Performance During Military Operations, February 2–3. Committee on Military Nutrition Research. Washington, DC.

MacRae, S. M. (2000). Supernormal vision, hypervision, and customized corneal ablation. *J Cataract Refract Surg, 26* (2), 154–157.

MacRae, S. M., Krueger, R. R., et al., eds. (2001). *Customized corneal ablation: The quest for supervision.* Thorofare, NJ: SLACK Incorporated.

Meijer, M. (2007). A human performance perspective on the ethical use of cogniceuticals: Commentary on "recommendations for the ethical use of pharmacologic fatigue countermeasures in the U.S. military." *Aviation, Space, and Environmental Medicine, 78* (5, Suppl.), B131–B133.

National Stock Number 8925–01–530–1219 "Stay Alert" caffeine supplement chewing gum made by Wrigley's www.stayalertgum.com

Nielsen, J. (2007). Danish perspective: Commentary on "recommendations for the ethical use of pharmacologic fatigue countermeasures in the U.S. military." *Aviation, Space, and Environmental Medicine, 78* (5, Suppl.), B134–B135.

Parens, E. (1998). The enhancement project—A special supplement to the hastings center report. Jan-Feb S1–S17.

President's Council on Bioethics. (2003). Beyond Therapy: Biotechnology and the Pursuit of Happiness. Washington, DC. http://bioethics.georgetown.edu/pcbe/reports/beyondtherapy/

Roedig, E. (2007). German perspective: Commentary on "recommendations for the ethical use of pharmacologic fatigue countermeasures in the U.S. military." *Aviation, Space, and Environmental Medicine, 78* (5, Suppl.), B136–B137.

Roehers, T., & Roth, T. (2008). Caffeine : Sleep and daytime sleepiness. *Sleep Medicine Reviews, 12*, 153–162.

Russo, M., Maher, C., & Campbell, W. (2005). Cosmetic neurology: The controversy over enhancing movement, mentation, and mood. Letter to the editor—The healthy human. *Neurology, 64*, April 12 (1 of 2), 1320–1321.

Russo, M. (2007). Recommendations for the ethical use of pharmacologic fatigue countermeasures in the U.S. military. *Aviation, Space, and Environmental Medicine, 78*(5, Suppl.), B119–B127.

Russo, M. B., Caldwell, J. A., & Thomas, M. L. (2007). Ethical considerations in the use of cogniceuticals as augmented cognition mitigation strategies: Alertness sustaining fatigue countermeasures. In D. D. Schmorrow, D. M. Nicholson, J. M. Drexler, & L.M. Reeves (Eds.), *Foundations of augmented cognition, 4th ed.* (pp. 112–116). Arlington, VA: Strategic Analysis, Inc.

Russo, M. B., Thomas, M. L., Caldwell, J. A., & Arnett, M. V. The ethical use of cogniceu-ticals in the militaries of democratic nations. *Am Journal Bioethics* (In press)

Sententia, W. (2004). Cognitive liberty and converging technologies for improving human cognition. *Ann. N.Y. Acad. Sci, 1013*: 221–228.

Shay, J. (1995). *Achilles in Vietnam*. New York: Simon and Shuster, pp. 208–209.

Swofford, A. (2003). Jarhead—*A marine's chronicle of the gulf war and other battles*. New York: Simon and Schuster.

Wilson J. (2004). An argument against biotechnical enhancements of service members in the armies of liberal democracies. *Ethical Perspectives, 11* (2–3), 189–197.

Woodcock, J. (1997). A shift in the regulatory approach. Available from www.fda.gov/Cder/present/diamontreal/regappr/sld001.htm (slide 5 of 17).

4

Marketing of Neuropsychiatric Illness and Enhancement

PETER CONRAD AND ALLAN HORWITZ

Over the past three decades a vast medicalization of behaviors and conditions has occurred, expanding the domains of what is referred to as neuropsychiatric illness. By "medicalization," we mean defining and/or treating a human condition as a medical (and in this case psychiatric) problem. Conditions such as attention deficit hyperactivity disorder (ADHD), conduct disorder, anorexia, posttraumatic stress syndrome (PTSD), panic disorder, premenstrual syndrome (PMS), social anxiety disorder, and mild depression, among many others, have become seen as medical or psychiatric disorders (Conrad, 2007; Horwitz, 2002). One manifestation of this has been the expansion of diagnoses in the *Diagnostic and Statistical Manual of Mental Disorders* (DSM), which grew from the 106 diagnostic categories of the DSM-I in 1952 to the 297 diagnoses in the DSM-IV in 1994 (Mayes & Horwitz, 2005). Medicalization studies have recently noted that factors including biotechnology (which includes the pharmaceutical industry), managed care, and consumer self-assessments all serve as "engines of medicalization" (Conrad, 2005). While medicalization can be bidirectional, there is strong evidence for an overwhelming trend toward medicalization rather than demedicalization (despite important contrary cases such as masturbation and homosexuality); thus, most critical attention is given to the overmedicalization of human problems.

In recent years medicalization researchers have begun to examine biomedical enhancement as well as the expansion of disease categories. It is likely that humans have sought enhancements for themselves or their children as long as they have recognized that improvements in individuals are possible. These might include strategies, techniques, or potions to make humans stronger, smarter, faster, more attractive, or longer-lived (Conrad & Potter, 2004). A marked increase in biomedical enhancements, including surgery, drugs, and other medical interventions, has emerged in recent years. Examples include human growth hormone for idiopathic shortness, steroids for athletic performance, cosmetic surgery for bodily improvements, and various methods of changing the effects of aging. Enhancement interventions are not meant to treat or cure a disorder, but rather to improve someone's performance or condition. Neuropsychiatric enhancements attempt to improve the individual's abilities or mental acuity. Examples include medications that help one sleep less (e.g., Provigil), focus better (e.g., Adderall), or improve an individual's "normal" mood or identity (e.g., SSRI antidepressants; Elliot, 2003). Peter Kramer's notion of taking medications like Prozac to become "better than well," popularized in his best-selling book *Listening to Prozac* (1993) is one well-known instance.

The lines between treatment and enhancement are often unclear. Medications that are originally developed for treating an accepted disorder can also be used for enhancement. One recent example is Viagra and erectile dysfunction (ED). Viagra was initially introduced in 1998 for individuals with ED from prostate surgery or other physiological maladies; soon it was also advertised for men who had ED from aging or other sources, and next it was touted as a remedy for men who were anxious about their sexual performance. Recently it and other similar drugs have been marketed as enhancements to one's sex life (Loe, 2004; Conrad & Leiter, 2008). For example, a recent television ad for another ED drug boldly states that "Cialis is ready when you are," indicating that people who take the medication will experience improved sexual performance. To take another example, if a drug existed that could substantially and selectively improve the memory of people with Alzheimer's disease, we have little doubt that the same drug would soon become very attractive as an enhancement (or perhaps used for a newly described illness, "memory deficit disorder").

The merging of treatment and enhancement has implications for clinicians. Pharmaceutical companies promote new uses for drugs that help improve patients' lives, some of which may be forms of enhancement. In part because of direct-to-consumer advertising, an increasing number of patients will come to their doctors asking, "is [name of drug] right for me?"—which could refer to an illness or an enhancement (Kravitz et al., 2005). Moreover, physicians will face pressures—sometimes from the pharmaceutical industry, sometimes from professional sources, and sometimes from consumers themselves—to lower the thresholds when patients are deemed suitable to receive treatments for specific symptoms or conditions. These pressures can lead clinicians to become more cavalier about prescribing powerful medications for mildly distressing conditions or as enhancements. Many of these medications are of questionable or unknown effectiveness for the condition they are expected to treat or enhance, and some also come with considerably increased risk from their adverse effects.

HISTORICAL PRIMER AND RECENT TRENDS

The use of psychotropic drugs to treat everyday psychosocial problems is not new. Through the first half of the twentieth century, physicians and psychiatrists frequently used drugs including morphine and opium to treat a wide variety of common psychological ills (Shorter, 1997). They were particularly likely to prescribe barbiturates for the range of common symptoms such as nerves, tension, or insomnia. In 1955 the development of meprobamate (Miltown) dramatically changed the nature of treatment for widespread distressing conditions. Miltown became the most popular prescription drug in history; by 1965 physicians and psychiatrists had written 500 million prescriptions for it (Smith, 1985). By the late 1960s the spectacular success of Librium, which was introduced in 1960, displaced Miltown. Valium, in turn, succeeded Librium by 1970 as the newest blockbuster tranquillizer; by 1981 Valium was the single most prescribed drug of any sort. At this time, fully 20% of all women and 8% of all men reported using a minor tranquillizer (sometimes dubbed "mother's little helper") over the past year (Parry et al., 1973).

The popularity of these tranquillizers was especially due to their usefulness in treating the diffuse kinds of problems seen in general medical practice—general physicians wrote between 70 and 80% of prescriptions for Miltown, Librium, and Valium (Smith, 1985). Studies during the 1950s and 1960s found that only about a third of the minor tranquilizers were prescribed for mental, psychoneurotic, or personality disorders, while the rest were given as a response to more diffuse complaints and psychosocial problems (e.g., Shapiro & Baron, 1961; Cooperstock & Lennard, 1979). A sharp backlash against the tranquillizing drugs began during the 1970s. Stimulated by hostile congressional hearings, the Food and Drug Administration (FDA) and Bureau of Narcotics began a crusade against this category of medications. This backlash resulted in their classification in 1975 as schedule 4 drugs, which required physicians to report all prescriptions written for them and limited the number of refills a patient could obtain. As well, press coverage of these medications changed dramatically from their highly positive mid-1950s reception to very unfavorable coverage in the 1970s that emphasized their addictive qualities. After twenty years of steadily rising sales since their introduction in the mid-1950s, consumption of tranquillizers plunged. From a peak of 104.5 million prescriptions in 1973, their use sharply dropped to 71.4 million prescriptions by 1980 (Smith, 1985).

A new era in the treatment of nonpsychotic conditions began when the Selective Serotonin Reuptake Inhibitors (SSRIs) came on the market in 1988. The SSRIs act very generally to raise levels of serotonin in the brain and treat a wide variety of depressive and anxious conditions, among others; they are not specific to any single disorder. By the 1980s, however, non-SSRI antianxiety medications were linked to dependency and many negative side effects. When the SSRIs were introduced it made much more marketing sense to promote them as antidepressants (Healy, 1997). The publication of Peter Kramer's wildly popular *Listening to Prozac* cemented the coupling of the SSRIs with the treatment of depression as well as inculcating the conception that these were the newest class of wonder drugs.

SSRI use increased spectacularly over the 1990s. From 1996 to 2001 alone, the number of users of SSRIs increased, from 7.9 million to 15.4 million. From 1996 and 2001, overall spending on antidepressants rose from $3.4 billion to $7.9 billion (Zuvekas, 2005). By 2000, three SSRIs—Prozac, Zoloft, and Paxil—were among the eight most prescribed drugs of any sort; the antidepressants were the best-selling category of drugs in the United States.

The FDA approval of direct-to-consumer (DTC) drug advertisements in the late 1990s both enhanced the popularity of the SSRIs and linked them to the treatment of common psychosocial problems. Advertising directly to consumers does not simply attempt to sell particular products, but strives to reshape consumers' understanding of their problems into conditions that should be treated by medications. It presents a worldview about the nature, cause, and remedy for psychosocial problems. Ads typically link the most common symptoms of psychiatric diagnoses—sadness, anxiety, fatigue, sleeplessness, and the like—with very common psychosocial situations involving problems with interpersonal relationships, employment, and achieving valued goals (Conrad & Leiter, 2008).

The marketing of social anxiety disorder provides perhaps the best example of how DTC advertising in particular, and a broader marketing campaign in general, can create a widespread psychiatric condition. Psychiatric diagnostic manuals did not mention this condition until 1980. When it first appeared, the manual noted that "the disorder is apparently relatively rare" (American Psychiatric Association 1980, 228). Initial studies of the disorder in the early 1980s indicated that about 1 to 2% of the population reported this condition. Yet the most recent studies indicate that over 13%, or one of eight people, has had a social phobia at some point in their lives. Indeed, by the early 2000s, social phobias were one of the two most common mental disorders (Kessler et al., 2005). How did the number of people with this condition more than quintuple in such a short period of time?

In 1999 the antidepressant Paxil was approved for the specific treatment of social anxiety disorder. Its manufacturer, GlaxoSmithKline (GSK), mounted a huge advertising campaign, spending over $90 million dollars on a barrage of print and television ads with the message "imagine being allergic to other people." Other ads asked the question: "Is it shyness or social anxiety disorder?" Because about 90% of people self-report feeling shy, these ads can easily exploit the ambiguous lines between shyness and an anxiety disorder (Zimbardo, 1977). In the pharmaceutical industry this is known as "condition branding," promoting a condition just like one would a product (Angelmar, Angelmar, & Kane, 2007). These ads were just the tip of the iceberg of a gigantic public relations campaign by the firm Cohn & Wolfe aimed at fundamentally reshaping public perceptions of social anxiety from being shy and uneasy in social situations to having a mental disorder treatable with drugs (Moynihan & Cassells, 2005). Other aspects of the campaign involved placing stories in the news media, often using celebrity figures, psychiatric experts, and testimonies from members of consumer advocacy groups about the pervasiveness of and disabilities from social anxiety disorder. The immense public relations campaign was highly successful—Paxil became the largest selling antidepressant at the time with sales of $3 billion a year, and social anxiety is now a common and well-recognized mental illness. Whether using Paxil to improve interactions in public is a treatment or enhancement depends on one's view of shyness and unease in social situations.

While psychotropic drug use has grown enormously among adults, its use among young people has increased to an even greater extent. The period from 1994–2001 witnessed a 250% increase in the number of visits to physicians that resulted in a prescription for psychotropic medication among adolescents (Thomas et al., 2006). Antidepressant use in this group has grown even faster, swelling from three- to five-fold since the early 1990s with especially accelerated rates after 1999 (Zito et al., 2003). The numbers of persons twenty years old and younger who received a prescription for an antipsychotic medication jumped from about 200,000 in 1993–1995 to about 1,225,000 in 2002, a more than six-fold increase in less than a decade (Olfson et al., 2006b).

ADHD has become a particularly prevalent psychiatric condition among youth. Just twenty years ago less than 1% of children were diagnosed and treated with ADHD (Olfson et al., 2003). By 2003, however, 7.8% of youth aged four to 17 years received an ADHD diagnosis, and 4.3% took medication for this disorder (Visser, Lesesne, & Perou, 2007). Most treatment for ADHD involves the use of stimulant drugs, in particular, amphetamines. Over two million children now receive these drugs each year (Zuvekas, Vitiello, & Norquist, 2006).

Until the 1990s ADHD was generally seen as a disorder of childhood. In the 1990s we began to see the emergence of "adult ADHD": adults who had never been diagnosed as ADHD but were deemed to have symptoms similar to childhood ADHD. The new diagnosis was popularized in the media and by the book *Driven to Distraction* (Hallowell & Ratey, 1995). One of the hallmarks of childhood ADHD is that a person other than the child (parent, school, physician) usually identifies it. Adult ADHD, however, is often self-identified with the individual going to a physician with complaints like "I'm disorganized" or "I just don't think I'm doing as well as I should be" and stating that they think they may have (adult) ADHD. When asked how she or he knew, the patient might well answer "I heard about it on TV," or "my child was diagnosed and I was just like him," and "I wonder if medications would help me." There is very little research on adult ADHD, but estimates indicate that there are millions of people now so diagnosed and treated with stimulant medications. It is very difficult to draw a line where a drug is given as a treatment or as an enhancement. The medications may improve the individual's focus and concentration, but that does not mean it is treating an illness. Adult ADHD can often be a "medicalization of underperformance" (Conrad & Potter, 2000).

One emerging trend that may promote medicalization is the screening of healthy individuals to uncover unrecognized depression, anxiety, and other common problems, especially among youth (Horwitz & Wakefield, 2009). Mental health professionals have launched large-scale efforts that attempt to screen, in principle, every child and adolescent in the United States for signs of mental illness. In 2003 a presidential commission recommended that "every child should be screened for mental illness once in their youth in order to identify mental illness and prevent suicide among youth" (New Freedom Commission, 2003). A number of state legislatures have adopted measures that aim to implement this goal. Pharmaceutical companies have sponsored the development of many screening measures, which promise to open previously untapped markets of potential drug users.

The screening movement has developed short scales to administer in school classrooms to identify early symptoms of mental disorder in order to evaluate children for psychiatric treatment and prevent more serious disorders in later life. But questions such as "In the past six months, were there times when you were very sad?" are common enough so that about a quarter of all students screened—which rises to nearly half in some schools—are considered to have problems that warrant a follow-up interview. The broad nature of the questions on these scales insures that many adolescents who are experiencing normal emotions will mistakenly be considered to suffer from a mental disorder. Remarkably, not a single study demonstrates that screening programs actually improve mental health outcomes or prevent suicide. No evidence currently exists that their possible benefits override such possible risks as inappropriate diagnosis, unnecessary and possibly harmful treatment, and stigma.

Monitoring the varying emotions of adolescents—and especially the sadness, moodiness, distress, despair, feelings of inadequacy, and extreme negative thoughts ("That was so embarrassing, I could just kill myself" or "If he doesn't come back to me, I will kill myself") to which normal adolescents are regularly subject—may or may not be a good idea. But it is surely an idea that deserves serious and honest discussion before it is wholeheartedly embraced, especially because it is likely to lead to an increase in the number of nondisordered adolescent who are prescribed medications. However, current screening efforts medicalize normal unhappiness and lability of adolescent emotions without distinguishing them from mental disorders. Screening instruments that specifically aim at identifying mental disorders could be valuable but do not yet exist. As currently constituted, screening programs are—and ought to be truthfully labeled as—methods of identifying teenage distress, normal and pathological, and likely mostly normal reactions to serious environmental stressors.

A second trend that has the potential to increase medicalization is the creation of a category of people who are considered to be at-risk for developing a mental illness even when they have never actually had the condition. Genetic tests are now available that can identify the genes that presumably raise the probability that someone will become mentally ill. Such people can then be placed on long-term regimes of drug therapies that might prevent them from ever developing symptoms. The market for such products is potentially huge—for example, two-thirds of the population has genetic markers that place them at risk for developing just the condition of depression (Caspi et al., 2003). If it is successfully employed, this new concept will provide a vast new market for a class of more refined, genetically tailored, antidepressant medications to replace the SSRIs whose patents are now expiring (Rose, 2007). The concept of being at-risk for some mental illness expands the concept of mental disorder to incorporate virtually the entire population.

ETHICAL ISSUES

Observers have pointed out that in recent decades Americans have become less tolerant of mild symptoms and relatively benign problems (Barsky & Boros, 1995).

The use of medication for an increasing range of psychosocial problems creates a number of problematic issues. One concerns the increasing tendency to define psychosocial problems as illnesses in need of chemical corrections. DTC advertisements, in particular, present antidepressant consumption as the solution to family, work, motivational, and behavioral problems, which they portray as forms of mental disorders. The thresholds for diagnosis and drug treatment of disorders like ADHD have decreased (Mayes et al., 2009), leading to medication regimens for an increasingly large number of children with mild behavior problems. There is some evidence that stimulant medications have also been prescribed as cognitive enhancements for adults (Conrad & Potter, 2000) and college students (Forlini & Racine, 2009) who have no existing ADHD diagnosis.

Attaching labels such as "mild depressive disorder," "social anxiety disorder," or "attention deficit hyperactivity disorder" to such widespread problematic conditions prejudges the sorts of treatment that is necessary. These labels, in combination with limitations on what health services or insurance plans will pay for, may lead to the neglect of responses such as cognitive or psychotherapies or nonprofessional remedies including changes in diet, exercise, or living situations. The often-reflexive use of drugs to treat minor symptoms and nondisordered conditions not only can come at the expense of the use of safer, cheaper, and more effective alternatives, but also ignores the risks that drug treatments often pose. In contrast to the general acceptance of the benefits of early treatment for conditions such as hypertension or hypercholesterolemia, the current evidence shows little advantage of antidepressant medications compared to placebo for conditions that are not severe (Kirsch et al., 2008).

Prescribing powerful psychoactive medications to children and adolescents presents another ethical issue, namely safety (see Singh & Kelleher, chapter 2 in this volume). One example of this is the controversy over whether SSRIs may actually increase the suicide risk of children and adolescents. There have been continuing reports of increased suicidality and suicides among adolescents prescribed SSRIs (Olfson et al., 2006a). This has led the FDA to place a black box warning (its highest warning short of banning a medication) on the prescription of SSRIs for adolescents. In the United Kingdom, the government health authority has banned prescribing all but one SSRI drug for patients under the age of eighteen. To cite another example, a recent study reported that children prescribed Ritalin had a four- to five-fold greater chance of a sudden unexplained cardiac arrest (Gould et al., 2009). While the numbers are small, the comparative risk is significant. When a medication is given for severe or intractable cases of ADHD, then the risk might be justified, but when it is given for mild difficulties or as an enhancement the acceptable risk of medication usage requires a higher bar and raises new ethical issues about appropriate thresholds for treatment. In addition, given the potential adverse effects and risks of psychoactive drugs, the risk-benefit ratio for enhancement medications should be held to a higher standard than when the drugs are used to treat a well-established psychiatric disorder.

When medications are prescribed as enhancements there may be additional ethical concerns. For example, if enhancements induce permanent changes, as with human growth hormone for children with idiopathic short stature, there are questions about whether such medications should be given to children who cannot legally consent to their use. Furthermore, enhancement medications can raise issues of fairness; prescribing medications to increase student or work performance raises concerns of what such enhancements will do to notions of fair play and comparison. This has already become a major issue in sports in terms of performance enhancing drugs, but with more psychoactive medications prescribed for enhancements it will become an issue with mental activities as well.

POSSIBLE SOLUTIONS

The potential expansion of concepts of mental disorder to more and more kinds of behaviors and populations is not a new phenomenon, although the increase in the past three decades is unprecedented. During the 1960s and early 1970s government regulatory agencies actively opposed the use of pharmaceuticals for an ever-widening range of conditions. The National Institute of Mental Health took the position that drug treatments should be reserved for responding to genuine disorders and not taken for psychosocial problems. For example, at a Senate hearing in 1967, the director of the National Institute of Mental Health (NIMH), Stanley Yolles, expressed his concerns: "To what extent would Western culture be altered by widespread use of tranquilizers? Would Yankee initiative disappear? Is the chemical deadening of anxiety harmful? . . . I feel that myself and my colleagues—or so it seems to be evident from the sales statistics—are so easily seduced by the clever advertisement" (Smith 1985, 179 9).

During this period the FDA also took a strong adversarial role toward the pharmaceutical industry, carefully scrutinized drug advertisements, and sanctioned companies that attempted to expand the boundaries of pathology beyond well-recognized disease entities. In 1971 the commissioner of the FDA, Charles Edwards, explicitly distinguished the frustrations of daily living from mental disorders, noting in testimony before a congressional committee that the drugs were only intended for patients suffering from mental conditions that were psychiatric disorders and not for the "ordinary frustrations of daily living" (Smith 1985, 187). Moreover, the psychiatric profession did not promote widespread drug use for nonsevere conditions in this era (Herzberg, 2009).

In recent decades, however, no regulatory agency has opposed the widespread use of pharmaceutical treatments for common psychosocial problems. For example, the NIMH and prominent psychiatrists have joined to create educational campaigns to stimulate treatment and promote the use of screening programs to expand the net of coverage for people who might use these drugs and other treatments (Olfson et al., 2000). In the 1980s during the Reagan and G. H.W. Bush presidencies, the FDA dropped its adversarial stance toward the pharmaceutical companies. The agency provided weak oversight over the industry's research, did not seek to uncover suppressed data, or to question the optimistic spin put on findings. Concern with the safety of new products was minimal (Shorter, 2009).

In contrast to the 1960s and 1970s, no federal agency currently serves to counter the expanding boundaries of conditions in need of drug treatments.

The weak governmental oversight of drug companies might be changing. In 2006 the FDA uncommonly stood up to the pharmaceutical industry and adopted a black-box warning label, the most serious type of warning in prescription drug labeling, indicating that antidepressants may increase the risk of suicidal thinking and behavior in some children and adolescents. In addition, stimulated by media reports of conflicts of interest between psychiatric researchers and the pharmaceutical industry, the NIMH has begun to pay more attention to the potential of these relationships to compromise the integrity of scientific research. Stronger federal control of the drug industry and its advertising, which would reveal the results of unpublished as well as published drug trials and provide more honest reporting of adverse reactions to drug treatments, will help ensure the appropriate use of medication for the treatment of mental disorders.

In addition to a stronger federal regulatory role, the psychiatric profession can potentially help insure that the conditions it treats are mental disorders and not problems in living. The American Psychiatric Association's current process of revising the DSM should strive to distinguish genuine dysfunctions from distressing and disturbing, but normal, emotions and behaviors. It is likely, however, that the DSM-V is instead likely to bring about a further expansion of the range of minor or subthreshold conditions that it defines as mental disorders (Carey, 2008). Managed care organizations also have a role in limiting reimbursements for drug treatments to those that meet the standards of evidence based medicine. Probably the single most effective way to slow the growth of medicating conditions such as pediatric bipolar disorder or minor depression, where there is no evidence that drug treatments are successful, would be for managed care to stop paying for them. These organizations should also pay more attention to compensating treatments that do not involve drugs and that in some cases can be more effective than drug treatments.

CONCLUSION

The recent marketing of neuropsychiatric illnesses to encompass a broader range of conditions and enhancements has been almost completely associated with a growing use of drug treatments. Indeed, the use of cognitive and psychotherapeutic alternatives to drug treatments has declined in recent years (Majtabai & Olfson, 2008). Stronger countervailing powers to the pharmaceutical industry's influence over clinical treatment, research, media reports, and advocacy groups could help limit the unwarranted use and spread of drugs. Greater attention to psychosocial treatments and lifestyle changes as alternatives to medication could also have beneficial impacts. Medication can often be an effective response to mental disorders and even to troubling conditions that are not disorders. It is important, however, that consumers receive accurate information about the risks as well as benefits of these treatments and be fully informed about what treatments can lead to the best outcomes for their problematic conditions.

REFERENCES

American Psychiatric Association. (1980). *Diagnostic and Statistical Manual of Mental Disorders* (3rd ed). Washington, DC: American Psychiatric Association.

Barsky, A. J., & Boros, J. F. 1995. Somatization and medicalization in the era of managed care. *Journal of the American Medical Asssociation, 274*, 131–134.

Carey, B. (2008). Psychiatrists revise the book of human troubles." *New York Times.* December 17. A1.

Caspi, A., Sugden, K., Moffitt, T. E., et al. (2003). Influence of life stress on depression: Moderation by a polymorphism in the 5-HTT gene. *Science, 301*, 386–389.

Conrad, P. 2005. The shifting engines of medicalization. *Journal of Health and Social Behavior, 46*, 3–12.

Conrad, P. (2007). *The medicalization of society: On the transformation of human conditions into medical disorders.* Baltimore: Johns Hopkins University Press.

Conrad, P., & Potter, D. (2000). From hyperactive children to ADHD adults: Some observations on the expansion of medical categories. *Social Problems, 247*, 559–582.

Conrad, P., & Leiter, V. (2008). From Lydia Pinkham to Queen Levitra: DTCA and medicalization. *Sociology of Health and Illness, 30*, 825–838.

Cooperstock, R., & Lennard, H. (1979). *Some social meanings of tranquillizer use. Sociology of Health and Illness,* 1, 331–347.

Elliott, C. (2003). *Better than well: American medicine meets the American dream.* New York: Norton, 2003.

Forlini, C., & Racine, E. (2009). Autonomy and coercion in academic 'cognitive enhancement' using methylphenidate: Perspectives of key stakeholder. *Neuroethics.* www.springerlink.com/content/m2pq750581113nj1/fulltext.pdf.

Gould, M. S., Walsh, B. T., Munfakh, J. L., Kleinman, M., Buan, N., Olfson, M., et al. (2009). Sudden death and use of stimulant medications in youths. *American Journal of Psychiatry, 166*, 992–1001.

Hallowell, E. M., & Ratey, J. J. (1995). *Driven to distraction.* New York: Pantheon Books.

Healy, D. (1997). *The antidepressant era.* Cambridge, MA: Harvard University Press.

Horwitz, A. V. (2002). *Creating mental illness.* Chicago: University of Chicago Press.

Horwitz, A. V., & Wakefield, J. C. (2007). *The loss of sadness.* New York: Oxford University Press.

Horwitz, A. V., & Wakefield, J. C. (2009). Should screening for depression among children and adolescents be demedicalized? *Journal of the American Academy of Child and Adolescent Psychiatry, 48*, 683–687.

Kessler, R. C., Chiu, W. T., Demler, O., & Walters, E. E. (2005). Prevalence, severity, and comorbidity of 12-month DSM-IV disorders in the national comorbidity survey replication. *Archives of General Psychiatry, 62*, 617–627.

Kirsch, I., Deacon, B. J., Huedo-Medina, T. B., Scoboria, A., Moore, T. J., et al. (2008). Initial severity and antidepressant benefits: A meta-analysis of data submitted to the food and drug administration. *PLoS Med, 5*(2), e45.

Kramer, P. D. (1993). *Listening to prozac: A psychiatrist explores antidepressant drugs and the remaking of the self.* New York: Viking.

Kravitz, R. L., Epstein, R. M., Feldman, M. D., Franz, C. E., Azari, A., Wilkes, M. S., et al. (2005). Influence of patients' requests for direct-to-consumer advertised antidepressants: A randomized controlled trial, *JAMA, 293* (16), 1995–2002.

Loe, M. (2004). *The rise of viagra: How the little blue pill changed sex in America.* New York: New York University Press.

Majtabai, R., & Olfson, M. (2008). National trends in psychotherapy by office-based psychiatrists. *Archives of General Psychiatry, 65,* 962–970.

Mayes, R., Bagwell, C., & Erkulwater, J. (2009). *Medicating children: ADHD and pediatric mental health.* Cambridge, MA: Harvard University Press.

Mayes, R., & Horwitz, A. V. (2005). DSM-III and the revolution in the classification of mental illness. *Journal of the History of Behavioral Sciences, 41,* 249–267.

Moynihan, R., & Cassells, A. (2005). *Selling sickness: How the world's biggest pharmaceutical companies are turning us all into patients.* New York: Nation Books.

New Freedom Commission on Mental Health. (2003). *Achieving the promise: Transforming mental health care in America.* Rockville, MD: US Department of Health and Human Services.

Olfson, M., Blaco, C., Linxu, L., Moreno, C., & Laje, G. (2006b). National trends in the outpatient treatment of children and adolescents with antipsychotic drugs. *Archives of General Psychiatry, 63,* 679–685.

Olfson, M,, Gameroff, M. J., Marcus, S. C., & Jensen, P. S. (2003). National trends in the treatment of attention deficit hyperactivity disorder. *American Journal of Psychiatry, 160,* 1071–1077.

Olfson, M., Guardino, M., Struening, E., Schneier, F. R., Hellman, F., & Klein, D. F. (2000). Barriers to the treatment of social anxiety. *American Journal of Psychiatry, 157,* 521–527.

Olfson, M., Marcus, S. C., & Shaffer, D. (2006a). Antidepressant drug therapy and suicide in severely depressed children and adults. *Archives of General Psychiatry, 63,* 865–872.

Parry, H., Balter, M., Mellinger, G., Cisin, I., & Manheimer, D. (1973). National patterns of psychotherapeutic drug use. *Archives of General Psychiatry, 28,* 769–783.

Rose, N. (2007). *The politics of life itself.* Princeton, NJ: Princeton University Press.

Shapiro, S., & Baron, S. (1961). Prescriptions for psychotropic drugs in a noninstitutional population. *Public Health Reports, 76,* 481–488.

Shorter, E. (1997). *A history of psychiatry: From the era of the asylum to the age of prozac.* New York: Wiley.

Shorter, E. (2008). *Before prozac: The troubled history of mood disorders in psychiatry.* New York: Oxford University Press.

Smith, M. C. (1985). *A social history of the minor tranquillizers.* New York: Pharmaceutical Products Press.

Thomas, C., Conrad, P., Casler, R., & Goodman. E. (2006). Trends in the use of psychotropic medications among adolescents, 1994–2001. *Psychiatric Services, 57,* 63–69.

Visser, S. N., Lesesne, C., & Perou, R. (2007). National estimates and factors associated with medication treatment for childhood attention-deficit/hyperactivity disorder. *Pediatrics, 119* (Supp. 1), S99–S106.

Zimbardo, P. (1977). *Shyness: What it is: What to do about it.* Reading MA: Addison Wesley.

Zito, J. M., Safer, D. J., dosReis, S., et al. (2003). Psychotropic practice patterns for youth. *Archives of Pediatric and Adolescent Medicine, 157,* 17–25

Zuvekas, S. H. (2005). Prescription drugs and the changing patterns of treatment for mental disorders, 1996–2001. *Health Affairs, 24,* 195–205.

Zuvekas, S. H., Vitiello, B., & Norquist, G. S. (2006). Recent trends in stimulant medication use among U.S. children. *American Journal of Psychiatry, 163,* 579–585.

5

Brain Training

BREEHAN CHANCELLOR AND ANJAN CHATTERJEE

An elderly woman on a limited fixed income seeks help because she has problems remembering little things. She is concerned that she might be developing Alzheimer's disease. Wishing to be proactive in keeping her mind as sharp as possible for as long as possible, she inquires about brain training programs. She would need to buy a computer and learn how to use it. But she thinks that classes in the library would help her with that. Her questions are: Which brain game should she buy, how much should she pay for them, and will they prevent her from getting Alzheimer's disease? Clinicians encounter variations of this scenario frequently. What advice should they give such patients?

The commerce of brain training games is one example of a large trend in the interface of neuroscience and the market. Commercial products increasingly align themselves with basic and clinical neuroscience research (Lynch, 2009). Recognizing this trend, Neurotechnology Industry Organization, a recently created trade group, "is working on programs that could translate into millions of dollars to your company's bottom-line and billions of dollars for commercial neuroscience" (NIO, 2010). We should rejoice in the fact that neuroscience is coming of age and delivering on the promise of basic science to help diagnose and treat disease. However, we might also be wary of promises that are likely to fall short.

In the context of clinical neuroscience, commercial values can sometimes undermine scientific and social values. The concerns are by no means unique to clinical neuroscience. In some form, this concern exists in all divisions of medicine. However, the tremendous growth of neuroscience knowledge, the hold that neuroscience has on the public imagination, and the vulnerability of the population in need of neuroscience treatments give "brain brands" a special halo in this evolving market.

We start with the principle that therapeutic interventions should conform to professional standards of care, and ideally they should be grounded in clinical and scientific evidence. Products sold in violation of this principle are suspect. As we have reviewed elsewhere, such violations arouse three kinds of concern

(Chancellor & Chatterjee, 2011). One concern is a general unease about the fact that blurring boundaries between academia and industry can insidiously compromise scientific and clinical standards. The evolving relationship between research universities and pharmaceutical corporations is a prime example of this category. The second concern is that marketing can get ahead of the science; that is, neuroscience is used prematurely as a marketing tool to sell products. The third concern is that products can be sold by misusing neuroscience. Some brain imaging practices in psychiatry cross over into this third unsavory category (see Imaging chapter). Brain training training falls in the middle. There are good scientific reasons to think that brain training might be helpful. But the commerce of the games has outpaced data from which a clinician can assuredly offer advice to patients.

POTENTIAL PROMISE OF BRAIN TRAINING

Pharmaceutical companies are familiar targets when we bemoan how profit motives undermine clinical scientific ones. However, the development and deployment of pharmaceuticals is highly regulated. By contrast, there is relatively little oversight of other health care products introduced into the marketplace. These can be introduced before evidence of efficacy has been established. Brain training programs target clinical populations, but are not subject to regulations and standards associated with traditional medical interventions.

Currently, about 30 million people have dementia worldwide. Given that older people are deeply concerned about cognitive decline (Connell, Scott Roberts, & McLaughlin, 2007) and treatments for dementia have limited efficacy, anything that could stave off cognitive decline is desirable. Any such intervention would be perfectly positioned to make a substantial profit. This is where brain training programs come in.

The rationale for brain training programs rests on observations that active mental engagement is associated with decreased cognitive decline (Verghese et al., 2003) and that adult brains are more plastic than previously appreciated (Lie, Song, Colamarino, Ming, & Gage, 2004). For example, the hippocampus has neurogenerative capabilities well into adulthood suggesting that memory systems are subject to ongoing modulation (Kitabatake, Sailor, Ming, & Song, 2007). Thus, it is plausible that behavioral interventions that promote such modulation would benefit people with cognitive decline.

Despite the plausibility of the hypothesis that brain training interventions might halt or reverse cognitive decline, direct evidence in support of the efficacy of specific commercial programs is sparse. One of the best studies supportive of this hypothesis is a very large trial on cognitive training, with 2,832 participants living independently in the United States. This study used instructors engaging participants in memory, speed of processing, and reasoning training with live instructors (Willis et al., 2006). Memory training involved teaching participants organization, visualization, and association strategies. Speed of processing involved teaching participants visual search and divided attention tasks at increasingly brief exposure times. Reasoning training involved teaching them strategies for finding patterns in letter or word series. Memory training and speed of processing training provided some benefits within the domains trained. However, the benefits did not consistently generalize to other domains. Reasoning training did generalize to improve instrumental activities of daily living. Notably, some of the benefits in instrumental activities of daily living were sustained five years after the training. Identifying characteristics (such as higher education and age) that predict response to training is critical and active research on this question is underway (Langbaum, Rebok, Bandeen-Roche, & Carlson, 2009). One hopes that soon we will know which interventions work for which subpopulations and the extent to which these interventions are efficacious. In the meantime, however, the games are sold and marketed widely.

POTENTIAL PROBLEMS OF THE MARKETING OF BRAIN TRAINING

Most research reports on brain training programs are sponsored by the companies themselves and are conducted by authors with a financial stake in the outcome of the studies reported. For example, one of the first studies using Posit Science software reported modest improvements with trained auditory memory among older nondemented participants. The authors reported that the effects generalized to nontrained neuropsychological tests and was sustained for three months. All ten authors held stock or options in the company and most of them worked for the company (Mahncke et al., 2006). The paper was published in PNAS as a communication by the senior author, who is a founder of the company. Such papers are not subject to the rigors of the traditional peer review process. As of January 1, 2009, PNAS no longer consider submissions through this route if the academy member or coauthors disclose a significant financial conflict of interest.

A subsequent larger study (n=487) also reported modest improvement in auditory memory (Smith et al., 2009). To its credit, this study included an active control group. The two principle coinvestigators were not employed by and did not hold equity in the company. However, Posit Science funded the study and three authors received consulting fees and a fourth held stock options and was employed by Posit Science, again raising the specter of bias in the study. The program currently sells commercially for $395 for a one-user version and $495 for a two-user version. Of course it does not necessarily follow that people with financial conflicts of interest will produce biased reports. However, in assessing efficacy it would be helpful to have more studies conducted with specific brain training programs by investigators without financial conflicts of interest in the commercial success of the products.

A recent meta-analysis of the existing data on brain training programs points out typical shortcomings in brain games research (Papp, Walsh, & Snyder, 2009). These shortcomings are a lack of consensus on what constitutes adequate improvement, limited follow-up times, absent active matched control conditions, and outcome measures that do not show changes in peoples' daily activities. In the largest study to date, over 11 thousand participants enrolled in a six-week online brain-training program. The game was designed to improve reasoning, memory, planning, visuospatial skills, and attention. The authors found that participants improved on each of these cognitive areas, but there was no generalization to untrained tasks, even those that were closely related to the cognitive domains being trained (Owen et al., 2010). It is difficult to know why the effects of this study differed from the Willis study (2006). One possibility is that the unsupervised online nature of the Owen (2010) study, a useful feature if it had worked, was its shortcoming. Perhaps, for brain training programs to work, supervised and extensive training with booster sessions are needed to have a meaningful clinical impact.

In the absence of solid evidence for clinical efficacy, brain training companies use several tactics to give their products credibility. They align their programs with related, even if not directly applicable, neuroscience findings. MindSparke Brain Fitness, sold by Mind Evolve, founded by a philosopher rather than a scientists or a clinician, aligns its product with research showing that software-based training improves working memory and "fluid intelligence"(Jaeggi, Buschkuehl, Jonides, & Perrig, 2008; Mind-Evolve, 2009b). They claim that this is the only brain training product proven to improve intelligence. The original study included a very demanding working memory task that minimized the ability of participants to develop automatic and task specific strategies. Mind Evolve claims that training for 30 minutes per day with its program increases working memory by 50% to 80%, and increases fluid intelligence by more than 40% in less than a month (Mind-Evolve, 2009a). The software being sold is an adaptation of the training used in the original study (Jaeggi et al., 2008). The original study used young healthy individuals, and whether the results would generalize to older people, the target population for most brain games, is not clear. One of the studies sponsored by Posit Science assessing the impact of 40 hours of training with Brain Fitness in healthy elderly adults found improvements in auditory processing speed and attention (Smith et al., 2009). From these findings, the company inferred that the program produces "an improvement in memory equivalent to approximately 10 years."(Posit-Science, 2009). Petersen notes that Posit Science has not thus far tested the longevity of the Brain Fitness improvements past three months and questions the real-world significance of their outcomes (quoted in Ellison, 2007).

Other companies cite peer-reviewed research involving company software, but their claims do not always match the actual findings. Cognifit, producer of "personalized brain fitness," states that "99.9% of 972 users showed improvement in at least one cognitive ability after eight weeks training" and that subjects improved an average of eight abilities (CogniFit, 2009). The study in which the software was featured tested the impact of insomnia on cognitive capacity and did not assess whether the software improved cognition (Haimov, Hanuka, & Horowitz, 2008). Sometimes studies cited on websites are in abstract form and not published in peer review journals. As such, they have not been subject to peer scientific scrutiny that is helpful in determining the usefulness of such interventions.

Companies that sell brain training software programs enhance their credibility by relying on the authority of scientists that design the products or others that have company affiliation. Lumosity has a scientific advisory board of academicians from Stanford and UCSF. In 2011, the company raised 32.5 million dollars of investments in its brain training programs. The company claims that its training improves memory and attention, but little scientific data using the actual software is offered (Lumos, 2009). One study mentions improvement in children that survived childhood cancer. As of this writing, no peer-reviewed studies of the efficacy of their specific program in elderly populations at risk for dementia have been published. Nintendo uses a similar approach to great effect. According to their website, more than 10 million people worldwide include playing Brain Age games in their daily lives (Nintendo, 2009). Ryuta Kawashima, a neuroscientist who specializes in brain imaging, inspired Brain Age and is featured in the game. He claims that daily training in these games helps prevent cognitive decline because it enhances blood flow to the prefrontal cortex. These claims are derived from a study of 32 people with Alzheimer's disease. Half the participants were given reading and math tasks and the other half no intervention. The training group, unlike the control group, showed improvements on assessments of frontal lobe functions and were observed to be more communicative than before the intervention (Kawashima et al., 2005). However, the authors acknowledge that they could not determine the role of social and emotional factors involved in the special attention given to the treatment group over the six months of the study. The same treatment has since been introduced in over three hundred nursing homes in Japan (Fuyuno, 2007). Kawashima says that he uses the royalties from the game on his research and is not interested in conducting more detailed studies on the effects of the game (Fuyuno, 2007).

In summary, brain training programs are designed with plausible scientific rationales. However, its commerce has moved ahead of its science. Hypotheses to be tested with these programs are portrayed as foregone conclusions. The marketing of these products often exaggerates or misrepresents the science motivating their use. For credibility, the companies rely on the authority of scientists and embellish their claims with anecdotal testimonials. Some of these products may very well turn out to be effective. However, they are sold in the market before their efficacy has been established.

Many academic scientists might be uneasy by the fact that some scientists are willing to leave academia and become entrepreneurs. Underlying this unease is the concern that they might forego basic tenets of the scientific methods and clinical practice in order to promote their wares. Despite this discomfort, one might reasonably ask the following question: Even if these products are not efficacious, where is the harm in them (and why shouldn't scientists make money)? Brain training programs keep older individuals occupied and can give them a sense of agency in so far as they are being active in maintaining their health. The risks seem trivial compared to potential side effects of medications. It seems reasonable that the sales of these programs should not be subject to the same regulatory scrutiny levied on pharmaceutical companies.

We suggest that this lack of scrutiny is what opens the way for companies to make claims that are not always justifiable. Clinicians should be sensitive to the possibility that for older people with fixed incomes and limited resources, several hundred dollars may not be a trivial expense. In addition, the fact that these products sequester the elderly away from more socially engaging activities might itself be cause for concern. Ongoing positive social contact is an important feature of aging well (Rohr & Leang, 2009), which contributes to older people's sense of emotional well-being (Charles & Carstensen, 2010).

MARKETING SUCCESS

Why do people buy into brain branding claims? We suggest that two factors contribute to peoples' susceptibility to the marketing of brain games: the allure of neuroscience and their own vulnerability.

The Allure of Neuroscience

People tend to believe explanations couched in the language of neuroscience. The persuasive power of neuroscience is evident in political commentary, marketing, and even in court deliberations (Aguirre, 2008; Farah, 2009; Mast & Zaltman, 2005; Morse, 2006). Weisberg and colleagues refers to this phenomenon as "the seductive allure of neuroscience." They reported that people find behavioral explanations with neuroscience language more satisfying than those without such language, even when the neuroscience information adds nothing substantial to the explanation (Weisberg, Frank, Joshua, Elizabeth, & Jeremy, 2008). In fact, the neuroscience language disproportionately impairs people's ability to detect logical flaws in bad explanations.

People find neuroscience explanations alluring for many reasons. Neuroscience language is reductionist (Trout, 2008) and is appealing because it is concrete and appears technical and objective. Neuroscience descriptions are easily misconstrued as providing causal explanations, which people prefer over descriptions of covariation when faced with complex phenomena (Brem & Rips, 2000). People are also not good at judging the validity of complicated explanations (Evans, 1993; Keil, 2005). They accept inappropriate teleological explanations, are distracted by seductive details, and fail to recognize logical circularities (Harp & Mayer, 1998; Lombrozo & Carey, 2006; Rips, 2002). Despite not being good at judging the quality of explanations, people overestimate their own abilities to explain complex phenomena (Keil, 2003).

Thus, it appears that the appeal of seemingly causal explanations, combined with an overconfidence in the validity of complex explanations, probably contribute to the allure of neuroscience. This allure confers brain branded products with a sense of credibility.

Vulnerability of Patients

While the allure of neuroscience can bias most people's ability to judge complex explanations, these factors take on added force with physicians, patients, and caregivers that are confronted with difficult health care decisions. In this context, lack of control and desires for a particular outcome can further push patients into believing claims made by brain brands.

People tend to evaluate evidence in a way that confirms their prior beliefs (Nickerson, 1998). Their attitudes can become more extreme in the direction of their initial point of view when presented with mixed evidence (Miller, McHoskey, Bane, & Dowd, 1993). These observations raise the possibility that physicians and patients who want to believe in the efficacy of brain games might be inclined to trust supporting claims even when they are inadequately supported or the evidence is mixed.

People with chronic diseases have limited control over their illnesses. Their symptoms may be unpredictable and treatments limited. Regardless of the condition of their health, people who lack a sense of control are disposed to infer causes where there are none. Whitson and Galinsky (2008) showed that healthy participants who lacked control were more likely to form false correlations from stock market data, develop conspiracy beliefs, and make superstitious causal connections. Patients with limited control over the course of their diseases might be similarly disposed to believe unwarranted causal claims when facing the promise of brain brands.

CONCLUSIONS

As neuroscience advances, the gap between scientific and clinical knowledge and public understanding widens. Scientists have a responsibility to communicate science to the public and offer antidotes to the presses' tendencies to simplify, exaggerate, and dramatize findings. Clinicians have a responsibility to guide patients making important health care decisions that involve a biological understanding of mind. These responsibilities are particularly important because the allure of neuroscience and patients' conditions and desires make them especially vulnerable to the claims of the brain brands. The issues raised by brain training programs are complicated. The scientific rational for these programs helping patients with few other choices is in principle plausible. However, in practice, as of this writing, the data are not quite convincing. Regardless of whether the data on brain games catches up with or ultimately undermines its marketing in the near future, brain training programs are a model case for when marketing outstrips its scientific foundations.

Clinicians should be vigilant when scientific and commercial enterprises are conflated. They should be sensitive to situations when neuroscience is used prematurely and when it is misused to sell diagnostic or therapeutic products. Independently conducted peer reviewed research in support of therapeutic efficacy remains the gold standard. Considering that many brain brand products fall short of this standard, what might clinicians suggest that buyers bear in mind? Buyers should be wary of brain brands that rely on the scientific authority of individuals with financial interests in the company. They should be wary when products emphasize anecdotes, testimonials, and press releases. When such marketing practices are evident, products deserve further scrutiny. Does the seller frame unproven hypotheses as foregone conclusions? Do they misrepresent the science behind the product? Sensitivity to such questions will help physicians better advocate for the responsible care of their patients and will help patients be better and thoughtful consumers of health care technologies.

REFERENCES

Aguirre, G. K. (2008). The political brain. *Cerebrum*, September 12.

Brem, S. K., & Rips, L. J. (2000). Explanation and evidence in informal argument. *Cognitive Science: A multidisciplinary journal, 24*(4), 573–604.

Chancellor, B., & Chatterjee, A. (2011). Brain branding: When neuroscience and commerce collide. *AJOB Neuroscience, 2*(4), 18–27.

Charles, S. T., & Carstensen, L. L. (2010). Social and emotional aging. *Annual Review of Psychology, 61*(1), 383–409.

CogniFit. (2009). CogniFit data tracks improved brain fitness for 99.9% of users. October 5. www.cognifit.com/press-releases/cognifit-data-tracks-improved-brain-fitness-999-users.

Connell, C. M., Scott Roberts, J., & McLaughlin, S. J. (2007). Public opinion about Alzheimer disease among blacks, hispanics, and whites: Results from a national survey. *Alzheimer Disease & Associated Disorders, 21*(3), 232–240.

Ellison, K. (2007). Video games vs. the aging brain. *Discover Magazine,* May.

Evans, J. S. B. T. N., SE. Byrne, RMJ. (1993). *Human reasoning: The psychology of deduction.* Hove, UK: Psychology Press.

Farah, M. J. (2009). A picture is worth a thousand dollars. *Journal of Cognitive Neuroscience, 21*(4), 623–624.

Fuyuno, I. (2007). Brain craze. *Nature, 447*(7140), 18–20.

Haimov, I., Hanuka, E., & Horowitz, Y. (2008). Chronic insomnia and cognitive functioning among older adults. *Behavioral Sleep Medicine, 6*(1), 32–54.

Harp, S. F., & Mayer, R. E. (1998). How seductive details do their damage: A theory of cognitive interest in science learning. *Journal of Educational Psychology, 90*(3), 414–434.

Jaeggi, S. M., Buschkuehl, M., Jonides, J., & Perrig, W. J. (2008). Improving fluid intelligence with training on working memory. *Proceedings of the National Academy of Sciences, 105*(19), 6829–6833.

Kawashima, R., Okita, K., Yamazaki, R., Tajima, N., Yoshida, H., Taira, M., et al. (2005). Reading aloud and arithmetic calculation improve frontal function of people with dementia. *The Journals of Gerontology Series A: Biological Sciences and Medical Sciences, 60*(3), 380–384.

Keil, F. C. (2003). Folkscience: Coarse interpretations of a complex reality. *Trends in Cognitive Sciences, 7*(8), 368–373.

Keil, F. C. (2005). Explanation and understanding. *Annual Review of Psychology, 57*(1), 227–254.

Kitabatake, Y., Sailor, K. A., Ming, G.-l., & Song, H. (2007). Adult neurogenesis and hippocampal memory function: New cells, more plasticity, new memories? *Neurosurgery Clinics of North America, 18*(1), 105–113.

Langbaum, J. B. S., Rebok, G. W., Bandeen-Roche, K., & Carlson, M. C. (2009). Predicting memory training response patterns: Results from ACTIVE. *The Journals of Gerontology Series B: Psychological Sciences and Social Sciences, 64B*(1), 14–23.

Lie, D. C., Song, H., Colamarino, S. A., Ming, G.-l., & Gage, F. H. (2004). Neurogenesis in the adult brain: New strategies for central nervous system diseases. *Annual Review of Pharmacology and Toxicology, 44*(1), 399–421.

Lombrozo, T., & Carey, S. (2006). Functional explanation and the function of explanation. *Cognition, 99*(2), 167–204.

Lumos. (2009). Brain games and exercises—lumosity.

Lynch, Z. (2009). Neurotechnology industry 2009 report. www.researchandmarkets. com/reports/997270.

Mahncke, H. W., Connor, B. B., Appelman, J., Ahsanuddin, O. N., Hardy, J. L., Wood, R. A., et al. (2006). Memory enhancement in healthy older adults using a brain plasticity-based training program: A randomized, controlled study. *Proceedings of the National Academy of Sciences, 103*(33), 12523–12528.

Mast, F. W., & Zaltman, G. (2005). A behavioral window on the mind of the market: An application of the response time paradigm. *Brain Research Bulletin, 67*(5), 422–427.

Miller, A. G., McHoskey, J. W., Bane, C. M., & Dowd, T. G. (1993). The attitude polarization phenomenon: Role of response measure, attitude extremity and behavioral consequences of reported attitude change. *Journal of Personality and Social Psychology, 64*(4), 561–574.

Mind-Evolve, L. (2009a). Mind sparke brain fitness pro—Benefits, features, purchase. http://mindsparke.com/brain_fitness_pro.php?id=tour.

Mind-Evolve, L. (2009b). Mind sparke brain fitness pro—science. www.mindsparke. com/science.php.

Morse, S. (2006). Brain overclaim syndrome and criminal responsibility: A diagnostic note. *Ohio State Journal of Criminal Law, 3*, 397–412.

Neutotechnology Industry Organization (NIO). (2010). www.neurotechindustry.org/ home.html.

Nickerson, R. S. (1998). Confirmation bias: A ubiquitous phenomenon in many guises. *Review of General Psychology, 2*(2), 175–220.

Nintendo. (2009). Nintendo Financial Results Briefing for Fiscal Year Ended March, 2009 o. (Document Number)

Owen, A. M., Hampshire, A., Grahn, J. A., Stenton, R., Dajani, S., Burns, A. S., et al. (2010). Putting brain training to the test. *Nature, 465*, 775–778.

Papp, K. V., Walsh, S. J., & Snyder, P. J. (2009). Immediate and delayed effects of cognitive interventions in healthy elderly: A review of current literature and future directions. *Alzheimer's and Dementia, 5*(1), 50–60.

Posit-Science. (2009). The impact study.

Rips, L. J. (2002). Circular reasoning. *Cognitive Science, 26*, 767–795.

Rohr, M. K., & Leang, F. R. (2009). Aging well together—A mini review. *Gerentology, 55*, 333–343.

Smith, G. E., Patricia, H., Kristine, Y., Ronald, R., Robert, F. K., Henry, W. M., et al. (2009). A cognitive training program based on principles of brain plasticity: Results from the improvement in memory with plasticity-based adaptive cognitive training (IMPACT) study. *Journal of the American Geriatrics Society, 57*(4), 594–603.

Trout, J. D. (2008). Seduction without cause: uncovering explanatory neurophilia. *Trends in Cognitive Neuroscince, 12*(8), 281–282.

Verghese, J., Lipton, R. B., Katz, M. J., Hall, C. B., Derby, C. A., Kuslansky, G., et al. (2003). Leisure activities and the risk of dementia in the elderly. *The New England Journal of Medicine, 348*(25), 2508–2516.

Weisberg, D. S., Frank, C. K., Joshua, G., Elizabeth, R., & Jeremy, R. G. (2008). The seductive allure of neuroscience explanations. *Journal of Cognitive Neuroscience, 20*(3), 470–477.

Whitson, J. A., & Galinsky, A. D. (2008). Lacking control increases illusory pattern perception. *Science, 322*(5898), 115–117.

Willis, S. L., Tennstedt, S. L., Marsiske, M., Ball, K., Elias, J., Koepke, K. M., et al. (2006). Long-term effects of cognitive training on everyday functional outcomes in older adults. *JAMA, 296*(23), 2805–2814.

PART TWO

Competence and Responsibility

6

Competence and Autonomy

The Cases of Driving, Voting, and Financial Independence

JASON KARLAWISH

WHAT IS COMPETENCE?

The bioethical principles of respect for autonomy, nonmaleficence, and beneficence guide ethical decision making in medicine (Beauchamp & Childress, 2001).
Competence is a tool we use on a case-by-case basis to negotiate the boundaries
between these principles. For example, when the principle of respect for autonomy conflicts with the principles of beneficence and nonmaleficence, competence
becomes an important instrument for determining how to resolve this dilemma.
We normally respect the choice a competent person makes. In contrast, we choose
for a noncompetent person in order to, at a minimum, protect them from harms
and, in some cases, maximize potential benefits (Grisso & Appelbaum, 1998).
In short, competency does substantial work to mediate some of the most vexing
dilemmas in bioethics, such as whether to respect a patient's refusal of treatment
or a cognitively impaired older adult's decision to continue driving. And not only
medicine uses this instrument. Competency guides the practices of a variety of
disciplines including nursing, social work, law, and financial management. We
live in the age of competency.

Competency is an instrument with a history (Grisso, 2003). Its story is part of
the histories of individual freedom and rights, and also the ascending influence
of the disciplines of neurology, psychiatry, and psychology. As a result, what we
mean when we say someone is "not competent" or "not capable of making a decision" has, over time, taken on and shed different meanings.

In the era before liberal democracy and the modern sciences and practices of
neurology, psychiatry, and psychology, a person's status did much of the work that,
today, competency does for us. That is, a person's choice was respected based on who
he was, or his "station in life." The barbarian, the slave, the lunatic, the senile, the

imbecile, and women were examples of kinds of persons who, by virtue of their lack of status, were deemed not capable of making many, if not all, decisions about their well-being. In contrast, the high-minded, property-holding man was free to make decisions. Of course, he needed to remain within the social and cultural norms that dictated what such a man ought to do. Indeed, this ethic was so embedded in culture and practice that these decisions came to define what a "reasonable"—hence competent—person should do. Within this conceptual framework, competency took on a global status; that is, a person was either competent or not competent.

The enlightenment project slowly toppled status as one of the principle arbiters of who could and who could not make decisions for themselves. As a result, the modern era has seen the rapid retreat of status and tradition, and the reasonableness that flowed from these features, as relevant to judging whether someone is competent. In their stead, we have developed the concept of capacity (Grisso & Appelbaum, 1998). This term describes a person's decisional abilities. A person is competent if he or she has sufficient abilities to make a decision. Hence, the label of, for example, "mentally ill" does not necessarily define someone with a mental illness as not competent. Instead, that person must be found to lack decisional capacity. The history of the last century can be said to be the fall of the reasonable man and the rise of the person who has adequate decisional capacity.

This transformation from status to capacity has exerted two significant consequences. Politically, it has empowered the individual. After childhood, the relevant consideration in judging competency is not *who you are* but *what you are capable of doing*—that is, your abilities. This, in turn, has led to the scientific transformation of competency from a judgment of whether a person fits within a class, such as "persons of unsound mind," to a measurement of a person's decisional abilities. Competency has moved from a global judgment to a task-specific assessment of ability. As a result, in the last thirty years, the law and psychology have developed of a host of task specific capacities and, in the case of some of these capacities, instruments to measure them (Grisso, 2003). They include, for example, the capacity to make medical decisions (Grisso & Appelbaum, 1998), enroll in research (Appelbaum & Grisso, 2000), execute a will or a contract, complete an advanced directive (Molloy et al., 1996), consent to sexual relations, and even to vote (Appelbaum et al., 2005).

Competency has thus acquired the qualities of science. For example, the state of Maine once excluded persons under guardianship for reasons of mental illness from voting. Now, a resident of Maine is competent to vote if he or she "understands the nature and effect of voting" (Doe v. Rowe, 2001). This definition has in turn served as the foundation for an instrument to measure the construct (Appelbaum et al., 2005), just as we have instruments to measure constructs such as short-term memory, depression, or neuroticism. A person can't vote if a capacity assessment shows he or she lacks decisional ability, and investigators have validated a set of questions that assess whether someone understands the nature and the effect of voting and is, therefore, capable to vote. We have transformed from the age of status to the age of assessment. Capacity assessment seems acultural or objective, but later, as we consider putting science into practice, we will see how the assessment is deeply grounded in context and culture.

Capacity is a continuum described by performance on measures of one or more of the decisional abilities, but as a continuum, it is a weak instrument. In order to do the work of mediating the balance among the principles of bioethics, that continuum must be split into two groups: the capable and the not capable. This division along a continuum is no different than how modern medicine diagnoses diseases such as hypertension.

Blood pressure is a continuum. How do physicians divide this continuum into the normal versus the abnormal; that is, the normotensive versus the hypertensive? They use a variety of data, such as the association between blood pressure values and the risk of future bad outcomes such as a stroke or heart failure. This, in turn, allows them to draw a line that thereby divides a normal from an abnormal blood pressure. The same approach applies to capacity. It follows then that how we measure capacity and the norms we apply to interpret that measure grant us substantial power.

The goal of this chapter is to examine these issues at the intersection of ethics and measurement. I will use three distinct decisions as case studies: driving, voting, and financial management. The point of this chapter is that while we have cast off the chains of the tyranny of status, we hazard shackling ourselves with a new tyranny: measurement. If competency is a numbers game, then whoever controls the numbers controls capacity and, in turn, autonomy.

ASSESSING COMPETENCY

The following two cases illustrate the scientific and technical issues at stake in assessing competency.

> Mrs. P. is an 80-year-old woman who fell and sustained a compound fracture of her right wrist and a head injury. After several days in hospital, she returns home. She continues to have trouble concentrating and has substantial limitations using her right hand, which is her dominant hand. She ceases driving, orders a cab when she needs to leave the house, and chooses her daughter to manage her checkbook. A few days before election day, her daughter asks her if she wants to vote. She says she does, but in conversation with her daughter, she cannot name the candidate for her preferred party. Her daughter decides not to drive her mother to the polls.
>
> Mr. X. is an 80-year-old man who has suffered the same injuries as Mrs. P. Although he too has trouble concentrating and using his right hand, he still drives and manages his checkbook. He has had one fender-bender and had double-paid at least one bill. His children offer to help him, but he insists everything is fine and declines their help. On election day, he drives to the polls. He asks a poll worker to help him mark his ballot as his hand is in a cast. The poll worker reads him the ballot. He does not recognize the names of the candidates for governor but he does recognize the parties, and he asks the worker to mark the candidate running on behalf of Mr. X's preferred party.

Both Mrs. P and Mr. X have functional impairments. That is, they cannot perform their usual and everyday activities as well as they once did: driving, managing money, and voting. Both persons made choices about how to manage these impairments. In the case of Mrs. P., she has chosen to have other people help her with her money and her transportation. In contrast, Mr. X. wants to continue doing what he has always done. And he is having some problems with those tasks. Together, these cases present two kinds of competencies—functional and decisional—and highlight the differences in those competencies and how we assess them.

Functional Competency Defined

Functional refers to activities of daily living (ADL) or independent or instrumental activities of daily living (IADL). Within this category of competency are activities such as managing finances, using transportation, managing medications, cooking, shopping, and cleaning. The other category of activities of daily living is the basic activities of daily living (BADL). These are the ability to transfer, toilet, dress and groom, bathe, and eat. Collectively, the BADLs and IADLs are a set of competencies a person needs in order to function as an autonomous adult.

The science of measuring activities of daily living began in the 1960s when Powell Lawton and Elaine Brody published their landmark article "Assessment of Older People" (Lawton & Brody, 1969). Subsequently, researchers have developed multiple scales that quantify the degree of impairment or dependence a person has in performing their activities of daily living. These measures have a substantial role in deciding who is entitled for assistance and how much assistance they require, and for measuring the benefits of assistance—that is, how well an intervention moves someone from being dependent to independent.

Both Mrs. P. and Mr. X. have functional problems. The social worker who visited them after they returned to their homes checked off a series of boxes on a form that assessed each of the IADLs; for example, the social worker rated them "partially dependent for financial management." These data help to frame a care plan and the resources needed to achieve that plan. In all likelihood, this social worker would not think of his functional assessments as competency assessments. But these assessments do in fact determine whether someone is capable of managing, or, we could say, is competent to manage a task.

Decisional Capacity Defined

The finding that a person is not capable or competent to perform a functional task often necessitates an assessment of the second kind of competency: decisional competency. This assessment may be necessary in order to decide whether the person who cannot *do* a task (such as finances) is capable of deciding how best to manage that disability (such as having someone else take care of their bills).

The science of measuring decisional abilities is comparatively recent compared to the measurement of functional status. It began when law, psychiatry, and ethics settled on a conceptual model of the four decisional abilities: understanding, appreciation, reasoning, and choice (Grisso & Appelbaum, 1998). The law uses at least one of these abilities to define decisional capacity for a task. For example, one element of the capacity to write a will, also called testamentary capacity, is the ability to understand the nature and extent of one's property.

Researchers have developed instruments to measure these abilities (Karlawish, 2008). The general strategy to measure a decision-making ability is to ask a patient a series of questions that assess that ability and to score those answers using criteria (for example, an adequate answer=2, marginal answer=1, and inadequate answer=0). The sum of scores for the questions then represents a score on the measure of ability. In the case of the element of testamentary capacity described above, a question such as "Can you tell me in your own words what property you have?" would measure the person's understanding of their property. The person's answer could be scored using a scale such as "2=adequate understanding," "1=marginal understanding," and "0=inadequate understanding."

SOCIAL AND ETHICAL CONTEXT OF COMPETENCY ASSESSMENT

Suppose we talk to Mrs. P. and Mr. X. about the functional problems we have measured. Specifically, suppose we engage each of them in a conversation about their capacity to solve their everyday functional problems—that is, to get help with transportation or managing their money or casting a ballot. Why do we want to have such a conversation? We want to talk with them because we sense, in the wings, that a collision looms, a collision between the principles of beneficence and autonomy, between taking care of these older adults versus letting them live as they choose to live. And this collision is complicated by contextual considerations and measurement uncertainties. I shall focus on one of these tasks: managing money.

Mrs. P. appreciates her problems with double-paying a bill; she understands a range of solutions to solve this, such as having someone else pay her bills; she appreciates that she has double paid some bills and that having someone take care of them for her could benefit her by preventing the loss of money-; and she reasons through how this solution would affect her daily life because, she explains, her daughter would come over to get her bills and that means she would need to organize them for her daughter.

Mr. X. can also make a choice. He will continue to manage his money. But when we measure his other decisional abilities, we discover deficits beyond the relatively simply ability to express a choice. Although he understands the options to manage the kinds of day-to-day problems he is experiencing and he can reason through these options, he does not recognize the problems he is having with money management. In other words, he has adequate understanding and reasoning but he lacks appreciation of, or awareness or insight into, his problems (Lai & Karlawish, 2007).

The ethical issues at stake here are the following. When we assess his decisional ability, what facts about a finding of impairment in a functional task do we want to examine: the full range of facts about his problems with financial management, or are there some key facts to focus on? We also need to decide what decisional abilities we will measure. Do we care only that he understands the facts, or should he appreciate them as well?

After we agree on the facts and the relevant abilities, we face a third consideration. How well do people like Mr. X. and Mrs. P. have to perform on the questions we use to assess these abilities? The extremes of performance are easy. If a person cannot answer any question with even marginal correctness, he or she is not competent to decide how to manage their functional problems. In contrast, if he or she answers the questions fully and accurately, we would say the person is competent to manage their functional problems. The latter statement deserves reinforcing, particularly in the case of Mr. X., the man who is double-paying bills and still chooses to manage his own money. We should all agree that he is free to do this if we found that he fully understood, appreciated, and reasoned through this functional problem. The financial risks are his and because he has capacity, they are his risks to manage.

The more vexing cases are the middle performers—that is, the persons who have some impairment in their abilities. To what norm or standard should we compare this intermediate performance? This is among the central questions in measurement science.

The field of capacity assessment has adopted two kinds of norms. One relies upon population norms. That is, the distribution of a comparison group's performance on the capacity measure describes normal performance and the degree of deviation from this norm describes the abnormal. For example, the capacity to consent to treatment instrument uses normative data from age-matched controls to define abnormal as performance on an ability that is more than two standard deviations below the norm (Marson et al., 1995). This kind of standard draws upon the concepts and methods of measurement science in psychology. These define what is normal cognitive function, such as memory and language, based on the distribution of performance of age and education matched controls.

The ethical issue at stake in this method to norm capacity scores is the choice of group whose performance will define the norm. We could include persons like Mr. X. who have had a head injury and, as a result, some mild cognitive problems. In contrast, we might exclude from the normal group all persons with diagnoses that might impair cognition, such as recent head injury. Either choice has merits and demerits, or, we would say in measurement, its errors.

If that group includes people like Mr. X., then we permit false negative labels. That is, we forgive or permit some degree of disability, thereby we can potentially label his performance as "normal" when, in fact, it is abnormal. In contrast, a sample of persons without any diagnoses that might impair cognition could set a standard for normal that is over and above what would be thought necessary. In other words, we permit false-positive labels. That is, we mistakenly label as abnormal a performance that is, in fact, normal.

The selection of the normal group becomes particularly vexing among persons who are elderly. Cognitive function declines with age. Hence, an age-matched group of normal controls will permit some degree of performance on measures of cognition that in younger persons would be abnormal. It could be sufficiently abnormal that it impairs decisional capacity.

A second and related reason why selecting among elderly persons who is normal is vexing is that the standards to define what are the diseases of cognitive aging

are changing. In the last ten years, psychologists and physicians have revolution-
ized what is normal cognitive aging. The concept of mild cognitive impairment
has gained increasing traction as a legitimate clinical entity (Roberts et al., 2010).
It defines persons who have cognitive complaints and cognitive performance that
do not meet criteria for dementia, but fall below the normal, for example, performance on a memory test that is 1.5 standard deviations below the norm (Petersen,
2003). It follows that if concepts of normal and disease shift, then who is in the
normal group and who is not in that group will shift as well.

A second standard used to create norms is an expert judgment of performance
(Kim et al., 2001; Karlawish et al., 2008). Here, experts in capacity assessment
review a set of capacity interviews and render their judgment of whether the person is or is not capable of making the decision. For a given score on the measure
of capacity, we can then use regression techniques to calculate the odds of being
judged not capable of consent. Inspection of how well each score classifies people
into the category of the capable or the not capable defines the optimal point on
the scale to define what is abnormal. This is similar to the method clinical medicine uses to define what level of a test best classifies persons as diseased versus not
diseased—that is, what level has the best sensitivity and specificity.

Unlike population norms, this standard integrates expert judgment of what I
call *extracompetency considerations*—that is, information other than the capacity
data, such as information about the risks and benefits of the decision at hand, as
well as social norms we judge relevant to the specific decision. Below, I discuss
these considerations more generally. Here, I focus on how they impact on generating norms.

Risk considerations do influence judgments of competency to consent.
Physicians who were experts in capacity assessments viewed one of two interviews
of capacity to consent to research (Kim et al., 2006). In both interviews, the same
person had the same scores on each of the measures of decisional ability. The only
difference between the two interviews was the risk of the research: one study was
a low-risk study of a promising drug, while the other was a high-risk study of a
neurosurgical procedure. Although the patient had the same scores on the measures of decisional ability for each interview, the experts who viewed the high-risk
interview were significantly more likely to judge that person not capable of consent than were their colleagues who viewed the low-risk capacity interview.

The implication of this finding for norming capacity scores is that expert consensus will yield different classification of what level of performance is normal
than will norms based on the distribution of population performance. Why? The
degree of risk, and other considerations as well, changes the cut point for judging
normal capacity *even when the capacity scores are exactly the same*. It follows that
the decision about which norm to use to decide what is a normal capacity score
(population distribution or expert consensus) cannot be answered using science.
Instead, this decision is a matter of ethics and politics.

Extracompetency considerations, also referred to as contextual considerations
or values, have other influences upon competency. Below, I discuss how such matters influence each of the competencies at stake in the cases of Mrs. P. and Mr. X.

Managing Money

Suppose Mr. X. said clearly and repeatedly to his children: "I know my head's a little foggy since the fall and I've had some trouble with the bills, but I got to keep active and it's my money, so I'll take my chances." We might judge that he has adequate appreciation of his problem and also that he appreciates one of the downsides of the solutions.—namely, the loss of independence. As a result, we might let him take his risks. It is his money. If he is capable of showing he has the capacity to decide what he is doing—where "what" refers not only to his successful transactions, but his *unsuccessful* ones as well—then his money is his to manage as he sees fit.

He is, to a degree, a free agent. The degree that might cause society to intrude is the harm Mr. X. suffers from his unsuccessful financial decisions. If his very house and home were threatened, we might demand a greater threshold of performance on the measures of decisional ability, or we might require he show decisional ability for a greater range of facts than simply his bill paying, such as the finances of home ownership, or we might cast aside capacity and simply intervene. Such considerations of risk and harm are at the heart of all capacity assessments. If the decision posed little harm—*chicken or beef tonight?*—we would not care about his capacity to arrive at the choice other than that he can express a choice. But when the decision poses real threat of harm, we are just like the experts who reviewed the interviews about enrolling in one of the two research studies. We ought to raise the bar for how well the person needs to perform at his choice.

Managing Transportation, Specifically Driving

Extracompetency considerations have considerable influence upon what Mr. X. can choose and, in some jurisdictions, what others such as physicians are obligated to do. Mr. X. may have the capacity to decide about his driving. He may fully understand and appreciate he is not a good driver—*Look at how wide I took that last turn! Wow, I'm way off*—but that appreciation and understanding are of little importance. His failure to adequately function is of such importance to society that we don't care what he has to say about it. If he's making mistakes, he needs to prove he can drive adequately. In some states of the United States, physicians are required to report such a person to state driving authorities. In other words, for some kinds of tasks, competence is function. Of course, this judgment of function engenders all of the considerations of norming measurement that I discussed above.

Managing Voting

Much like driving, extracompetency considerations have considerable influence upon what Mr. X. can do. But, these considerations operate in a manner quite different from the case of driving. For example, if Mr. X. lived in Canada, the question of whether he is competent to vote would not be germane as that country has no standard for competency to vote (Karlawish, 2007). If a person is otherwise qualified to vote (for example, he or she is a citizen), then he or she can vote. And if the person requires assistance to vote, then he or she must receive it. Under such a system, not only is autonomy fully respected—if you want to vote, you can vote—but beneficence is as well—if you need assistance, we will provide it. This is autonomy fully realized in the liberal state.

A historical perspective on the right to vote illustrates the transformation from competence constructed along lines of status, to a reconstruction along lines of capacity, either as decisional capacity, or, as in the case of Canada, functional capacity alone—that is, the ability to express a choice. In many states, who can vote was based on a status-based concept of respect for persons. In the eighteenth century, females and males without property were disenfranchised, and in the nineteenth century, states began to disenfranchise persons with cognitive and psychiatric problems who fit into categories such as imbeciles and the insane. In short, once you were deemed an imbecile, the state took no consideration of your capacity to vote. You simply could not vote. In recent years, some states, such as Illinois, dropped any consideration of capacity to vote or, like Maine, they adopted a task-specific standard, such as understanding the nature and effect of voting (Karlawish et al., 2004).

These examples of finances, driving, and voting illustrate how competency is an instrument of social control. Its history has been a transformation from a global judgment based on status labels, to a task specific assessment based on measurement of the decisional abilities and judgments about risks and benefit. Hence, that control is in the hands of a professional class with authority to assess competency using scales that, at least in some cases, incorporate norms and values about what risks are permissible and what benefits are sufficient.

NEUROETHICAL CHALLENGES FOR THE CONCEPTUALIZATION AND ASSESSMENT OF COMPETENCE

Competency as a task-specific and measured capacity is not a passing trend. It is part of the fabric of our social structure of cognition and autonomy in this age of assessment. Hence, we need to pay much better attention to how we measure it, who measures it, and what we do with those data.

What norm should we use? I would submit that this question matters if we believe that judgments of performance on a measure of capacity would materially change if we applied a different standard to norm our scores. That is, if either different population norms or expert criteria instead of population norms would change whether someone is classified as having adequate capacity, then we should be up front and quite specific and transparent about what norm we selected and why we selected it.

In deciding what norm we use, we ought to attend to the shifting concepts of what is normal cognitive aging. Of particular relevance is whether the concept of abnormal cognitive aging is defined based on the future loss of function or only on present disability. If it is defined based on future loss of function, like hypertension, it is arguably irrelevant to norm capacity. Capacity is about a here and now decision, not the future risk of loss of function. Hence, excluding persons with mild cognitive impairment from the category of normal makes sense, as these persons have been shown to have impairments in IADLs (Jefferson et al., 2007). They are not cognitively normal. But excluding persons who have neuroimaging findings that put them at risk of future loss of cognition but have normal function makes little sense. Again, capacity is about the here and the now, not the future loss of capacity.

There are two aspects of normal cognitive aging that might impact on capacity assessments and, if they do, require us to reconceptualize capacity among the elderly. Cognitive psychology has shown that as people age, two events occur not as a result of disease but of aging in general. First, their emotions exert a unique influence on cognition (Carstensen et al., 1999; Kensinger et al., 2007). Specifically, older adults perceive their time as limited. This perception of time changes cognitive processes. In particular, older adults (as well as other persons who perceive their time as limited, such as persons with a terminal illness) show a preferential effect for learning and recalling information with a positive emotional valence over and above information with a negative emotional valence. For example, they are more likely to remember a photo of a cute infant than one of roaches crawling about on a pizza slice. They are also more likely than persons who do not perceive their time as limited to prefer messages that emphasize achieving emotional goals (Fung & Carstensen, 2003).

How might these age-related cognitive changes affect capacity? To the extent that information has an emotional valence, we would expect that a person's ability to understand that information would vary as a function of that emotional valence. That is, they would be more likely to understand and appreciate positive information than they would negative information. This, in turn, may mean that compared to younger persons, older adults will be less likely to understand and appreciate risks.

A second feature of normal cognitive aging is changes in executive function, specifically executive function mediated by the prefrontal cortex. Lesion studies of this region of the brain show that it affects how humans process risk, delay, ambiguity, and emotions such as regret. The region has been described as "the accountant and the executive" of the brain. Studies of cognitively normal elderly persons show that some have impairments in the cognitive processes this brain region mediates. Specifically, when they are put through a task that measures risk-reward processing, such as the Iowa Gambling Task, their profile resembles what would be seen in persons with a lesion to their prefrontal cortex (Denburg et al., 2007). They are insensitive to long-term punishment. How may this kind of aberrant risk-reward processing affect capacity? It is entirely possible that it will affect the abilities to appreciate and reason through risks and benefits. Specifically, within the category of older adults there may be persons who do not appreciate the net disadvantage of a low-reward, high-risk situation and instead, they assert that such a condition benefits them.

If we find that changes in the emotional processing of facts and risk-reward processing impact upon capacity, then we need to consider whether we would accept this performance as part of normal aging and thus normal capacity, or as abnormal and thus not to be included in our capacity norms. If we choose the latter and argue it is abnormal, then we have reshaped what is acceptable behavior for the (formerly) autonomous older adult.

A second proposal to address the ethical challenges of competency in the age of assessment is to be more transparent in the role that values necessarily serve in how we judge whether a particular level of capacity is adequate. The study of the physicians who viewed the interviews of an older adult reviewing one of two research risk scenarios raises the question of what if we had different kinds of raters. What if, instead of physicians trained in psychiatry, we had researchers who studied older adults with cognitive problems? Or institutional review board members who reviewed the ethics of research that enrolls these older adults?

These questions are not theoretical. A National Institute of Mental Health-sponsored clinical trial that enrolled persons with schizophrenia found that the study site, more so than cognition or the severity of psychiatric symptoms, explained most of the variance in scores on a measure of the capacity to consent to enroll in that clinical trial (Stroup et al., 2005).

The point is that we need to scrutinize the values of the judges who norm capacity measures. If that judge is liberal, we may get a very different decision than, all other things being equal, the judge who is conservative. A reasonable approach here is to measure the values of our judges. These values are data that is as relevant as both the capacity of the person and the risks of the decision that the judge is called to judge.

In tough, ethically controversial cases, we might want to separate the assessment of capacity from the judgment of whether that capacity is adequate. This is, of course, the model for guardianship decisions where an appropriate expert testifies about the person's capacity, and a judge reviews this evidence in a court of law. This may be a model for other kinds of close-call decisions, such as financial capacity.

IMPLICATIONS FOR POLICY AND PRACTICE

Competency is and always has been an instrument of social control. In its present form, it is largely conceptualized as a measurement of function such as the IADLs and the decisional abilities. I have argued that our ethics and values influence how we conceptualize it, measure it, and interpret those measures. Among these three points, I think the second—that ethics influences how we measure decisional abilities—is perhaps the most controversial, as I am asserting that our concepts of normal are hidebound with our values.

The present status quo relies on robust consensus about what is normal function (and therefore abnormal). To the extent then that advances in neurosciences reconceptualize what is a normal brain, we should be attentive to how this may change our status quo. Of particular concern is how functional and structural neuroimaging could supplant assessments of the decisional abilities. If, for example, we find positron emission tomography (PET) imaging showing hypometabolism in the prefrontal cortex signifies poor prefrontal function, we are one step toward using PET imaging of the prefrontal cortex to assess the ability to process risk and reward. Or we could use magnetic resonance imaging (MRI) volumes of regions of interest relevant to driving skills to determine who is not capable of driving. And so on.

Such measures may be even more precise than our present methods of measuring function such as the IADLs and the decisional abilities using interviews and pencil and paper tasks. And yet, our present culture argues that everyone should have a fair chance to prove their merit and ability. Every adult should have a shot at managing their own money, casting a ballot, and perhaps even getting behind the wheel of a car. In other words, we value respect for autonomy over and above how precisely we measure that autonomy.

Setting aside these conceptual issues, what does the internist, neurologist, geriatrician, or psychiatrist do when faced with a patient in the office in whom capacities such as driving or financial management are being questioned?

At present, practitioners do have a robust set of tools that give them a coherent language to assess capacity. By "coherent" I mean a language that makes sense from one practitioner to the other. These tools include instruments that contain validated questions to measure the decisional abilities (understanding, appreciation, and reasoning) and access to formal assessments of functional capacity, such as a driving assessment, to provide concrete and vivid data that inform whether a person retains the ability to drive.

In addition to having a toolkit of standard methods to assess the decisional abilities and functional capacities, practitioners should be informed about which abilities and capacities they ought to assess. For example, a practitioner who is asked to determine in a guardianship hearing whether a person is capable of voting should only focus on the abilities that jurisdiction deems relevant to determine whether someone is capable of voting. In Maine, for example, that would mean focusing on the person's ability to understand the nature and effect of voting. In the case of driving capacity, a practitioner ought to have access to a driver's rehabilitation facility where patients can receive a structured and third party assessment of their driving ability.

Finally, practitioners ought to have a network of expert colleagues to whom they can refer tough cases. Such a network is a valuable resource to get a second opinion, test out different approaches, or mediate disagreements.

REFERENCES

Appelbaum, P. S., Bonnie, R. J., et al. (2005). The capacity to vote of persons with Alzheimer's disease. *American Journal of Psychiatry, 162* (11), 2094–2100.

Appelbaum, P. S., & Grisso, T. (2000). *The MacArthur competence assessment tool—Clinical research*. Sarasota, FL: Professional Resources Press.

Beauchamp, T. L., & Childress, J. F. 2001. *Principles of biomedical ethics*. 5th ed. New York: Oxford University Press.

Carstensen, L. L., Isaacowitz, D. M., et al. (1999). Taking time seriously. A theory of socioemotional selectivity [see comment]. *American Psychologist, 54* (3), 165–181.

Denburg, N. L., Cole, C. A., et al. (2007). The orbitofrontal cortex, real-world decision making, and normal aging. *Annals of the New York Academy of Sciences, 1121,* 480–498.

Doe v. Rowe, Docket No. 00-CV-206-B-S. (2001). United States District Court For the District of Maine 156 F. Supp. 2d 35; 2001 US Dist. Lexis 11963.

Fung, H. H., & Carstensen, L. L. (2003). Sending memorable messages to the old: Age differences in preferences and memory for advertisements. *Journal of Personality & Social Psychology, 85* (1), 163–178.

Grisso, T. (2003). *Evaluating competencies: Forensic assessments and instruments.* New York: Kluwer Academic/Plenum Publishers.

Grisso, T., & Appelbaum, P. S. (1998a). Abilities related to competence. In xxx, *Assessing competence to consent to treatment. A guide for physicians and other health professionals* (pp. 31–60). New York: Oxford University Press.

Grisso, T., & Appelbaum, P. S. (1998b). Using the MacArthur competence assessment tool—treatment. In *Assessing competence to consent to treatment. A guide for physicians and other health professionals* (pp. 101–126). New York: Oxford University Press.

Grisso, T., & Appelbaum, P. S. (1998c). Why competence is important: The doctrine of informed consent. In *Assessing competence to consent to treatment. A guide for physicians and other health professionals* (pp. 1–15). New York: Oxford University Press.

Jefferson A.L., Lambe, S., et al. (2007). Decisional capacity for research participation among individuals with mild cognitive impairment." *Journal of the American Geriatrics Society, 56,* 1236–1243.

Karlawish, J. (2008). Measuring decision-making capacity in cognitively impaired individuals. *Neurosignals, 16* (1), 91–98.

Karlawish, J. H., & Bonnie, R.J. (2007). Voting by elderly persons with cognitive impairment: Lessons from other democratic nations. *University of the Pacific McGeorge Law Review, 38* (4), 880–916.

Karlawish, J. H., Bonnie, R. J., et al. (2004). Addressing the ethical, legal, and social issues raised by voting by persons with dementia. *JAMA, 292* (11), 1345–1350.

Karlawish J. H., Kim, S.Y., et al. (2008). Interpreting the clinical significance of capacity scores for informed consent in Alzheimer disease clinical trials. *American Journal of Geriatric Psychiatry, 16* (7), 568–574.

Kensinger, E. A., Garoff-Eaton, R. J., et al. (2007). Effects of emotion on memory specificity in young and older adults. *Journals of Gerontology Series B-Psychological Sciences & Social Sciences, 62* (4), P208–P215.

Kim, S. Y., Caine, E. D., et al. (2001). Assessing the competence of persons with Alzheimer's disease in providing informed consent for participation in research. *American Journal of Psychiatry 158* (5), 712–717.

Kim, S. Y., Caine, E. D., et al. (2006). Do clinicians follow a risk-sensitive model of capacity-determination? An experimental video survey. *Psychosomatics, 47* (4), 325–329.

Lai, J. M., & Karlawish, J. (2007). Assessing the capacity to make everyday decisions: A guide for clinicians and an agenda for future research.*American Journal of Geriatric Psychiatry, 15* (2), 101–111.

Lawton, M. P., & Brody, E. M. (1969). Assessment of older people: Self-maintaining and instrumental activities of daily living. *Gerontologist, 9,* 179–186.

Marson, D. C., Ingram, K. K., et al. (1995). Assessing the competency of patients with Alzheimer's disease under different legal standards: A prototype instrument. *Archives of Neurology, 52* (10), 949–954.

Molloy, D. W., Silberfeld, M., et al. (1996). Measuring capacity to complete an advance directive. *Journal of the American Geriatrics Society, 44,* 660–664.

Petersen, R. C. (2003). *Mild cognitive impairment: Aging to Alzheimer's disease.* Oxford: Oxford University Press.

Roberts, J. S., Karlawish, J. H., et al. (2010). Mild cognitive impairment in clinical care: A survey of American Academy of Neurology members. *Neurology, 75* (5), 425–431.

Stroup, S., Appelbaum P., et al. (2005). Decision-making capacity for research participation among individuals in the CATIE schizophrenia trial. *Schizophrenia Research 80* (1), 1–8.

7

Competence for Informed Consent for Treatment and Research

SCOTT Y. H. KIM

INFORMED CONSENT AND DECISION-MAKING CAPACITY

As the populations of western nations age, greater numbers of their citizens will experience the effects of normal and abnormal brain aging. Abnormal brain aging due to neurodegenerative disease is common and often impairs the cognitive processes needed to give informed consent for medical treatment and participation in research. In addition, other neuropsychiatric diseases and neurological injuries can impair a person's capacity to give informed consent. This chapter will review the clinical and ethical issues involved in the definition and assessment of competence for informed consent.

There are three elements required for valid informed consent, whether it be for treatment or research (Berg, Appelbaum, Lidz, & Parker, 2001). First, there needs to be adequate disclosure of information. In the treatment setting, this includes the patient's condition or diagnosis, the nature of the treatment in question, its risks and potential benefits, and any alternatives and their risks and benefits, including no treatment at all. Second, the patient must not be coerced or unduly influenced, so that the decision is a free and voluntary choice. Third, the patient must be capable of actually using the information provided to reach a free decision. This capacity has been called by various names, including decision-making capacity (DMC), capacity, and competence. Most modern laws use the term "decision-making capacity" and provide for clinicians to make the determination. If the DMC determination is made by the courts, it is simply qualified with "adjudicated."

Ordinarily, adults are presumed to have DMC. But this presumption may be challenged when an individual has a condition that impairs cognitive abilities necessary for making decisions.

The need for DMC assessment has increased over the past several decades with the rise of the informed consent doctrine. This doctrine has led to less reliance on broad paternalistic labels, such as "unsound mind." Patients are being called upon to provide formal informed consent for an increasing number of medical and surgical procedures. Also, due to rising health care costs, hospitals have an incentive to discharge patients quickly, often creating the need for impaired patients to make decisions that previously could be avoided or at least delayed until the patients regained their faculties.

CRITERIA FOR CAPACITY: THE FOUR ABILITIES MODEL

The modern concept of DMC focuses on the patient's functional abilities to use the disclosed information to arrive at a choice, rather than on some feature of the person such as diagnosis, age, legal status, or a quasi-psychological label that functions as a proxy for "normal" (e.g., "being of sound mind"). In other words, competence is defined as a functional concept that depends on what a person can actually do or not do, rather than the person's clinical or legal status (Grisso, 2003). A comprehensive review of state laws, court cases, commission reports, and other ethicolegal literature reveals, with slight variations, four standards or abilities that are commonly cited (Berg, Appelbaum, & Grisso, 1996). These four abilities are communicating a choice, understanding, appreciation, and reasoning. This four-abilities model has been further delineated (Grisso et al., 1998; Appelbaum, 2007), along with considerable amount of empirical application of the model over the years, making it the most commonly cited framework for assessing DMC (Kim, 2010).

Communicating a Choice

The ability to communicate a choice simply requires that the patient evidence a decision regarding a treatment or procedure. The decision must be stable enough to carry out the patient's decision. Communicating a choice is a necessary but insufficient basis for competence.

Understanding

The ability to understand the information relevant to decision making is the most intuitive standard; some version of it is present in all discussions of competency standards and in all legal definitions of capacity (Berg et al., 1996). The ability to understand, however, is broader than a mere retention and regurgitation; the patient must be able to "grasp the fundamental meaning"(Appelbaum, 2007) of the disclosed information. One can get at this by asking, "Can you tell me in your own words what the doctors have told you so far?" and asking the patient to explain the relevant concepts in his answer ("What's involved in the surgery?" "What do you mean by 'the treatment will work'?").

Appreciation

Appreciation is the ability to apply the disclosed facts to one's own situation. Thus appreciation can be assessed only if understanding is intact. One can assess this ability by first confirming the patient's understanding of the facts ("What have the doctors told you about your condition and what are they recommending?") and then probing the patient's ability to apply those facts to his or her situation. This typically involves probing the patient's *beliefs* regarding the facts conveyed to him or her, assessing whether they meet three conditions: substantially erroneous, directly affecting the ability to appreciate, and caused by a neuropsychiatric pathology (Kim, 2010). A delusional patient (due to a psychotic disorder), for example, may be able to convey a perfect factual understanding of what the doctors said, but deny that those facts apply to him or her because she believes that the doctors are in fact CIA agents.

Reasoning

Reasoning is a constellation of abilities that are involved in processing information that leads to a decision. This involves being able to compare options, make logical inferences, and weigh evidence. Reasoning emphasizes the formal features of processes leading to a choice, rather than the rational content of the choice. Thus, the standard does *not* refer to the reasonableness of the decision made by the patient. Although a very unconventional decision (say, refusing a high-benefit but no-burden intervention) may trigger an evaluation, the reasonableness of the content of a choice cannot be the sole basis for judging someone incompetent. The reasoning standard is not always included in laws or in court decisions, and it is never used alone but always in conjunction with other standards (Berg et al., 1996).

Although the four abilities model of communicating a choice, understanding, appreciation, and reasoning are a commonly used framework for assessing DMC, certain jurisdictions may not use all four standards, and the clinician needs to be aware of the specific requirements in his or her jurisdiction (Grisso et al., 1998).

PREVALENCE AND RISK FACTORS FOR LOSS OF CAPACITY

The neurobiological systems involved in decision-making capacities are often affected by disease. Generally, decision making involves assigning a subjective value to a situation or thing, and then making choices based on the values assigned. Research over the last decade (see Kable & Glimcher, 2009; Rangel et al., 2008) has demonstrated that subjective values are assigned by a complex interplay of striatal and ventromedial circuits. Striatal neurons modulated by dopamine are involved in predicting and learning from how well predictions of rewards match the actual outcomes. Frontoparietal circuits are then involved in actually choosing among possible actions based on the values that have been assigned. Thus, diseases that damage striatal, ventromedial frontal, and parietal neuronal circuitry could impair decision-making capacities.

Prevalence in Various Clinical Settings

The lack of treatment consent capacity is quite common, especially among those admitted to acute care hospitals. Recent studies indicate that 37% to 44% of such patients may lack DMC for treatment consent (Raymont et al., 2004; Etchells et al., 1999). In nursing homes, lack of DMC ranges from 44% to 69% (Pruchno, Smyer, Rose, Hartman-Stein, & Henderson-Laribee, 1995; Barton, Mallik, Orr, & Janofsky, 1996). Another treatment setting with high prevalence of incapacity is the acute care psychiatric hospital where approximately 30% to 60% of patients lack the capacity to consent to medication treatment or hospital admission (Okai et al., 2007; Owen et al., 2008).

Neuropsychiatric Conditions that Affect Capacity

Delirium (often in patients with dementia) is the major cause of incapacity in general hospitals (Umapathy, Ramchandani, Lamdan, Kishel, & Schindler, 1999; Farnsworth, 1990). Although delirium is a global, acute dysfunction of cognitive abilities (such as attention, memory, visual-spatial and other functions), there are instances where some psychotic symptoms are present without a similar degree of cognitive impairment (Meagher et al., 2007). For these patients, brief cognitive screens (such as the mini mental state [Folstein, Folstein, & McHugh, 1975]) may be misleading. A highly educated, delirious patient, for instance, could have a score that is in the normal range and yet have an underlying delusion that prevents the person from making a competent medical decision.

A recent population-based study found that nearly 14% of adults over the age of 70 suffer from dementia in the United States. Of these, 70% have Alzheimer's disease and another 17% suffer from vascular dementia (Plassman et al., 2007). Not unexpectedly, persons with dementia or cognitive impairment are more likely to be incompetent than their elderly counterparts without these diagnoses (Bassett, 1999; Dymek, Atchison, Harrell, & Marson, 2001; Kim, Caine, Currier, Leibovici, & Ryan, 2001; Marson, Ingram, Cody, & Harrell, 1995; Marson, Annis, McInturff, Bartolucci, & Harrell, 1999; Marson, Earnst, Jamil, Bartolucci, & Harrell, 2000; Fazel, Hope, & Jacoby, 1999).

Even among those with known diagnoses of dementia (such as Alzheimer's disease), there is sufficient heterogeneity that one cannot simply equate a diagnosis of dementia with incapacity (Marson et al., 1995; Stanley et al., 1988). For example, we found that 34% of mild to mild-moderate AD patients (mean MMSE 22.9) performed above a threshold validated by a clinician panel on all four standards of decision-making ability (Kim et al., 2001). Nevertheless, despite the heterogeneity, dementing illnesses in general do have a major impact on treatment consent capacity, even when the disease is in the early stages (Okonkwo et al., 2007). Other neurodegenerative disorders such as Parkinson's disease, when accompanied by cognitive dysfunction, also lead to impairment in decision-making capacity (Dymek et al., 2001).

The influence of psychotic disorders on treatment consent capacity has been extensively studied, and the results can be summarized as follows. First, chronic psychotic disorders (schizophrenia, schizoaffective disorder, and chronic debilitating cases of bipolar disorder) are a risk factor for impaired consent capacity. A multicenter study found that about 25% of the persons with schizophrenia failed a psychometric threshold for capacity, with 52% failing on at least one standard of decision-making ability (Grisso & Appelbaum, 1995). Second, although chronic psychotic disorders are an important risk factor, there is still a wide range of DMC performance among persons with such diagnoses. For example, among stable outpatients in assisted living, one study (Palmer, Dunn, Appelbaum, & Jeste, 2004) showed that only the measure of understanding significantly differed between controls and patients and that their other abilities (such as reasoning and appreciation) were relatively preserved (Palmer et al., 2004). Third, performance of chronically psychotic patients on abilities related to consent capacity is more a function of cognitive symptoms (and negative symptoms such a lack of volition, slowness of thought, etc.) than of classic positive psychotic symptoms such as delusions and hallucinations (Palmer & Savla, 2007; Carpenter, Jr. et al., 2000; Moser et al., 2002; Palmer et al., 2004). Finally, several studies have shown that the factual comprehension (understanding) aspect of DMC can be improved in these patients by a variety of methods (Dunn et al., 2002; Carpenter, Jr. et al., 2000; Moser et al., 2002).

A manic episode, the hallmark of bipolar disorder (sometimes referred to as manic depressive illness), is accompanied by impulsivity, grandiose thinking, distractibility, rapid speech and "racing thoughts," increased activity, and lack of need for sleep. It is often accompanied by psychotic beliefs and poor judgment in personal interactions, in spending money, and in risky activity. Although some research has shown that repeated disclosures can improve factual comprehension in this group (Misra, Socherman, Park, Hauser, & Ganzini, 2008), mania is very strongly associated with loss of DMC for treatment decisions (Owen et al., 2008; Beckett & Chaplin, 2006).

Mild to moderate depression has little effect on the abilities relevant for consent capacity (Vollmann, Bauer, Danker-Hopfe, & Helmchen, 2003; Appelbaum, Grisso, Frank, O'Donnell, & Kupfer, 1999; Stiles, Poythress, Hall, Falkenbach, & Williams, 2001; Grisso et al., 1995). However, some persons with severe depression, if accompanied by psychotic symptoms or cognitive impairment, may evidence significant loss of decision-making abilities (Lapid et al., 2003; Bean, Nishisato, Rector, & Glancy, 1994).

Traumatic brain injury (TBI) affects 5.3 million Americans with an annual societal cost of $48.3 billion (National Center for Injury Prevention and Control, 2008). The leading causes of TBI are falls, motor vehicle accidents, moving injuries (such as sports), and assault. TBI disability is becoming more common, especially with the increasing number of combat veterans who survive head injuries. The issue of decision-making capacity looms large in the brain injury rehabilitation setting (Mukherjee & McDonough, 2006; Marson et al., 2005). In a study of moderate to severe TBI patients 6 months after acute hospitalization, 25% to 34% were marginally capable or incapable on key measures of DMC abilities (Marson et al., 2005). Frontal lobe injuries are common in TBI, with attendant impairment of executive functioning (Reid-Proctor, Galin, & Cumming, 2001) and decreased consent related abilities (Holzer, Gansler, Moczynski, & Folstein, 1997; Royall et al., 1997; Marson, Chatterjee, Ingram, & Harrell, 1996). Because executive dysfunctions are often difficult to measure without formal neuropsychological evaluation (Reid-Proctor et al., 2001), evaluation of such patients requires corroborating evidence regarding events and behavior outside the hospital.

PRINCIPLES OF ASSESSMENT: EVALUATING FUNCTION IN CONTEXT

Although courts and state laws have delineated the broad criteria for capacity, how they are applied in practice depend on principles of practice that have evolved over the years. These principles are not found in laws (Grisso et al., 1998), but they are widely endorsed by various reports and guidelines nationally (President's Commission for the Study of Ethical Problems in Medicine and Biomedical and Behavioral Research, 1982; National Bioethics Advisory Commission, 1998) and internationally (WHO, 2005).

Capacity is Task-Specific

Just because a patient is capable of consenting to a medical procedure, it does not mean that she is also capable of consenting to a research protocol. Thus, evaluation of these different DMCs must be task and decision-specific.

This also means that *capacity is not defined by diagnosis or condition*. The evaluation must focus on the function, not a diagnosis. Further, the testing of the patient's functioning itself has to be keyed into the decision-making abilities themselves. Thus, *capacity must be assessed by measuring the abilities or criteria directly, not by using generic tests of cognitive functioning*. However, cognitive tests can be useful in establishing the degree of cognitive impairment, alerting the clinician to the need for a more thorough examination of the person's DMC.

Risk-benefit Analysis Matters in Setting Thresholds for Capacity

How impaired must a person be to be deemed lacking the capacity to provide informed consent? The most important contextual consideration is the risk-benefit profile of the patient's choice. Although caution must be exercised so as not to return to the days of unacceptable medical paternalism, risk-sensitivity of thresholds is a widely accepted principle (National Bioethics Advisory Commission, 1998; President's Commission for the Study of Ethical Problems in Medicine and Biomedical and Behavioral Research, 1982).

Capacity Should Be Enhanced When Possible

As noted above there is a strong body of evidence that educational interventions can improve comprehension in persons with bipolar disorder, even in a manic state (Misra et al., 2008) and schizophrenia (Dunn, Lindamer, Palmer, Schneiderman, & Jeste, 2001; Wirshing, Wirshing, Marder, Liberman, & Mintz, 1998; Stiles et al., 200; Carpenter, Jr. et al., 2000; Dunn & Jeste, 2001). It would appear that as long as the neuropsychiatric impairment does not itself severely impair the ability to learn, there is the hope of improving the patient's treatment consent capacity.

Whether one can improve appreciation and reasoning through interventions has not been well studied.

Avoiding a Capacity Evaluation May Sometimes Be the Best Course

An elderly woman with mild impairment in her cognition may be at risk if she goes home with no social or other supports. In such a case, it is better to make her life safe (if resources are available), rather than focusing on her capacity. Or when a patient with a prominent personality disorder whose treatment team's frustration with him leads to a request for a capacity evaluation ("Is he competent to refuse...?"), the consult request is best interpreted as a request for help in managing the situation.

GOING TO COURT

In many instances the loss of patient decision-making capacity requires adjudication by the courts. The reasons for going to court may vary from case to case.

Inability to Care for Oneself

Perhaps the most common reason for going to court is when an elderly patient with dementia who had been living alone but no longer can do so safely, and needs to be placed in a living facility. When such a patient objects to the recommended plan, doctors must refer to the courts.

Special Medical Treatments or Procedures

Some medical interventions require court authorization. Such medical interventions are controversial because there is the risk of exposing the patient to harm, burden, or indignity, not for the patient's own welfare, but for some other purpose. For psychiatric interventions—such as antipsychotics, electroconvulsive therapy, and psychosurgery—a primary concern has been the issue of using medical procedures for social control (Appelbaum, 1983; Valenstein, 1986). Another extraordinary intervention with a history of spectacular abuses is sterilization (Committee on Bioethics, 1999; Dubler & White, 1995).

Patient Disagrees with Clinical Determination of Incompetence

Most health care proxy laws do not authorize a proxy to override the active objection of a patient. This remains true even if that patient has been deemed incapacitated by a physician who has conducted a formal capacity evaluation. In such a situation, the court is the proper recourse to help guide health care workers on how to proceed.

Surrogate Decision Maker(s) Not Available, Unqualified, or in Conflict

When a surrogate is not present (or if the incompetent patient objects to the surrogate's decision) and the situation is urgent, an emergency court hearing might be necessary. Factors that favor proceeding to court include (a) availability of urgent court hearing in the jurisdiction, (b) the relative invasiveness of the medical procedure in question, and (c) a benefit to risk ratio in favor of intervention being less clear cut.

When the surrogate decision maker is not able to carry out his or her duties—whether this person is a de facto surrogate, a health care proxy or durable power of attorney for health care, or even a guardian—the courts may need to appoint an alternative decision maker. For example, the surrogate may himself be impaired or may have a conflict of interest.

When there are intractable conflicts among potential surrogates, courts may have to decide who will have the final decision-making authority for the patient. For patients without a formally designated surrogate such as a health care proxy, most states have some type of de facto surrogate treatment laws which guide practice (American Bar Association Commission on Law and Aging, 2008).

Guardianship should be sought when other mechanisms for surrogate decision making are not available for an incapacitated patient who is likely to face a series of major medical decisions. It does not serve such a patient well if one has to engage in a slow administrative or legal process for each future decision.

SPECIAL ISSUES IN RESEARCH CONSENT CAPACITY

Research involving the decisionally impaired remains controversial. Most of this controversy has centered around the ethics of surrogate consent, and the permissibility of exposing subjects to research risks when they cannot consent to such research themselves (Kim, Appelbaum, Jeste, & Olin, 2004). The best way to determine DMC to provide informed consent for research is relatively neglected in policy discussions. From a practical point of view, how subjects are categorized as either competent or not profoundly affects the implementation of policy regulating surrogate consent for research.

The overall approach to assessing DMC for research consent is similar to that for treatment. Capacity for research consent should be assessed in a task specific way, rather than using a generic cognitive test or inferring from a diagnosis. The four abilities model can also be used in the research context (Kim et al., 2001; Kim & Caine, 2002). But there are some notable differences between treatment and research.

First, the risk-benefit analysis is different for research than it is for treatment. The individual subjects' welfare is not the primary goal of the research. Indeed, in most situations of research, the subject forgoes some advantage in order to enhance the goals of science (Lidz & Appelbaum, 2002). The implication for capacity assessment is that the threshold for competence must take into account this different risk-benefit context. Second, the elements of disclosure for informed consent for research are obviously different from the treatment situation, and are spelled out in the Federal regulations (45 CFR 46.116). Third, the assessment of capacity in research provides an opportunity for a more standardized assessment since the subjects are being enrolled in the same study, and thus are making the same decision.

This may allow for more direct incorporation of relevant evidence in setting cutoff thresholds, for example, since standardization is possible (Kim et al., 2007).

The need for conducting research with persons afflicted with neuropsychiatric disorders is increasing. Alzheimer's disease, the most common form of dementia, will afflict 80.1 million people worldwide by 2040 (Ferri et al., 2005). The disease leads to early decisional incapacity (Okonkwo et al., 2007; Kim et al., 2001; Warner, McCarney, Griffin, Hill, & Fisher, 2008). Research on Alzheimer's disease can involve invasive procedures with unpredictable risks (Orgogozo et al., 2003; Tuszynski et al., 2005). In addition, scientists are using invasive methods to target other nervous system conditions such as tetraplegia (Hochberg et al., 2006), disorders of consciousness (Schiff et al., 2007), refractory mood and anxiety disorders (Mayberg et al., 2005; Greenberg et al., 2006), Huntington's disease (Dunnett & Rosser, 2004), and other conditions (Damier et al., 2007; Maciunas et al., 2007). Given the pace of progress in neuroscience research and the impetus toward implementing those advances in the clinic, it is likely that the assessment of capacity for informed consent for clinical neuroscience and interventions research will continue to challenge researchers, policymakers, and clinicians for some time to come.

REFERENCES

American Bar Association Commission on Law and Aging (2008). *Surrogate Consent in the Absence of an Advance Directive—January 2008.*

Appelbaum, P. S. (1983). Refusing treatment: the uncertainty continues. *Hospital & Community Psychiatry, 34,* 11–12.

Appelbaum, P. S. (2007). Assessment of patients' competence to consent to treatment. *New England Journal of Medicine, 357,* 1834–1840.

Appelbaum, P. S., Grisso, T., Frank, E., O'Donnell, S., & Kupfer, D. (1999). Competence of depressed patients for consent to research. *American Journal of Psychiatry, 156,* 1380–1384.

Barton, C. D. J., Mallik, H. S., Orr, W. B., & Janofsky, J. S. (1996). Clinicians' judgement of capacity of nursing home patients to give informed consent. *Psychiatric Services, 47,* 956–960.

Bassett, S. S. (1999). Attention: Neuropsychological predictor of competency in Alzheimer's disease. *Journal of Geriatric Psychiatry & Neurology, 12,* 200–205.

Bean, G., Nishisato, S., Rector, N. A., & Glancy, G. (1994). The psychometric properties of the competency interview schedule. *Canadian Journal of Psychiatry—Revue Canadienne de Psychiatrie, 39,* 368–376.

Beckett, J. & Chaplin, R. (2006). Capacity to consent to treatment in patients with acute mania. *Psychiatric Bulletin, 30,* 419–422.

Berg, J. W., Appelbaum, P. S., & Grisso, T. (1996). Constructing Competence: Formulating Standards of Legal Competence to Make Medical Decisions. *Rutgers Law Review, 48,* 345–396.

Berg, J. W., Appelbaum, P. S., Lidz, C. W., & Parker, L. S. (2001). *Informed consent: Legal theory and clinical practice* (2nd ed.) New York: Oxford University Press.

Carpenter, W. T., Jr., Gold, J., Lahti, A., Queern, C., Conley, R., Bartko, J. et al. (2000). Decisional capacity for informed consent in schizophrenia research. *Archives of General Psychiatry, 57,* 533–538.

Committee on Bioethics (1999). Sterilization of minors with developmental disabilities. *Pediatrics*, *104*, 337–340.

Damier, P., Thobois, S., Witjas, T., Cuny, E., Derost, P., Raoul, S. et al. (2007). Bilateral deep brain stimulation of the globus pallidus to treat tardive dyskinesia. *Archives of General Psychiatry*, *64*, 170–176.

Dubler, N. & White, A. (1995). Fertility control: legal and regulatory issues. In W. T. Reich (Ed.), *Encyclopedia of Bioethics* (pp. 839–847). New York: Simon & Shuster Macmillan.

Dunn, L. & Jeste, D. V. (2001). Enhancing informed consent: A review. *Neuropsychopharmacology*, *24*, 595–607.

Dunn, L., Lindamer, L., Palmer, B. W., Golshan, S., Schneiderman, L., & Jeste, D. V. (2002). Improving understanding of research consent in middle-aged and elderly patients with psychotic disorders. *American Journal of Geriatric Psychiatry*, *10*, 142–150.

Dunn, L. B., Lindamer, L. A., Palmer, B. W., Schneiderman, L. J., & Jeste, D. V. (2001). Enhancing comprehension of consent for research in older patients with psychosis: A randomized study of a novel consent procedure. *American Journal of Psychiatry*, *158*, 1911–1913.

Dunnett, S. B. & Rosser, A. E. (2004). Cell therapy in Huntington's disease. *NeuroRx*, *1*, 394–405.

Dymek, M., Atchison, P., Harrell, L., & Marson, D. C. (2001). Competency to consent to medical treatment in cognitively impaired patients with Parkinson's disease. *Neurology*, *56*, 17–24.

Etchells, E., Darzins, P., Silberfeld, M., Singer, P. A., McKenny, J., Naglie, G. et al. (1999). Assessment of patient capacity to consent to treatment. *Journal of General Internal Medicine*, *14*, 27–34.

Farnsworth, M. G. (1990). Competency evaluations in a general hospital. *Psychosomatics*, *31*, 60–66.

Fazel, S., Hope, T., & Jacoby, R. (1999). Assessment of competence to complete advance directives: validation of a patient centred approach. *British Medical Journal*, *318*, 493–497.

Ferri, C., Prince, M., Brayne, C., Brodaty, H., Fratiglioni, L., Ganguli, M. et al. (2005). Global prevalence of dementia: a Delphi consensus study. *Lancet*, *366*, 2112–2117.

Folstein, M. F., Folstein, S. E., & McHugh, P. (1975). Mini-mental state. A practical guide for grading the cognitive state of patients for the clinician. *Journal of Psychiatric Research*, *12*, 189–198.

Greenberg, B. D., Malone, D. A., Friehs, G. M., Rezai, A. R., Kubu, C. S., Malloy, P. F. et al. (2006). Three-year outcomes in deep brain stimulation for highly resistant obsessive-compulsive disorder. *Neuropsychopharmacology*, *31*, 2384–2393.

Grisso, T. (2003). *Evaluating competencies* (2nd ed.) New York: Kluwer/Plenum.

Grisso, T. & Appelbaum, P. S. (1995). The MacArthur treatment competence study. III: Abilities of patients to consent to psychiatric and medical treatments. *Law & Human Behavior*, *19*, 149–174.

Grisso, T. & Appelbaum, P. S. (1998). *Assessing competence to consent to treatment: A guide for physicians and other health professionals*. New York: Oxford University Press.

Hochberg, L. R., Serruya, M. D., Friehs, G. M., Mukand, J. A., Saleh, M., Caplan, A. H. et al. (2006). Neuronal ensemble control of prosthetic devices by a human with tetraplegia. *Nature*, *442*, 164–171.

Holzer, J. C., Gansler, D. A., Moczynski, N. P., & Folstein, M. F. (1997). Cognitive functions in the informed consent evaluation process: a pilot study. *Journal of the American Academy of Psychiatry & the Law, 25*, 531–540.

Kable, J.W., Glimcher, P.W. (2009). The neurobiology of decision: consensus and controversy. *Neuron, 63*, 733–745.

Kim, S. & Caine, E. D. (2002). Utility and limits of the mini mental state examination in evaluating consent capacity in Alzheimer's disease. *Psychiatric Services, 53*, 1322–1324.

Kim, S. Y. H. (2010). *Evaluation of capacity to consent to treatment and research.* New York: Oxford University Press.

Kim, S. Y. H., Appelbaum, P. S., Jeste, D. V., & Olin, J. T. (2004). Proxy and surrogate consent in geriatric neuropsychiatric research: Update and recommendations. *American Journal of Psychiatry, 161*, 797–806.

Kim, S. Y. H., Appelbaum, P. S., Swan, J., Stroup, T. S., McEvoy, J. P., Goff, D. C. et al. (2007). Determining when impairment constitutes incapacity for informed consent in schizophrenia research. *The British Journal of Psychiatry, 191*, 38–43.

Kim, S. Y. H., Caine, E. D., Currier, G. W., Leibovici, A., & Ryan, J. M. (2001). Assessing the competence of persons with Alzheimer's disease in providing informed consent for participation in research. *American Journal of Psychiatry, 158*, 712–717.

Lapid, M., Rummans, T., Poole, K., Pankratz, S., Maurer, M., Rasmussen, K. et al. (2003). Decisional capacity of severely depressed patients requiring electroconvulsive therapy. *Journal of ECT, 19*, 67–72.

Lidz, C. W. & Appelbaum, P. S. (2002). The therapeutic misconception: Problems and solutions. *Medical Care, 40*, V55–V63.

Maciunas, R. J., Maddux, B. N., Riley, D. E., Whitney, C. M., Schoenberg, M. R., Ogrocki, P. J. et al. (2007). Prospective randomized double-blind trial of bilateral thalamic deep brain stimulation in adults with Tourette syndrome. *J Neurosurg, 107*, 1004–1014.

Marson, D. C., Annis, S. M., McInturff, B., Bartolucci, A., & Harrell, L. E. (1999). Error behaviors associated with loss of competency in Alzheimer's disease. *Neurology, 53*, 1983–1992.

Marson, D. C., Chatterjee, A., Ingram, K. K., & Harrell, L. E. (1996). Toward a neurologic model of competency: Cognitive predictors of capacity to consent in Alzheimer's disease using three different legal standards. *Neurology, 46*, 666–672.

Marson, D. C., Dreer, L. E., Krzywanski, S., Huthwaite, J. S., Devivo, M. J., & Novack, T. A. (2005). Impairment and partial recovery of medical decision-making capacity in traumatic brain injury: a 6-month longitudinal study. *Arch Phys.Med Rehabil., 86*, 889–895.

Marson, D. C., Earnst, K., Jamil, F., Bartolucci, A., & Harrell, L. (2000). Consistency of physicians' legal standard and personal judgments of competency in patients with Alzheimer's disease. *Journal of the American Geriatrics Society, 48*, 911–918.

Marson, D. C., Ingram, K. K., Cody, H. A., & Harrell, L. E. (1995). Assessing the competency of patients with Alzheimer's disease under different legal standards. A prototype instrument. *Archives of Neurology, 52*, 949–954.

Mayberg, H., Lozano, A., Voon, V., McNeely, H., Seminowicz, D., Hamani, C. et al. (2005). Deep brain stimulation for treatment-resistant depression. *Neuron, 45*, 651–660.

Meagher, DJ., Moran, M., Raju, B., Gibbons, D., Donnelly, S., Saunders, J. et al. (2007). Phenomenology of delirium: Assessment of 100 adult cases using standardised measures. *The British Journal of Psychiatry, 190*, 135–141.

Misra, S., Socherman, R., Park, B. S., Hauser, P., & Ganzini, L. (2008). Influence of mood state on capacity to consent to research in patients with bipolar disorder. *Bipolar. Disord., 10,* 303–309.

Moser, D. J., Schultz, S. K., Arndt, S., Benjamin, M. L., Fleming, F. W., Brems, C. S. et al. (2002). Capacity to provide informed consent for participation in schizophrenia and HIV research. *American Journal of Psychiatry, 159,* 1201–1207.

Mukherjee, D. & McDonough, C. (2006). Clinician perspectives on decision-making capacity after acquired brain injury (ethics in practice). *Topics in Stroke Rehabilitation, 13,* 75–83.

National Bioethics Advisory Commission (1998). *Research involving persons with mental disorders that may affect decisionmaking capacity.* (Volume 1 ed.) Rockville, MD: National Bioethics Advisory Commision.

National Center for Injury Prevention and Control (2008). *Traumatic brain injury.* Atlanta: Centers for Disease Control and Prevention.

Okai, D., Owen, G., McGuire, H., Singh, S., Churchill, R., & Hotopf, M. (2007). Mental capacity in psychiatric patients: Systematic review. *The British Journal of Psychiatry, 191,* 291–297.

Okonkwo, O., Griffith, H. R., Belue, K., Lanza, S., Zamrini, E. Y., Harrell, L. E. et al. (2007). Medical decision-making capacity in patients with mild cognitive impairment. *Neurology, 69,* 1528–1535.

Or gogozo, J. M., Gilman, S., Dartigues, J. F., Laurent, B., Puel, M., Kirby, L. C. et al. (2003). Subacute meningoencephalitis in a subset of patients with AD after Aβ42 immunization. *Neurology, 61,* 46–54.

Owen, G., Richardson, G., David, AS., Szmukler, G., Hayward, P., & Hotopf, M. (2008). Mental capacity to make decisions on treatment in people admitted to psychiatric hospitals: cross sectional study. *BMJ, 337,* a448.

Palmer, B. W. & Savla, G. N. (2007). The association of specific neuropsychological deficits with capacity to consent to research or treatment. *J Int Neuropsychol.Soc, 13,* 1047–1059.

Palmer, B. W., Dunn, L. B., Appelbaum, P. S., & Jeste, D. V. (2004). Correlates of treatment-related decision-making capacity among middle-aged and older patients with schizophrenia. *Archives of General Psychiatry, 61,* 230–236.

Plassman, B. L., Langa, K. M., Fisher, G. G., Heeringa, S. G., Weir, D. R., Ofstedal, M. B. et al. (2007). Prevalence of dementia in the United States: the aging, demographics, and memory study. *Neuroepidemiology 29,* 125–132.

President's Commission for the Study of Ethical Problems in Medicine and Biomedical and Behavioral Research (1982). *Making health care decisions: The ethical and legal implications of informed consent in the patient-practitioner relationship* (Rep. No. One).

Pruchno, R. A., Smyer, M. A., Rose, M. S., Hartman-Stein, P. E., & Henderson-Laribee, D. L. (1995). Competence of long-term care residents to participate in decisions about their medical care: a brief, objective assessment. *Gerontologist, 35,* 622–629.

Rangel, A., Camerer, C., Montague, P.R. (2008). *Nature Reviews Neuroscience, 9,* 545–556.

Raymont, V., Bingley, W., Buchanan, A., David, A. S., Hayward, P., Wessely, S. et al. (2004). Prevalence of mental incapacity in medical inpatients and associated risk factors: cross-sectional study. *The Lancet, 364,* 1421–1427.

Reid-Proctor, G. M., Galin, K., & Cumming, M. A. (2001). Evaluation of legal competency in patients with frontal lobe injury. *Brain Injury, 15*, 377–386.

Royall, D. R., Cordes, J., & Polk, M. (1997). Executive control and the comprehension of medical information by elderly retirees. *Experimental Aging Research, 23*, 301–313.

Schiff, N. D., Giacino, J. T., Kalmar, K., Victor, J. D., Baker, K., Gerber, M. et al. (2007). Behavioural improvements with thalamic stimulation after severe traumatic brain injury. *Nature, 448*, 600–603.

Stanley, B., Stanley, M., Guido, J., & Garvin, L. (1988). The functional competency of elderly at risk. *Gerontologist, 28*, 53–58.

Stiles, PG., Poythress, NG., Hall, A., Falkenbach, D., & Williams, R. (2001). Improving understanding of research consent disclosures among persons with mental illness. *Psychiatric Services, 52*, 780–785.

Tuszynski, M. H., Thal, L., Pay, M., Salmon, D. P., Hoi, S., Bakay, R. et al. (2005). A phase 1 clinical trial of nerve growth factor gene therapy for Alzheimer disease. *Nature Medicine, 11*, 551–555.

Umapathy, C., Ramchandani, D., Lamdan, R., Kishel, L., & Schindler, B. (1999). Competency evaluations on the consultation-liaison service. *Psychosomatics, 40*, 28–33.

Valenstein, E. S. (1986). *Great and Desperate Cures: The Rise and Decline of Psychosurgery and Other Radical Treatments for Mental Illness*. New York: Basic Books.

Vollmann, J., Bauer, A., Danker-Hopfe, H., & Helmchen, H. (2003). Competence of mentally ill patients: a comparative empirical study. *Psychological Medicine, 33*, 1463–1471.

Warner, J., McCarney, R., Griffin, M., Hill, K., & Fisher, P. (2008). Participation in dementia research: rates and correlates of capacity to give informed consent. *Journal of Medical Ethics, 34*, 167–170.

World Health Organization (WHO). (2005). *WHO Resource book on mental health, human rights and legislation*. Geneva: World Health Organization.

Wirshing, D. A., Wirshing, W. C., Marder, S. R., Liberman, R. P., & Mintz, J. (1998). Informed consent: Assessment of comprehension. *American Journal of Psychiatry, 155*, 1508–1511.

8

Addiction and Responsibility

STEVEN E. HYMAN

Cognitive and social neuroscience and studies of the pathophysiologic processes underlying neuropsychiatric disorders have begun to probe the mechanisms by which human beings regulate their behavior in conformity with social conventions and in pursuit of chosen goals—and the circumstances under which such "cognitive control" may be eroded (Miller & Cohen, 2001; Miller & D'Esposito, 2005; Montague, Hyman, & Cohen, 2004). The resulting ideas call into question folk psychology views on the voluntary control of behavior; that is, for the most part, we regulate our actions based on conscious reasons. Even in health, critical processes that intervene between sensory inputs to the brain and the execution of actions, including processes that permit top-down or cognitive control of behavior, do not appear to depend on conscious exertion of will (Wenger, 2002). Challenges to folk psychology views of the voluntary control of behavior may be highlighted most vividly, however, by conditions such as addiction, in which the core symptoms reflect a failure of the underlying processes (Hyman, 2005; Kalivas & Volkow, 2005; Montague et al., 2004), which I refer to as *cognitive control*.

The major justification for demarcating neuroethics from the broader field of bioethics derives from the special status of the brain (Roskies, 2002), which is the causal underpinning of our conscious mental lives and of our behavior. This is not a reductionist claim. The structure and function of the brain is influenced not only by bottom-up factors such as genes, but also by top-down factors such lived experience and context. Moreover, neuroscience does not obviate the need for social and psychological level explanations intervening between the levels of cells, synapses, and circuits and that of ethical judgments. Indeed, modern cognitive and social neuroscience (Cacioppo, Berntson, Adolphs, & Carter, 2002; Gazzaniga, 2004) are, in no small measure, attempts to mediate between understandings of the functioning of neural networks in one regard and of sensation, thought, and action in another. What neuroscience contributes to ethical discourse is mechanistic insight that constrains our interpretations of psychological observations and that suggests new explanatory frameworks for thought and behavior.

Neuroscience should make it possible to ask how the nature of our brains shapes and constrains what we call *rationality*, and therefore, ethical principles themselves, and it should permit us to probe deeply into the nature of reason, emotion, and the control of behavior (Churchland, 2006). Having recently reviewed the neurobiology of addiction for clinicians (Hyman, 2005) and for neuroscientists (Hyman, Malenka, & Nestler, 2006), I would like to examine the implications of emerging ideas about reward, cognitive control, and the pathophysiology of addiction for insights into the voluntary control behavior.

ADDICTION AND RESPONSIBILITY

The question of whether and to what extent an addicted individual is responsible for his or her actions remains a matter of unsettled debate. One proxy (albeit imperfect) for this question is disagreement as to whether addiction is best conceptualized as a brain disease (Leshner, 1997; McLellan, Lewis, O'Brien, & Kleber, 2000), as a moral condition (Satel, 1999), or as some combination of the two (Morse, 2004b). Those who argue for the disease model not only believe it is justified by empirical data but also see virtue in the possibility that a disease model decreases the stigmatization of addicted people and increases their access to medical treatments. Those who argue that addiction is best conceptualized as a moral condition are struck by the observation that drug-seeking and drug-taking involve a series of voluntary acts that often require planning and flexible responses to changing conditions—not simply impulsive or robotic acts. They worry that medicalization will lead addicted people to fatalism about their condition and to excuses for their actions rather than full engagement with treatment and rehabilitation and an effort to conform to basic societal expectations.

Current definitions of addiction come from medical texts and thus, not surprisingly, favor a disease model. Indeed, addiction looks very much like a disease (admittedly definitions of "disease" remain somewhat fuzzy). Addiction has known risk factors (family history, male sex) and a typical course and outcome: often a chronic course punctuated by periods of abstinence followed by relapse (Hser, Hoffman, Grella, & Anglin, 2001; McLellan et al., 2000). True, the precise alterations in physiology that account for the symptoms and course are not yet known with certainty, but there is little doubt in the scientific community that such mechanisms will be found (Chao & Nestler, 2004). Similarly, the search for the precise genetic variants that confer familial risk is in its early days, but existing data from family, twin, and adoption studies convincingly argue that genes play a central role in vulnerability (Goldman, Oroszi, & Ducci, 2005).

What is more interesting is that modern definitions of addiction focus squarely on the issue of voluntary control. The current medical consensus is that the cardinal feature of addiction is compulsive drug use despite significant negative consequences (American Psychiatric Association, 1994). The term *compulsion* is imprecise, but at a minimum implies diminished ability to control drug use, even in the face of factors (e.g., illness, failure in life roles, loss of job, arrest) that should motivate cessation of drug use in a rational agent willing and able to exert control over behavior. The focus on loss of control is not derived primarily from a theory, but from extensive observation of the behavior of addicted individuals (O'Brien, Childress, Ehrman, & Robbins, 1998; Tiffany, 1990) and indeed recognition of the failure of previous definitions to capture clinical realities. The current focus on compulsive use as the defining features of addiction superseded previous views that focused on dependence and withdrawal. These previous views implied that addicted individuals take drugs to seek pleasure and avoid aversive withdrawal symptoms. Although the avoidance of withdrawal might create strong motivation to take drugs, this view does not imply a loss of voluntary control. This previous view failed on several counts. First, some highly addictive drugs such as cocaine and amphetamine may produce mild withdrawal symptoms and lack a physical withdrawal syndrome entirely. Moreover, the previous view does not explain the stubborn persistence of relapse risk long after detoxification, long after the last withdrawal symptom, if any, has passed, and despite incentives to avoid a resumption of drug use (Hyman, 2005).

Before discussing my views of the neural basis of addiction, I should stipulate that the science is in its early stages and that there is not yet a fully convincing theory of how addiction results from the interaction of risk factors, drugs, and the brain. Moreover, there are still disagreements at the theoretical level of what the existing data signifies for the mechanisms of addiction. (Compare, for example, Hyman, 2005; Koob & Le Moal, 2005; and Robinson & Berridge, 2003). This state of affairs invites skepticism from those wary of a disease model (Satel, 1999). Nonetheless, we cannot select models of human behavior based on desired social implications, but must rely on the scientific evidence we have. Despite somewhat different views of mechanism, all current mainstream formulations agree that addiction diminishes voluntary behavioral control. At the same time, none of the current views conceives of the addicted person to be devoid of all voluntary control and thus absolved of all responsibility for self-control.

NEURAL BASIS OF ADDICTION AND SELF-CONTROL

Short of being harshly coerced, severely psychotic, or significantly demented, what can it mean to say that a person cannot control his or her actions? An alcoholic must obtain money, go to the liquor store or otherwise obtain alcohol (perhaps carefully hidden from a spouse), and consume drinks. A heroin user may have to go to great lengths to obtain the drug, perhaps committing one of more crimes, before beginning the ritual that ends in self-injection. How can these extended chains of apparently voluntary acts be the result of compulsion? In my view, addictive drugs tap into and, in vulnerable individuals, usurp powerful mechanisms by which survival-relevant goals shape behavior (Hyman, 2005; Hyman et al., 2006).

Diverse organisms, including humans, pursue goals with positive survival value such as food, safety, and opportunities for mating; such goals act as rewards (Kelley & Berridge, 2002). Rewards are experienced as pleasurable and as motivating (they are desired). Environmental cues that predict their availability (e.g., the smell of baking bread) are rapidly learned and are imbued with incentive properties: They activate "wanting" and initiate behaviors aimed at obtaining the desired goal. Such goal-directed behaviors tend to increase in frequency over time (reinforcement) and to become highly efficient. Of course, rewarding goals for humans can vary enormously in immediacy, complexity, and motivational power, ranging from a well-liked food to seeing a favorite painting in a museum.

The brain has evolved several specialized mechanisms to maximize the ability of an organism to obtain rewards. There are mechanisms to provide internal representations of rewards and to assign them relative values compared with pursuing other possible goals; these mechanisms depend primarily on the orbital prefrontal cortex (Schoenbaum, Roesch, & Stalnaker, 2006). There are mechanisms that permit an organism to learn and to make relatively efficient and automatic sequences of actions to obtain specific rewards; these depend primarily on the dorsal striatum (Everitt & Robbins, 2005). Mechanisms of cognitive control support successful completion of goal-directed behaviors by maintaining the goal representation over time, suppressing distractions, and inhibiting impulsive actions that redirect the organism. Cognitive control is dependent on the prefrontal cortex and its connections to the striatum and thalamus. In humans, the capacity for cognitive control appears to be a relatively stable trait that is an important predictor of life success (Eigsti et al., 2006). Deficits in cognitive control play an important role in attention deficit–hyperactivity disorder (Vaidya et al., 2005) and may increase vulnerability to later substance misuse.

These circuits respond in a coordinated fashion to new information about rewards through the action of the neurotransmitter dopamine (Montague et al., 2004). Dopamine is released from neurons with cell bodies in the ventral tegmental area (VTA) and substantia nigra within the midbrain. These neurons project widely through the forebrain and can influence all of the circuits involved in reward-related learning, as well as in other aspects of cognition and emotion. Dopamine projections from the VTA to the nucleus accumbens bind the pleasurable (hedonic) response to a reward to desire and to goal-directed behavior (Berridge & Robinson, 1998; Everitt & Robbins, 2005). Dopamine projections from the VTA to the prefrontal cortex play a critical role in the assignment of value and in updating goal representations in response to the state of the organism (Montague et al., 2004). Dopamine projections from the substantia nigra to the dorsal striatum are critical for consolidating new behavioral responses so that reward-related cues come to activate efficient strategies to reach the relevant goal (Everitt & Robbins, 2005).

Addictive drugs are Trojan horses. Unlike natural rewards, addictive drugs have no nutritional, reproductive, or other survival value. However, all addictive drugs exert pharmacologic effects that cause release of dopamine. Moreover, the effects of addictive drugs on dopamine release are quantitatively greater than that produced by natural rewards under almost all circumstances.

Normally dopamine serves as a learning signal in the brain. Dopamine is released when a reward is new, better than expected, or unpredicted in a particular circumstance (Schultz, 2006; Schultz, Dayam, & Montague, 1997). When the world is exactly as expected, there is nothing new to learn; no new circumstances to connect either to desire or to action—and no increase in dopamine release. Because addictive drugs increase synaptic dopamine by direct pharmacologic action, they short circuit the normal controls over dopamine release that compare the current circumstance with prior experience. Thus, unlike natural rewards, addictive drugs always signal "better than expected." Neural circuits "overlearn" on an excessive and grossly distorted dopamine signal (Hyman, 2005; Hyman et al., 2006; Montague et al., 2004). Cues that predict drug availability such as persons, places, or certain bodily sensations gain profound incentive salience and the ability to motivate drug-seeking. Because of the excessive dopamine signal in the prefrontal cortex (Volkow & Fowler, 2000), drugs become overvalued compared with all other goals. Rational goals such as self-care, working, parenting, and obeying the law are devalued. In addition, normal aspects of cognitive control weaken; even if the addicted person wants to cut down, prepotent cue-initiated drug-seeking responses are extremely difficult to suppress. If the person is successful in delaying drug-seeking (or is for external reasons unable to seek drugs), intense craving may result (Tiffany, 1990). Because the changes in synaptic weight and synaptic structure that underlie memory are among the longest-lived alterations in biology, the ability of drug-related cues to cause relapses may persist for many years, even a lifetime.

CONCLUSION

There remains much to learn about the pathophysiology of addiction. Currently, much research is attempting to demonstrate that drug-induced changes in synaptic connectivity and drug-induced changes in the expression of neuronal genes and proteins are causally involved in addiction-related behaviors (Chao & Nestler, 2004; Hyman et al., 2006). This model of pathogenesis, and the research on reward-related learning on which it rests, suggest highly plausible mechanisms by which addicted individuals may lose control over drug-seeking and drug-taking (Hyman, 2005; Hyman et al., 2006; Kalivas & Volkow, 2005; Montague et al., 2004). Mechanisms that evolved to motivate survival behaviors, the pursuit of natural rewards, are usurped by the potent and abnormal dopamine signal produced by addictive drugs. The result is a brain in which drug cues powerfully activate drug-seeking, and in which attempts to suppress drug-seeking result in intense craving. This model does not, however, reduce addicted individuals to zombies who are permanently controlled by external cues. As overvalued as drugs become, as potent as the effects of drug cues on behavior, other goals are not extirpated. Perhaps in a drug-free context, perhaps with a good measure of initial coercion, perhaps with family, friends, and caregivers acting as external "prostheses" to strengthen and partially replace damaged frontal mechanisms of cognitive control, and often despite multiple relapses, addicts can cease drug use and regain a good measure of control over their drug-taking. Our current models help explain why recovery is difficult and why relapses occur even long after detoxification and rehabilitation. The long experience of humanity with addiction does not counsel fatalism, but implacable efforts to overcome the behavioral effects of neural circuits hijacked by drugs. Finally, views based on cognitive neuroscience and studies of addiction pathogenesis suggest that some apparently voluntary behaviors may not be as freely planned and executed as they first appear. Such cognitive views have not yet penetrated folk psychology, and it is premature for these views to have any place in the courtroom (Greene & Cohen, 2004; Morse, 2004a). Nonetheless, these cognitive views deserve a place in current ethical discussions of personal responsibility. For many reasons, it may be wise for societies to err on the side of holding addicted individuals responsible for their behavior and to act as if they are capable of exerting more control than perhaps they can; however, if the ideas expressed in this review are right, it should be with a view to rehabilitation of the addicted person and protection of society rather than moral opprobrium.

ACKNOWLEDGMENTS

This reading originally appeared in 2007 in the *American Journal of Bioethics-Neuroscience,* volume 7, pages 8–11, and is used with permission. Section headings were added by the editor.

REFERENCES

American Psychiatric Association. (1994). *Diagnostic and statistical manual of mental disorders,* 4th edition. Washington, DC: American Psychiatric Association.

Berridge, K. C., & Robinson, T. E. (1998). What is the role of dopamine in reward: Hedonic impact, reward learning, or incentive salience? *Brain Research Reviews, 28,* 309–369.

Cacioppo, J. T., Berntson, G. G., Adolphs, R., & Carter, C. S. (Eds.). (2002). *Foundations in social neuroscience.* Cambridge, MA: The MIT Press.

Chao, J., & Nestler, E. J. (2004). Molecular neurobiology of drug addiction. *Annual Review of Medicine, 55,* 113–132.

Churchland, P. S. (2006). Moral decision-making and the brain. In J. Illes (Ed.), *Neuroethics: Defining the issues in theory, practice, and policy* (pp. 3–16). Oxford: Oxford University Press.

Eigsti, I. M., Zayas, V., Mischel, W., Schoda, Y., Ayduk, O., Dadlani, M. B., et al. (2006). Predicting cognitive control from preschool to late adolescence and young adulthood. *Psychological Science, 17* (6), 478–484.

Everitt, B. J., & Robbins, T. W. (2005). Neural systems of reinforcement for drug addiction: From actions to habits to compulsion. *Nature Neuroscience, 8,* 1481–1489.

Gazzaniga, M. S. (Ed.). (2004). *The cognitive neurosciences III,* 4th edition. Cambridge, MA: The MIT Press.

Goldman, D., Oroszi, G., & Ducci, F. (2005). The genetics of addictions: Uncovering the genes. *Nature Reviews Genetics, 6,* 521–532.

Greene, J., & Cohen, J. (2004). For the law, neuroscience changes nothing and everything. *Philosophical Transactions of the Royal Society of London. Series B, Biological Sciences, 359,* 1775–1785.

Hser, Y. I., Hoffman, V., Grella, C. E., & Anglin, M. D. (2001). A 33-year follow-up of narcotics addicts. *Archives of General Psychiatry, 58* (5), 503–508.

Hyman, S. E. (2005). Addiction: A disease of learning and memory. *American Journal of Psychiatry, 162* (8), 1414–1422.

Hyman, S. E., Malenka, R. C., & Nestler, E. J. (2006). Neural mechanisms of addiction: The role of reward-related learning and memory. *Annual Review of Neuroscience 21* (29), 565–598.

Kalivas, P. W., & Volkow, N. D. (2005). The neural basis of addiction: A pathology of motivation and choice. *American Journal of Psychiatry, 162* (8), 1403–1413.

Kelley, A. E., & Berridge, K. C. (2002). The neuroscience of natural rewards: Relevance to addictive drugs. *Journal of Neuroscience, 22,* 3306–3311.

Koob, G. F., & Le Moal, M. (2005). *Neurobiology of addiction.* New York, NY: Academic Press.

Leshner, A. I. (1997). Addiction is a brain disease, and it matters. *Science, 278* (5335), 45–47.

McLellan, A. T., Lewis, D. C., O'Brien, C. P., & Kleber, H. D. (2000). Drug dependence, a chronic medical illness: Implications for treatment, insurance, and outcomes evaluation. *Journal of the American Medical Association, 284* (13), 1689–1689.

Miller, B. T., & D'Esposito, M. (2005). Searching for "the top" in top-down control. *Neuron, 48,* 535–538.

Miller, E. K., & Cohen, J. D. (2001). An integrative theory of prefrontal cortex function. *Annual Review of Neuroscience, 24,* 167–202.

Montague, P. R., Hyman, S. E., & Cohen, J. D. (2004). Computational roles for dopamine in behavioural control. *Nature, 431,* 760–767.

Morse, S. J. (2004a). New neuroscience, old problems: Legal implications of brain science. *Cerebrum, 6* (4), 81–90.

Morse, S. J. (2004b). Medicine and morals, craving and compulsion. *Substance Use & Misuse, 39* (3), 437–460.

O'Brien, C. P., Childress, A. R., Ehrman, R., & Robbins, S. J. (1998). Conditioning factors in drug abuse: Can they explain compulsion? *Journal of Psychopharmacology, 12,* 15–22.

Robinson, T. E., & Berridge, K. C. (2003). Addiction. *Annual Review of Psychology, 54,* 25–53.

Roskies, A. (2002). Neuroethics for the new millennium. *Neuron, 35* (1), 21–23.

Satel, S. L. (1999). What should we expect from drug abusers? *Psychiatric Services, 50* (7), 861.

Schoenbaum, G., Roesch, M. R., & Stalnaker, T. A. (2006). Orbitofrontal cortex, decision-making and drug addiction. *Trends in Neurosciences 29* (2), 116–124.

Schultz, W. (2006). Behavioral theories and the neurophysiology of reward. *Annual Review of Psychology, 57,* 87–115.

Schultz, W., Dayan, P., & Montague, P. R. (1997). A neural substrate of prediction and reward. *Science, 275,* 1593–1599.

Tiffany, S. T. (1990). A cognitive model of drug urges and drug-use behavior: Role of automatic and nonautomatic processes. *Psychological Review, 97,* 147–168.

Vaidya, C. J., Bunge, S. A., Dudukovic, N. M., Zalecki, C. A., Elliott, G. R., & Gabrieli, J. D. (2005). Altered neural substrates of cognitive control in childhood ADHD: Evidence from functional magnetic resonance imaging. *American Journal of Psychiatry, 162* (9), 1605–1613.

Volkow, N. D., & Fowler, J. S. (2000). Addiction, a disease of compulsion and drive: Involvement of the orbitofrontal cortex. *Cerebral Cortex, 10,* 318–325.

Wenger, D. M. (2002). *The illusion of conscious will.* Cambridge, MA: The MIT Press.

PART THREE

Brain Imaging

9

Medicolegal Issues in Neuroimaging

STACEY TOVINO

The use of neuroimaging technologies in the clinical setting raises a variety of criminal, civil, and administrative law issues. During the past five years, legal theorists have focused on the impact of advances in neuroimaging for issues in criminal law, including criminal responsibility, criminal procedure, capital punishment, and national security. The aim of this chapter is to identify and examine the multiplicity of ways in which clinical neuroimaging might intersect with civil and administrative law and to highlight those areas of the law that are most relevant for clinicians.

TORT LAW

A tort is a negligent or an intentional civil wrong that arises in a context other than contract or statute. Neurologists, neurosurgeons, psychiatrists, radiologists, and other clinicians who use structural and functional neuroimaging in the clinical setting may find themselves involved in a number of different types of tort claims, including negligent neuroimaging, negligent diagnosis, and negligent misdiagnosis, as well as lawsuits in which either the plaintiff or the defendant attempts to introduce neuroimaging evidence and a clinician is asked to testify regarding the meaning and import of the evidence.

Negligent Neuroimaging

Claims involving negligent neuroimaging may arise if a patient experiences an injury or dies as a result of a structural or functional neuroimaging procedure. In the context of magnetic resonance imaging (MRI), patient harm may result from the unexpected movement or displacement of a metal object, side effects of MRI contrast agents, and excessive noise.

MRI's permanent magnetic field can easily move coins, pens, watches, and other objects that contain metal and that are located within the imaging suite or on or within the patient. The Food and Drug Administration (FDA) has found several instances of lapses in human-controlled screening and safety measures in both hospitals and imaging centers that have resulted in patient injury and death, including one oft-cited case in which a patient died when her aneurysm clip moved during a clinical MRI scan and lacerated her middle cerebral artery (FDA, 1992). In a second well-known case, *Colombini v. Westchester County Healthcare Corporation*, a sedated six-year-old boy was undergoing an MRI scan when his anesthesiologist realized that he was not receiving sufficient oxygen (Colombini, 2005). A hospital nurse who happened to be passing by the imaging suite heard the anesthesiologist's call for oxygen and brought into the imaging suite an oxygen tank made of ferrous metal. The oxygen tank was immediately pulled into to the bore of the magnet, striking and killing the boy. The boy's parents sued the anesthesiologist, the hospital nurse, and other individual and institutional defendants on a number of theories of liability (Gilk, 2006).

A clinician, such as the anesthesiologist or the hospital nurse in the *Colombini* case, may be held civilly liable under tort law for a patient injury or death that results from the movement or displacement of a metal object during a neuroimaging procedure if the clinician fails to adhere to the appropriate standard of care and that failure actually and proximately causes the patient's injury or death. Clinicians who are involved in brain scanning procedures and wish to minimize their liability must be specifically trained regarding the ferromagnetic dangers of MRI and perform sufficiently detailed and redundant screening procedures (NIMH, 2005).

Although metal objects located in the imaging suite or on the subject's person are not too difficult to identify, less obvious are metal objects that lie within the subject's body, including pacemakers, aneurysm clips, surgical clips, other metal implants and prostheses, metallic shavings, dental and orthodontic apparatuses, and even metallic substances remaining around the eye due to the application of cosmetic eye shadow (NIMH, 2005). Although most patients will inform their clinicians regarding their metal exposure during the informed consent to treatment process, some patients, including vegetative and minimally conscious patients whose brains are scanned in an attempt to determine the extent of traumatic injury or to identify any remaining cognitive capacity (see Schiff et al., 2005; Owen et al., 2006; Di et al., 2007) may be unable to do so. In these cases, the patient's metal exposure would have to be revealed by a family member or other person familiar with the patient's medical and social history, a handheld metal detector, another method of body scanning or, preferably, a combination of all three (NIMH, 2005).

To minimize liability associated with ferromagnetic risks, clinicians must familiarize themselves with safe neuroimaging practices, including the American College of Radiology's recent guidance document that outlines safe MRI practices (Kanal, 2007). Among other things, MRI procedures should only be assigned to certified radiology technicians and other clinicians who (1) have been educated in MRI safety; (2) follow detailed and redundant screening policies and procedures designed to ensure that no metal objects are left in the imaging suite, are brought into the imaging suite during the scanning procedure (including during emergency situations), or are located on or within the patient; and (3) follow detailed policies and procedures, including cancellation of the imaging procedure when metal is found or a metal object becomes necessary for treatment or resuscitation of the patient (DOL, 2008–2009). Adherence to safe neuroimaging practices is equally important in the research setting. Although institutional review boards (IRBs) routinely approve research protocols involving MRI procedures, little attention may be given by the IRB to the qualifications of the principal investigators, research fellows, and graduate students who will be conducting the scanning procedures (Kulynych, 2007). IRBs may incorrectly assume that these individuals, who may be affiliated with departments of psychology or other nonmedical disciplines, have as much training in MRI safety as trained radiologists and radiology technicians (Kulynych, 2007).

A second potential area of negligent neuroimaging liability relates to the use of MRI contrast agents. Using exogenous contrast media, clinicians have significantly improved MRI's sensitivity of detection and delineation of pathological structures, including primary and metastatic brain tumors, inflammation, and ischemia (Roberts, 2000). Unfortunately, the FDA also has received dozens of reports of patients with renal conditions who developed nephrogenic systemic fibrosis after receiving certain gadolinium-based contrast agents designed to enhance their MRIs (FDA, 2007). Armed with this information, several plaintiffs' law firms now specialize in the representation of patients in contrast-based MRI lawsuits (Saiontz et al., 2008). Clinicians who conduct MRI procedures thus face a difficult decision. If they decide not to use contrast in patients who have renal conditions for safety reasons, they may fail to identify or accurately diagnose a pathology that otherwise may have been detected. If they use contrast, they may risk patient injury. The proper balancing of risks may depend on the patient's particular condition, the necessity of an accurate diagnosis, and the relative risks the patient is willing to accept through the process of informed consent to treatment.

A third potential area of liability relates to the noise of the MRI machine. The sound of the magnet working within the MRI machine can be quite loud, and litigious patients may attempt to hold an involved clinician liable for noise discomfort and related hearing damage. Noise discomfort and hearing damage claims in the context of MRI sound far-fetched until one realizes that an increasing number of individuals who voluntarily expose themselves to loud noises, including pop music, are suing the sources of those noises for civil damages. In one recent case, a Louisiana man sued iPod maker Apple for unspecified damages and an injunction that would require Apple to restrict iPod's output to 100 decibels (dB) (BBC, 2006).

To minimize liability associated with noise discomfort and hearing damage claims, clinicians should develop and follow policies and procedures designed to minimize the magnet's noise, including, perhaps, policies and procedures developed in research studies involving infants. Research studies involving infants find that an MRI's magnet noise can be minimized to 12 dB higher than quiet conversation, which is 18 dB lower than a lawn mower and 38 dB lower than a car horn, by covering the magnet tunnel with noise-protection foam and placing over each infant's head a noise-protection helmet. The combination of the foam applications and the helmet reduces noise and vibrations inside the tunnel and may provide evidence of adherence to the standard of care in noise minimization.

Negligent Diagnosis or Misdiagnosis

In the rare case of a clinician who uses structural or functional neuroimaging in an attempt to diagnose a neurologic or psychiatric condition, the clinician may risk civil liability if the clinician incorrectly diagnoses or fails to diagnose the patient's condition and the incorrect or missed diagnosis actually and proximately causes the patient's injury or death. At present these claims likely will be more successful in structural neuroimaging cases designed to detect traditional anatomical diseases and injuries. Consider a patient who has an undiagnosed basilar bifurcation aneurysm and whose brain is scanned using CT or MRI. If a clinician reviews the scan and diagnoses the aneurysm, the patient can be referred for surgical or endovascular treatment. If a clinician reviews the scan but fails to diagnose the aneurysm when a reasonably prudent clinician in the same or similar circumstances would have diagnosed the aneurysm, the patient may miss the opportunity for treatment and suffer injury or death if the aneurysm ruptures and bleeds. In the latter case, the clinician may be civilly liable if the patient or next of kin proves that the clinician failed to adhere to the appropriate standard of diagnostic care and that the breach of that standard actually and proximately caused the patient's injury or death.

Although incorrect or missed diagnosis is an unlikely problem in the context of clinical psychiatry, simply because neuroimaging currently plays such a limited role at present in psychiatric diagnosis (See Farah & Gillihan, chapter 11 this volume), the problem of missed incidental findings in research scans is a more realistic worry because of the frequent use of neuroimaging to investigate psychiatric and other psychological phenomena in the research setting (see Detre and Bockow, chapter 10 in this volume). In any event, until structural or functional neuroimaging becomes the standard of care for clinical diagnosis of psychiatric and behavioral conditions, a plaintiff would have difficulty proving that a particular clinician has a duty to order or correctly interpret a diagnostic brain scan. The standard of care may be defined as the level of psychiatric diagnostic care that a reasonably prudent clinician in the same or similar circumstances would have taken to diagnose the patient's condition. If structural or functional neuroimaging does become the standard of care for clinical diagnosis of psychiatric and behavioral conditions, a clinician may be civilly liable if she fails to order a brain scan that a reasonably prudent clinician in the same or similar circumstances would have ordered to arrive at a diagnosis, or if she orders a brain scan but incorrectly reads the scan or otherwise fails to identify a condition that a reasonably prudent clinician in the same or similar circumstances would have identified.

Evidence of Pain and Suffering in Torts Cases

The above discussion focused on ways in which clinicians through their own actions or omissions may incur tort liability vis-à-vis patients whose brains are scanned. A separate issue is the extent to which clinicians will be asked to testify regarding the meaning and import of neuroimaging evidence submitted in tort lawsuits filed by individuals seeking damages from nonclinician, third-party defendants, such as drunk drivers, physically or emotionally abusive domestic partners, or supermarket chains in which slip-and-falls occur.

Many torts, including the negligence cause of action, require proof of compensable damages. Depending on the state, compensable damages may include lost past and future earnings, damaged or destroyed personal property, harm to marital relations, medical and hospital bills, and physical and emotional pain and suffering. Given that pain and suffering damages can account for a significant portion of personal damage awards in tort cases (McCaffrey et al., 2005), the ability to prove or measure pain and suffering or the lack thereof could be invaluable (Kolber, 2007).

In the past, pain and suffering complaints have been based on the plaintiff's first-person reports of pain and suffering coupled with circumstantial evidence, such as evidence of missed work or other missed activities, and some expert and lay testimony, such as testimony by a physician that a reasonable patient in the same or similar circumstance would experience the pain and suffering claimed by the plaintiff, or testimony by a friend or family member that the patient complained of headaches for months after the injury. The inability of all plaintiffs to provide unassailable and readily available evidence of their physical and emotional pain and suffering allows for two undesirable tort results, including cases in which healthy (although deceitful or malingering) patients recover significant damages based on false evidence, as well as cases in which patients who have experienced significant pain and suffering recover no or insufficient damages due to a lack of persuasive evidence.

In the past decade, however, several groups of scientists have used functional neuroimaging technologies in an attempt to better understand the neural correlates of both physical and emotional pain (Coghill et al., 1999; Coghill et al., 2003). In some of these studies, scientists have found significantly greater activations in certain areas of the brain when subjects are exposed to painful stimuli, as well as a correlation between the amount of brain activation and the intensity of the painful stimulus (see Coghill et al., 1999; Coghill et al., 2003). In other studies, scientists have investigated the neural correlates of emotional pain (Eisenberger et al., 2003; Eisenberger & Lieberman, 2004; Eisenberger, 2006; Ochsner et al., 2006), including in subjects who suffer from posttraumatic stress disorder (Bremner, 2006; Bremner, 2007). These studies raise at least three important questions for tort law. First, what impact will functional neuroimaging have on plaintiffs' ability to prove their tort claims, especially their allegations of physical and emotional pain and suffering? Second, what impact will functional neuroimaging have on defendants' ability to impugn plaintiffs' pain and suffering claims (Kolber, 2007)? Third, what should clinicians bear in mind when they are requested to scan a plaintiffs' brain in search of evidence of physical or emotional pain and suffering, or to testify regarding the meaning and import of a scan ordered by another clinician?

The impact functional neuroimaging will have on tort claims likely will depend on the tort involved. In the negligence cause of action, a plaintiff's functional brain scan that is interpreted by a clinician to reveal the neural correlates of pain and suffering may help prove the existence of compensable damages, the fourth element of the negligence cause of action (Viens, 2007). However, plaintiffs seeking damages for negligence also must prove three other elements: (1) a duty of the defendant; (2) a breach of that duty by the defendant; and (3) proof that the defendant's breach of her duty actually and proximately caused the plaintiff's damages (Viens, 2007). The elements of duty and breach generally are matters of law and fact for the court and jury and are not matters to which a clinician would testify in cases involving nonclinician, third party defendants (Viens, 2007). A posttort brain scan also would not necessarily prove the third element, causation, due to most plaintiffs' lack of a pretort (or baseline) brain scan that would help refute a preexisting pain and suffering defense (Viens, 2007). Functional neuroimaging thus may provide additional evidentiary support for the damage element of the plaintiff's negligence cause of action, but not the duty, breach, or causation elements (Viens, 2007). Stated another way, even persuasive functional neuroimaging evidence would not be dispositive of a plaintiff's negligence cause of action (Viens, 2007).

Again, the impact functional neuroimaging will have on tort claims likely will depend on the tort involved. In the intentional infliction of emotional distress (IIED) cause of action, plaintiffs are required to prove that the defendant engaged in extreme and outrageous conduct; however, once that conduct is proved, the plaintiff generally is not required to provide expert medical testimony regarding the severity of her emotional distress or any physical symptoms thereof (Grey, 2007). Neuroimaging evidence of emotional distress thus may not significantly impact a plaintiff's ability to prove an IIED cause of action (Grey, 2007).

On the other hand, some jurisdictions require plaintiffs alleging the negligent infliction of emotional distress (NIED) cause of action to prove not only that most people would suffer severe emotional distress, but also that the plaintiff did in fact suffer severe emotional distress (Grey, 2007). Some jurisdictions also require NIED plaintiffs to prove a physical manifestation or symptom of their emotional distress (Grey, 2007). Neuroimaging evidence of emotional distress thus may impact the NIED cause of action in at least two different ways: (1) by providing evidence of the emotional distress experienced by a particular plaintiff; or, more broadly, (2) by changing the tort's proof requirements, including eliminating the need for proof of a physical manifestation or symptom of emotional distress (Grey, 2007).

In summary, clinicians may be requested by tort plaintiffs who allege physical and emotional pain and suffering to scan their brains in order to find proof of their pain and suffering (and to testify thereto), or to testify regarding the meaning and import of a scan ordered by another clinician. Before a clinician agrees to testify in a tort proceeding regarding the meaning and import of a brain scan, the clinician must be extremely familiar with the limitations of her neuroimaging findings, including limitations relating to the indirect nature of functional neuroimaging based on blood-oxygenation level dependent (BOLD) signal, the lack of a pretort (or baseline) functional neuroimage, and the possibility that the plaintiff's pain and suffering predated the tort. Opposing counsel will emphasize these limitations on cross-examination. Clinicians also must be carefully counseled regarding the content of their testimony and the ways in which opposing counsel will critique the functional neuroimages submitted as evidence, including the choice of colors used to represent different areas of BOLD activity and the prejudicial effect these colors may have on the judge and jury.

HEALTH INSURANCE CONTRACT DISPUTES

The discussion above focused on the tort implications of advances in structural and functional neuroimaging. The discussions below focuses on the multiplicity of ways in which clinical neuroimaging might intersect with other areas of civil and administrative law, beginning with the interpretation of health insurance contracts.

Advances in structural and functional neuroimaging have the potential to impact the interpretation of health insurance contractual provisions, including provisions that distinguish physical illnesses from mental illnesses and provide fewer insurance benefits for individuals whose illnesses are classified as mental in nature. The 1990 case of *Blake v. Union Mutual Stock Life Insurance Company* is illustrative (Blake, 1990). In *Blake*, a woman named Pam Blake sued her health insurance company when it classified her postpartum depression as a mental, rather than a physical, illness and paid only a small portion of the cost of Mrs. Blake's hospitalization and treatments. In deciding whether Blake had a mental or physical illness, the court reviewed the evidence provided by Blake's treating psychiatrists and psychologists about postpartum depression. Although Blake's clinicians testified regarding Blake's erratic postpartum behavior and thoughts, including her specific and stated desire to physically harm her baby within three days of the baby's birth, the court found that none of the clinicians were able to provide any physical test results or other measurements that could prove that Blake had an organic illness. The court thus held that the defendant insurer was not required to reimburse Blake for most of the costs of her postpartum hospitalizations and psychiatric treatments.

The potential of neuroimaging to provide neuroradiological evidence of postpartum mood disorders could impact health insurance contract disputes, including the type of dispute presented in *Blake*. Over the last decade, several groups of scientists have used neuroimaging in an attempt to provide neuroradiological evidence of the postpartum mood disorders. One small study published ten years ago, for example, found that the brains of women with postpartum psychoses looked structurally different than the brains of age-matched women with nonpostpartum psychoses, leading the study authors to conclude that they may have found evidence of subtle, unspecified neurostructural abnormalities in ill new mothers, and that these abnormalities might constitute an unspecific vulnerability factor (Lanczik et al., 1998).

More recently, in 2007, scientists published a study using fMRI to compare the brain function of women with postpartum depression compared to asymptomatic postpartum female control subject (Silverman et al., 2007). Although the scientists emphasized that it would be premature to conclude that postpartum depression has a unique depression phenotype, they stated that their findings suggest the potential of neuroimaging to identify an empirically based neural characterization of postpartum depression. And, in March 2008, scientists published a study using positron emission tomography (PET) to measure brain serotonin receptor binding potential in a small sample of both healthy and depressed postpartum women (Moses-Kolko et al., 2008). The study authors found that postsynaptic receptor binding in the depressed subjects was reduced 20 to 28% relative to controls.

Recall that in the *Blake* case, insured Pam Blake was unable to secure reimbursement for her postpartum hospitalizations and psychiatric treatments because her clinicians were unable to provide any tests or measurements that could prove to the court that Blake's postpartum depression had an organic basis. If neuroimaging advances to the point where clinicians are able to use neuroimaging to identify, for example, an unspecific neural vulnerability factor for postpartum depression, clinicians may be able to assist insured individuals in the more favorable resolution of their health insurance contract disputes. In the studies referenced in this section, the focus of scientific investigation was the effect, if any, of structural and functional differences on behavior, including postpartum behavior; however, clinicians also may be asked to testify or may wish to clarify the effects that psychological variables may have on the brain to provide a more complete picture of the interdependent relationship between the brain and behavior. As discussed in more detail in the conclusion of this chapter, clinicians who voluntarily testify or are requested or compelled to testify regarding their neuroimaging findings must be carefully counseled regarding the ways in which the content of their testimony may be understood or misunderstood by judges and jurors, as well as the ways in which opposing counsel may attempt to slant their testimony to serve opposing counsel's own purposes.

MENTAL HEALTH PARITY LAW

Advances in structural and functional neuroimaging also have the potential to impact the application of federal and state mental health parity laws. Although mental health parity laws vary widely in scope, they generally are designed to remedy health insurance plans' less comprehensive coverage (through fewer inpatient days, fewer outpatient visits, or lower cost-sharing amounts) of mental illnesses compared to physical illnesses. On October 3, 2008, President Bush signed the Paul Wellstone and Pete Domenici Mental Health Parity and Addiction Equity Act (Mental Health Parity Act) into law. The Mental Health Parity Act requires certain group health plans that provide both medical and surgical benefits as well as mental health and substance use disorder benefits to ensure that: (1) the financial requirements applicable to mental health or substance use disorder benefits are no more restrictive than the predominant financial requirements applied to substantially all medical and surgical benefits covered by the plan; and (2) the treatment limitations applicable to mental health or substance use disorder benefits are no more restrictive than the predominant treatment limitations applied to substantially all medical and surgical benefits covered by the plan (HR 1424, 2008). In March 2010, President Obama signed the Affordable Care Act (ACA) into law. ACA extended federal mental health parity law to all qualified health plans, including individual and small group health plans, that will be offered on state health insurance exchanges on or after January 1, 2014. In summer 2012, the United States Supreme Court upheld ACA's extension of mental health parity law, resulting in mental health parity protections for additional insured's.

As currently written, the Mental Health Parity Act does not define the terms "mental health" or "mental health benefits"; instead, the act refers to the underlying plan's definitions, which may be regulated by state law. Some states define the mental health conditions to which parity mandates apply in terms of whether the conditions have a biological basis. Consider New Jersey's mental health parity law, which mandates equal health insurance coverage for those mental conditions that are "caused by a biological disorder of the brain" (N.J. Stat. Ann. § 17B:27A-19.7). Further, consider Nebraska's mental health parity law, which mandates equal health insurance coverage of "any mental health condition that current medical science affirms is caused by a biological disorder of the brain" (Neb. Rev. Stat. § 44–792(5)(b)). When a state law expressly refers to the current state of medical science in its insurance parity mandate, insured individuals are encouraged to refer to scientific studies that support the classification of their mental health conditions as biological disorders of the brain in their health insurance parity disputes.

Stakeholders are beginning to refer to scientific studies to achieve their mental health parity goals. For example, several recent neuroimaging studies have found structural and functional differences in the brains of women with both active and recovered eating disorders (Lambe et al., 1997; Muhlau et al., 2007; Wagner et al., 2007). In some states, lobbyists have referenced these studies en route to successfully arguing that eating disorders should be included within the statutory list of mental health conditions that require equal insurance benefits (Cal. Health & Safety Code § 1374.72(d)(8) 2009), notwithstanding the fact that many of the referenced studies referenced do not conclude that brain differences cause eating disorders or preclude a particular woman's eating disorder from being caused by a factor unrelated to the structure or function of her brain. If neuroimaging advances to the point where clinicians can use neuroimaging to identify women who have current eating disorders, however, patients may begin to request clinical neuroimaging diagnoses of their eating disorders (or other mental health conditions) in order to secure equal health insurance benefits. Advances in structural and functional neuroimaging thus have the potential to assist patients (or clinicians testifying in lawsuits filed on behalf of patients) in the more favorable resolution of mental health benefit parity disputes.

DISABILITY BENEFIT LAW

Advances in structural and functional neuroimaging also have the potential to impact disputes regarding the availability of benefits under public and private disability, social security, and other benefit programs. To prevent healthy plaintiffs from receiving benefits when they do not have a disability, disability plans and programs tend to define disability in terms of an abnormality that is demonstrable by medically acceptable clinical and laboratory diagnostic techniques. Medical evidence generally is considered *the* cornerstone of disability status under disability programs and plans (SSA, 2007). US Social Security Disability Insurance (SSDI), for example, is only available to claimants who can furnish medical and other evidence of the existence of a disability, including "medical signs and findings, established by medically acceptable clinical or laboratory diagnostic techniques" (42 USC § 423(d)(5)(A)).

In the past, the Social Security Administration (SSA) has denied disability status to many individuals who lack medical evidence of their conditions, especially controversial conditions such as chronic fatigue syndrome and fibromyalgia. The 2003 case of *Bartyzel v. Commissioner* illustrates this point. In *Bartyzel*, the plaintiff sued the SSA in an effort to receive disability benefits for her symptoms of fatigue and pain, which were diagnosed by at least one physician as chronic fatigue syndrome or fibromyalgia (Bartyzel, 2003). Both the trial and appellate courts affirmed the SSA's decision to deny the plaintiff SSDI benefits, reasoning that the plaintiff did not provide sufficient evidence of her disability status because she did not submit objective, medical evidence of her fatigue and pain. In its opinion, the appellate court stated that the SSA would have considered findings from an "abnormal magnetic resonance imaging (MRI) brain scan" had the plaintiff presented them (Bartyzel, 2003).

At present, clinicians do not routinely use neuroimaging in the clinical setting to diagnose conditions such as chronic fatigue syndrome and fibromyalgia. However, a number of recent structural and functional neuroimaging studies have been designed to investigate the brains of individuals with these conditions. In one study published in 2006, scientists scanned the brains of seventeen patients with chronic fatigue syndrome and twelve healthy controls during the performance of a working memory task (Caseras et al., 2006). The research findings suggested that the brains of patients with chronic fatigue syndrome show both quantitative and qualitative differences in activation of the working memory network compared with healthy control subjects. A review article published in late 2006 summarizes more generally functional neuroimaging findings in the context of fibromyalgia (Williams & Gracely, 2006). According to the review article, fibromyalgia patients differ from healthy controls in baseline levels of neural activity, especially in the caudate nucleus, and utilize more extensive brain resources than do same-aged peers in order to achieve comparable performance on cognitive tasks.

The claimant in *Bartyzel* was unable to secure disability benefits because she did not submit sufficient objective evidence of her pain and fatigue. If neuroimaging advances to the point where clinicians can confirm the claims of fatigue and pain associated with chronic fatigue syndrome and fibromyalgia, as well as the symptoms of other controversial conditions, disability claimants likely will request clinicians to testify to, or otherwise provide diagnostic neuroimaging evidence of, such symptoms for use in disability benefit disputes.

CONSTITUTIONAL LAW

Clinical neuroimaging has the potential to intersect with several other areas of the law, including constitutional law, which has been interpreted by a number of courts to allow the withholding and withdrawal of life-sustaining treatment from incompetent patients in certain situations. In a 1976 New Jersey Supreme Court case, *In re Quinlan*, the father of twenty-one-year-old Karen Ann Quinlan brought a declaratory judgment action against a number of individual and institutional defendants, including a physician and a New Jersey hospital, requesting that the court authorize the discontinuance of life-sustaining techniques for Karen, who was diagnosed as vegetative (in re Quinlan, 1976). In deciding that Karen's constitutional right to privacy, including the right to withdraw medical treatment, could be asserted by a guardian, the court emphasized that Karen had "no realistic possibility of returning to any semblance of cognitive or sapient life" and that "the focal point of decision should be the prognosis as to the reasonable possibility of return to cognitive and sapient life" (in re Quinlan, 1976). The court concluded, "If that consultative body agrees that there is no reasonable possibility of Karen's ever emerging from her present comatose condition to a cognitive, sapient state, the present life-support system may be withdrawn and said action shall be without any civil or criminal liability therefor, on the part of any participant, whether guardian, physician, hospital or others" (in re Quinlan, 1976).

At present, clinicians know of no medications or surgical interventions that will definitively reduce the length of the impaired consciousness of individuals such as Karen Ann Quinlan. Over the past decade, however, scientists have used various functional neuroimaging technologies to better understand disorders of consciousness, improve differential diagnoses, and predict short-term improvement, such as improvement from the vegetative to the minimally conscious state. As discussed by Fins and Schiff (chapter 13 in this volume), scientists have recently made a number of important findings. In a 2005 study, several American scientists suggested that some individuals in the minimally conscious state may retain widely distributed cortical systems that have potential for cognitive and sensory function (Schiff et al., 2005). In a 2006 study, several European scientists suggested that functional neuroimaging might be a means by which individuals with disorders of consciousness can use their residual cognitive capabilities to communicate their thoughts to those around them (Owen et al., 2006). And, in a 2007 study, scientists from China and Belgium suggested that traditional behavioral assessments can miss cerebral processing that might herald short-term improvement (Di et al., 2007).

Although clinicians presently know of no medications or surgical interventions that will definitely reduce the length of patients' impaired consciousness, family members and other stakeholders who do not want life-sustaining treatment withheld or withdrawn from a patient may attempt to rely on studies such as those identified above in an attempt to persuade a court that their kin may have some cognitive or sapient activity for purposes of the *In re Quinlan* holding. Family members and other stakeholders who do (or do not) want life-sustaining treatment withheld or withdrawn from a patient also may request a clinician to perform a functional neuroimaging procedure or to testify that an existing brain scan shows that the patient has no cognitive function (or significant cognitive function), as appropriate. Given the number of lawsuits and other informal disputes involving the appropriateness of withholding and withdrawal of life-sustaining treatment, including the Terri Schiavo case, as well as the recent flurry of scientific studies investigating the various disorders of consciousness, it is reasonable to anticipate that family members and other stakeholders will rely on recent scientific findings or request functional neuroimaging procedures in an attempt to support their legal agendas, whatever they may be.

COMPETENCY DETERMINATIONS

A number of civil and administrative proceedings require evidence of competency or incompetency, as appropriate. In the law of trusts and estates, a court is permitted to invalidate a will if a contestant can prove that the testator was incompetent at the time of the will's execution. Under most States' health care power of attorney laws, an agent's power to make health care decisions on behalf of a principal becomes effective only when a clinician determines that the principal is incompetent (see, e.g., Tex. Health & Safety Code § 166.152(b)).

Advances in structural and functional neuroimaging have the potential to impact competency disputes under these and other civil and administrative laws (see Karlawish, chapter 6, this volume). Consider recent studies concluding that neuroimaging has the potential to become an important tool in the diagnosis of Alzheimer's disease and other sources of dementia at an early stage (Smith et al., 2007). If neuroimaging develops to the point where structural or functional brain scans become the standard for dementia diagnosis, stakeholders in disputes that require proof of competency or incompetency may request clinicians to order and interpret new brain scans, or testify regarding existing brain scans. The clinician's findings or testimony may be used by a party to validate or invalidate a will, to exercise or prohibit the exercise of duties under a health care power of attorney, or to support or refute another legal position or argument.

CONCLUSION

The use of neuroimaging technologies in the clinical setting raises a variety of criminal, civil, and administrative law issues. During the past five years, scholars have overwhelmingly focused on the impact of advances in neuroimaging for criminal law, especially criminal responsibility, criminal procedure, capital punishment, and national security, as well as related evidentiary and privacy issues. The aim of this chapter was to balance this literature by identifying and examining the multiplicity of ways in which clinical neuroimaging might intersect with civil and administrative doctrine, including tort law, health insurance contract law, mental health parity law, disability benefit law, constitutional law, trusts and estates law, and advance directive law. Clinical neuroimaging undoubtedly will intersect with other civil and administrative laws not discussed herein.

This chapter shows that clinical neuroimaging has the potential not only to intersect, but also to significantly impact, civil and administrative law. For example, clinical neuroimaging has the potential to impact health insurance contract disputes by providing evidence that a particular mental health condition has a biological basis and, therefore, is worthy of comprehensive health insurance coverage. Clinical neuroimaging has the potential to impact mental health parity law by providing evidence that a particular mental health condition meets a state's definition of biologically based illness to which parity mandates apply. Clinical neuroimaging also has the potential to impact the future of certain civil causes of action, including negligent infliction of emotional distress, by eliminating the need for proof of a physical manifestation of a plaintiff's emotional distress.

In order to limit liability, clinicians who are requested to perform or interpret brain scans for civil or administrative purposes must be aware of the ways in which their own actions or omissions may result in tort liability, including through negligent neuroimaging, negligent diagnosis, or negligent misdiagnosis. Because many neuroimaging findings will constitute protected health information under federal and state health information confidentiality laws, including the federal Health Insurance Portability and Accountability Act (HIPAA) privacy rule, clinicians must ensure that they have obtained a prior written authorization from the patient who is the subject of the neuroimage, or satisfy an exception to the authorization requirement, before disclosing a patient's neuroimaging findings to a court, an administrative tribunal, or other interested stakeholder.

Clinicians must ensure that they are thoroughly and accurately documenting both their neuroimaging findings and, if possible, relevant limitations, so that their findings are not inappropriately used by plaintiffs, defendants, lobbyists, or other stakeholders with contrary legal agendas. Before a clinician agrees to testify in a civil or administrative proceeding regarding the meaning and import of a neuroimage, the clinician must be extremely familiar with all of the limitations of her findings. Finally, clinicians who voluntarily testify or are requested or compelled to testify regarding their neuroimaging findings must be carefully counseled regarding the content of their testimony and the ways in which opposing counsel and others will attempt to slant their testimony.

REFERENCES

Bartyzel v. Commissioner. (2003). *Federal Appendix, 74,* 515–529.

BBC News. (2006). Man sues over iPod hearing risk. *BBC News.* http://news.bbc.co.uk/2/hi/technology/4673584.stm

Blake v. Union Mutual Stock Life Insurance Company. (1990). *Federal Reporter 2nd, 906,* 1525–1531.

Bremner, J. D. (2006). Traumatic stress: effects on the brain. *Dialogues in Clinical Neuroscience, 8* (4), 445–461.

Bremner, J. D. (2007). Functional neuroimaging in post-traumatic stress disorder. *Expert Review of Neurotherapeutics, 7* (4), 393–405.

California Health and Safety Code § 1374.72(d)(8). (2009).

Caseras, X., et al. (2006). Probing the working memory system in chronic fatigue syndrome: A functional magnetic resonance imaging study using the *n*-back task. *Psychosomatic Medicine, 68* (6), 947–955.

Coghill, R. C., et al. (1999). Pain intensity processing with the human brain: A bilateral, distributed mechanism. *Journal of Neurophysiology 82* (4), 1934–1943.

Coghill, R. C., et al. (2003). Neural correlates of interindividual differences in the subjective experience of pain. *Proceedings of the National Academy of Sciences, 100,* 8538–8542.

Colombini v. Westchester County Healthcare Corp. (2005). *N.Y.S.2d, 808,* 705–710.

Department of Labor Bureau of Labor Statistics (DOL). (2008–2009). Radiology technologists and technicians. *Occupational Outlook Handbook.* www.bls.gov/oco/ocos105.htm.

Di, H. B., et al. (2007). Cerebral response to patient's own name in the vegetative and minimally conscious states. *Neurology, 68,* 895–899.

Eisenberger, N. I. (2006). Identifying the neural correlates underlying social pain: Implications for developmental processes. *Human Development, 49* (5), 273–293.

Eisenberger, N. I., et al. (2003). Does rejection hurt? An fMRI study of social exclusion. *Science, 302* (5643), 290–292.

Eisenberger, N. I. & Lieberman, M. D. (2004). Why rejection hurts: A common neural alarm system for physical and social pain. *Trends in Cognitive Sciences, 8* (7), 294–300.

Food and Drug Administration (FDA). (1992). FDA safety alert: MRI related death of patient with aneurysm clip. www.fda.gov/downloads/MedicalDevices/Safety/AlertsandNotices/PublicHealthNotifications/ucm063104.pdf.

Food and Drug Administration (FDA). (2007). Information for healthcare professionals: gadolinium-based contrast agents for magnetic resonance imaging. *FDA Alerts.* www.fda.gov/Drugs/DrugSafety/PostmarketDrugSafetyInformationforPatientsandProviders/ucm142884.htm.

Gilk, T. (2006). MRI suites: Safety outside the bore. *Patient Safety & Quality HealthCare.* www.psqh.com/sepoct06/mrisuites.html.

Grey, B. J. (2007). Neuroscience, emotional harm, and emotional distress tort claims. *American Journal of Bioethics-Neuroscience 7* (9), 65–67.

H.R. 1424 (2008). Paul Wellstone and Pete Domini mental health parity and addiction equity act of 2008. *110th Congress, Regular Session, Public Law Number* 110–343.

Kanal, E., et al. (2007). ACR guidance document for safe MR practices. *American Journal of Radiology, 188,* 1–27.

Kolber, A. J. (2007). Pain detection and the privacy of subjective experience. *American Journal of Law & Medicine, 33,* 433–469.

Kulynych, J. (2007). The regulation of MR neuroimaging research: disentangling the gordian knot. *American Journal of Law and Medicine 33,* 295–317.

Lambe, E. K., et al. (1997). Cerebral gray matter volume deficits after recovery from anorexia nervosa. *Archives of General Psychiatry, 154* (6), 537–542.

Lanczik, M., et al. (1998). Ventricular abnormality in patients with postpartum psychoses. *Archives of Women's Mental Health, 1,* 45–47.

Lo v. Burke. (1995). *Virginia Reporter, 249*, 311–319.

In the Matter of Karen Quinlan. (1976). *Atlantic Reporter Second, 355*, 647–672.

McCaffrey, E. J., et al. (1995). Framing the jury: Cognitive perspectives on pain and suffering awards. *Virginia Law Review 1995*, 1341–1420.

Moses-Kolko, E. L., et al. (2008). Serotonin 1A receptor reductions in postpartum depression: A PET study. *Fertility & Sterility 89* (3), 685–692.

Muhlau, M., et al. (2007). Gray matter decrease in the anterior cingulate cortex of anorexia nervosa. *American Journal of Psychiatry, 164* (12), 1850–1857.

National Institute of Mental Health (NIMH) Council Workgroup on MRI Research Practices. (2005). MRI research safety and ethics: points to consider. www.nimh.nih.gov/about/advisory-boards-and-groups/namhc/reports/mri-research-safety-ethics.pdf.

Nebraska Revised Statutes § 44–792(5)(b). (2009).

New Jersey Statutes Annotated § 17B:27A-19.7. (2009).

Ochsner, K. N., et al. (2006). Neural correlates of individual differences in pain-related fear and anxiety. *Pain, 120*, 69–77.

Owen, A. M., et al. (2006). Detecting awareness in the vegetative state. *Science, 313*, 1402.

Roberts, T. P., et al. (2000). Neuroimaging: do we really need new contrast agents for MRI? *European Journal of Radiology 34* (3), 166–178.

Saiontz, K., & Miles, P. A. (2009). MRI lawsuits: gadolinium contrast agents. www.youhavealawyer.com/gadolinium/index.html.

Schiff, N. D., et al. (2005). fMRI reveals large-scale network activation in minimally conscious patients. *Neurology, 64*, 514–523.

Silverman, M. E., et al. (2007). Neural dysfunction in postpartum depression: An fMRI pilot study. *CNS Spectrums, 12* (11), 853–862.

Smith, C. D., et al. (2008). White matter diffusion alterations in normal women at risk of alzheimer's disease. *Neurobiology of Aging*, ePub ahead of print.

Social Security Administration (SSA). (2007). Disability evaluation under Social Security. *Blue Book*, Part II, Evidentiary Requirements. www.ssa.gov/disability/professionals/bluebook/.

Texas Health and Safety Code § 166.152(b). (2009).

United States Code, 42, § 423(d)(5)(A). (2007).

Viens, A. M. (2007). The use of functional neuroimaging technology in the assessment of loss and damages in tort law. *American Journal of Bioethics, 7* (9), 63–65.

Wagner, A., et al. (2007). Altered reward processing in women recovered from anorexia nervosa. *American Journal of Psychiatry, 164* (12), 1842–1849.

Williams, D. A., & Gracely, R. H. (2006). Functional magnetic resonance imaging findings in fibromyalgia. *Arthritis Research and Therapy, 8* (6), 224–232.

10

Incidental Findings in Magnetic Resonance Imaging Research

JOHN DETRE AND TAMARA B. BOCKOW

Magnetic Resonance Imaging (MRI) is a widely available and versatile imaging modality that provides unprecedented spatial resolution and a variety of image contrast mechanisms within a single examination. Magnetic resonance images are derived from radiofrequency signals that readily penetrate dense tissue such as bone, making MRI by far the most sensitive method for imaging human brain structure in vivo. With three-dimenstional imaging based on T1 contrast, isotropic image resolution of less than a millimeter through whole brain is readily achievable within a matter of minutes on most current commercial MRI scanners, and excellent contrast between gray matter, white matter, and CSF is provided. T1 contrast can also be used to image arteries with magnetic resonance angiography and to measure tissue perfusion with arterial spin labeling. T2 contrast adds sensitivity to inflammatory responses that occur in a broad range of neuropathology. A modification of T2-weighted MRI termed diffusion imaging enhances sensitivity to microscopic diffusion of water and can be used to detect changes in water diffusivity in conditions such as acute stroke or to image white matter tracts based on the anisotropic diffusivity of water in white matter. T2* contrast adds exquisite sensitivity to paramagnetic deoxyhemoglobin and methemoglobin that are deposited in hemorrhagic disorders, and this contrast mechanism is also exploited as a surrogate marker of changes in regional cerebral blood flow and metabolism that occur with regional brain activation in functional MRI (fMRI) with blood oxygenation level dependent (BOLD) contrast. A number of other contrast mechanisms such as magnetization transfer, T1rho, and quantification of metabolites using magnetic resonance spectroscopy are also available, but are used less commonly.

Brain imaging including all of the aforementioned contrasts can be routinely obtained within a single MRI scanning session of less than an hour. Because MRI uses no ionizing radiation and does not require exogenous contrast administration in routine use, it is considered entirely noninvasive. The only absolute contraindications to MRI are the presence of ferrous metal in the brain or orbit or the presence of certain electronic implants such as a pacemaker. In clinical applications, MRI is considered safe during pregnancy. In research applications, MRI is typically classified as an "insignificant risk device." The main source of morbidity associated with noncontrast MRI is a risk of injury from metallic projectiles, which has resulted in at least one death in clinical practice (Colletti, 2004). This risk is readily controlled using operational procedures to insure that no ferrous objects enter the magnet room.

The noninvasiveness and versatility of MRI has also made it an increasingly popular modality for brain research. The relative ease and low cost of imaging regional brain function with BOLD fMRI as compared to prior PET methods spurred an explosion in its use as a research tool in basic and clinical neuroscience for studying brain-behavior relationships. The development of advanced computational approaches for analyzing the complex brain morphometry of both gray matter and white matter in various populations has also rendered structural MRI an important neuroscience research tool. Because there are no known risks of cumulative exposure to MRI, it can be used for serial studies of brain structure and function during development, aging, pathological processes, and learning. A number of large database projects acquire cross-sectional or longitudinal MRI data from hundreds or thousands of individuals.

Although the MRI procedure itself carries little or no risk, MRI scanning is capable of revealing previously unsuspected neuropathology in subjects who are scanner for research. While detecting neuropathology is the expressed intent of clinical MRI examinations, it is typically not the objective of research MRI, particularly when applied to a healthy population. Even in clinical populations, MRI scanning may reveal neuropathology unrelated to the patient's diagnosis. Such findings are incidental to the purpose of the study. Although many incidental findings have little or no obvious clinical significance, certain incidental findings, such as unsuspected tumors or vascular malformations, may have immediate implications for the medical status of the person in whom they are found. While incidental findings have been commonly recognized in the setting of clinical testing and are therefore an accepted consequence of medical evaluation, the widespread use of sensitive medical testing such as MRI for research into normal brain function is unprecedented and has dramatically increased the likelihood of an incidental medical finding in otherwise healthy research subjects.

Although the exact incidence of incidental findings in an otherwise healthy population is unknown and depends on demographic variables such as age and the extent of diagnostically relevant imaging that is carried out, a number of studies now suggest that medically significant abnormalities may be found in 5 to 20% of diagnostic-quality MRI scans from normal subjects, and somewhat higher in elderly subjects (Yue et al., 1997; Katzman et al., 1999; Weber & Knopf, 2006; Vernooij et al., 2007), though many such findings may have no immediate diagnostic or therapeutic implications. Incidental findings requiring urgent management are less frequent, and include brain tumors (~0.5–2%), aneurysms (~0.1%), and other vascular lesions (~0.2%).

An important general concept concerning incidental findings is that their significance may be difficult or impossible to determine in isolation. As an example, the presence of a few subcortical white matter hyperintensities is typically insignificant in a subject with no history of neurological complaints, but may strongly support a diagnosis of multiple sclerosis in a patient with a clear history of transient neurological deficits. Accordingly, the significance of an incidental finding for any individual is optimally determined by their personal physician in the context of the their complete medical history. The risks associated with an incidental finding may also vary from individual to individual. One subject may accept the presence of a small aneurysm or tumor and decide to undergo periodic follow-up scanning to insure that it is stable, while another subject may want immediate surgery, despite a significant perioperative risk. The process of evaluating and managing an incidental finding may take months or years and is also optimally carried out under the direction of the individual's personal physician.

INCIDENTAL FINDINGS IN MRI: RISK OR BENEFIT?

While minor incidental findings have the potential to cause anxiety and stress, the more significant findings have the potential to alter life expectancy and insurability as well as mandate additional testing or interventions that carry associated morbidity. Accordingly, the possibility of an incidental finding must be considered as a potential risk of participating in research involving brain MRI. By the same token, if a life-threatening incidental abnormality is discovered and treated, it could also be considered as a benefit of participating in the research. Although most research MRI consent forms include specific verbiage indicating that the study is not intended for diagnostic purposes, existing data suggests that subjects participating in research MRI may nonetheless expect that clinically significant abnormalities will be found (Kirschen et al., 2006).

A basic tenet of human subjects research is a positive risk: benefit ratio (Emanuel et al., 2000). Neglecting for the moment the risk of an incidental finding, MRI research in subjects without pacemakers or ferrous metal foreign bodies studied in an environment that carefully excludes ferrous projectiles carries no significant risk, hence as long as the research has some merit, a positive risk: benefit ratio is virtually guaranteed. In clinical populations, research studies may also provide some direct benefit to the patient if study results are also shared with their physician. However, in most instances, the benefit of MRI research is not to the research subject, but rather to society or future patients based on the new knowledge that is anticipated to be derived from the study. The benefit to the individual research subject is usually limited to a small honorarium provided to offset the time and expense of participating in the study.

The riskbenefit ratio for participating in MRI research is clearly altered by the recognition that the discovery of an incidental finding represents a risk to the research subject of participating in the study. It has been argued that the requirement to minimize the riskbenefit ratio of human subjects research implies that MRI researchers must also maximize the potential benefit of discovering and disclosing a clinically-significant incidental finding, while at the same time minimize the risk of causing unnecessary anxiety by disclosing an insignificant finding (Wolf et al., 2008). The problem with this argument is that many findings have uncertain clinical significance, or, as noted above, an assessment of the significance cannot be made based on the imaging data alone. In these instances, disclosure can actually increase the risk of unnecessary anxiety due to a false-positive finding (Kumra et al., 2006). Additionally, maximizing the risk-benefit ratio of an MRI study from the standpoint of determining the presence or absence of clinically significant abnormalities may require obtaining additional imaging data that are not needed for research purposes, and even then a conclusive determination may not be possible. While this type of screening could conceivably be explicitly built into the study design as an incentive for participating, as a general matter it does not seem necessary or even desirable to significantly alter the data acquired to try to derive clinical benefits that are unrelated to the intent of a research study.

DIAGNOSTIC REVIEW OF RESEARCH MRI STUDIES

Of course, MRI scanning per se does not lead to incidental findings. Rather, incidental findings are discovered through visual inspection or other forms of image analysis. The likelihood of discovering an incidental finding depends on the extent of diagnostic quality imaging that is carried out as well as the diagnostic skills of the individuals who are involved in the data acquisition and analysis. In neuroimaging research, there is considerable variability in both the extent and diagnostic quality of the imaging data that is obtained and in the diagnostic acumen of the research team. A cognitive fMRI study carried out by a psychologist may include only a single clinical-quality T1-weighted image series used for spatial normalization of the imaging data to a standard template and all of the data may be acquired and analyzed by a graduate student with no medical knowledge or training, in some cases in an institution with no medical personnel. In contrast, an fMRI study of motor activation in a patient recovering from stroke may include extensive structural imaging to define and characterize the ischemic lesion and may be reviewed by a physician skilled in clinical image interpretation. This variability in the quality of diagnostically useful MRI data and the clinical skills of the research team make it difficult to use a uniform approach for handling incidental findings in all MRI research studies.

One possible approach to dealing with the heterogeneity in research MRI is to require a basic set of diagnostically useful imaging data for each research study and to mandate the review of these data by a skilled interpreter, typically a neuroradiologist. This approach is operationally appealing in that each study can be managed in the same way from the standpoint of detecting incidental findings. It is also much easier and faster for a skilled interpreter to review a standard set of data than to try to make clinical interpretations from a variable set of data acquired for other purposes. However, this approach is also likely to yield the highest number of incidental findings, many of which may have uncertain clinical significance. In a retrospective review of structural MRI data sets acquired during fMRI studies in normal volunteers, skilled neuroradiologists were able to identify reportable abnormalities approximately half of the cases, with 4% requiring urgent referral (Illes et al., 2004). This approach would make the most sense under the scenario that maximizing the ability to identify incidental findings is a benefit of participating in the MRI study from the research subject's standpoint, or to minimize the medical and medicolegal consequences from missing a clinical disorder that presents later. This approach is also best suited to facilities where the effort of skilled interpreters is available at a reasonable cost.

An alternative approach links the extent of data acquisition and scrutiny for incidental findings to the intent of the research MRI study, and more naturally applies to research MRI carried out in nonclinical settings or for nonclinical purposes. In this approach, studies carried out in healthy volunteers with no clinical hypothesis or intent do not require any formal review for incidental findings, whereas studies involving clinical hypotheses and clinical populations must be reviewed more systematically. In neither case are additional imaging data required for clinical interpretation, but studies in clinical populations may already include imaging sequences to further characterize the pathological condition being examined as part of the research design. For example, an fMRI study of cognition in Alzheimer's patients may include T2-weighted MRI to exclude clinically unsuspected cerebrovascular disease. Such data are typically already acquired and reviewed with a clinical intent in mind, and hence it is a natural extension of this process to develop a more formal clinical interpretation of the imaging data. Indeed, a doctor-patient relationship is essentially implied in the study design, even if the investigator is not the research subject's personal physician.

This dichotomized approach is also consistent with research subjects' reasonable expectations. As long as healthy volunteers for a nonclinical study truly understand that a research MRI study is not intended for diagnostic purposes and will not be reviewed as such, there should be no expectation of any medical value to the study, particularly if it is being carried out in a nonclinical environment such as a department of psychology. In contrast, clinical populations studied in a clinical setting and recruited because of their known neuropathology might reasonably expect their diagnosis to be at least confirmed, and any medically significant deviations to be detected. Control populations scanned under a clinical research protocol should be managed as if they were patients, particularly if they are matched to the patient population in other ways.

Even if no special scrutiny is given to MRI data from healthy volunteers acquired for nonclinical purposes, there remains the possibility that study personnel will notice an incidental abnormality, and this finding will need to be reviewed by someone with appropriate clinical skills. Accordingly, all research MRI facilities must include some mechanism for timely access to this expertise. However, this expertise will likely only be required sporadically for nonclinical studies.

DISCLOSING INCIDENTAL FINDINGS

Research subjects are entitled to be informed of incidental findings that carry a significant risk to their health (Emanuel et al., 2000), and the possibility of such notification should be described as part of the consent procedure for all research MRI studies. The principal investigator is ultimately responsible for insuring that this disclosure is made in a timely fashion, though this responsibility may be deferred to another study investigator with greater clinical expertise. It is less clear what actions should be taken in response to incidental findings with questionable or unlikely clinical significance, since the stress and anxiety generated by disclosure may outweigh any potential heath benefits.

Some existing recommendations regarding disclosure conceptually stratify incidental findings into categories (Illes et al., 2008; Wolf et al., 2008). One category includes the minority of findings with obvious and urgent health consequences such as tumors and aneurysms, where the need for disclosure is clear-cut. Another category includes findings with possible health significance that may or may not require further evaluation or management, where the benefits of disclosure is less clear-cut, and disclosure is optional. An example of this is the finding of an occult venous malformation deep in the brain that carries a small risk of hemorrhage but for which no therapy is available. A third category includes findings of no health consequence for which disclosure is not required, though if a subject was interested in the findings there is no good reason why such inconsequential information should not be disclosed. An example of this would be a cavum septum pellucidum, a congenital variation in the ventricular system. A shortcoming of this stratification scheme is that, as noted previously, the health consequences of an incidental finding cannot be completely assessed in the absence of additional knowledge about the subject's medical history and risk tolerance. An additional complication for determining whether to disclose an incidental finding of uncertain significance is the limited imaging data that may be available in research MRI, which may preclude an accurate characterization lesion based on that study alone.

A less well recognized issue is how the transition is made from research data to clinical care in those subjects in whom the decision is made to disclose an incidental finding. Although several publications on the management of incidental findings recommend making the research study images demonstrating the abnormality available to the subject and their physician, the standard currency of diagnostic studies in the clinical realm is not raw images but rather a written radiological report describing the abnormality and its differential diagnosis. Uninterpreted brain images are of little use to primary care physicians, and even for neurologists and neurosurgeons it may be difficult to interpret data acquired using nonclinical protocols. Furthermore, if an incidental finding is to be the basis for additional medical evaluation or therapy under the subject's health insurance, it is necessary for this finding to be formally incorporated into their medical record. For these reasons, it may be preferable to again consider two options for incidental findings based on the intent of the study and the diagnostic quality of the imaging obtained in the research protocol. For studies in healthy volunteers with more limited diagnostic data, if a potentially significant incidental finding is detected, a recommendation to obtain a clinical quality study to confirm the findings would allow a formal and radiological interpretation to be generated that is not qualified by limited imaging. On the other hand, studies in clinical populations with more extensive diagnostic imaging may provide sufficient information for a formal clinical report to be generated without an additional clinical study. To respect the privacy of research subjects, any such recommendation or report should be given directly to the subject or their guardian. It should be up to the subject and their guardian to determine whether to pursue it further through their health care provider, though the investigator should endeavor to assist in this process to whatever extent possible.

Several recent articles on the management of incidental findings suggest that research subjects should be offered the opportunity to opt out of being notified of incidental findings (Illes et al., 2008; Wolf et al., 2008). In the event of a potentially life-threatening finding, the recommendation is then for the investigator to recontact the subject to confirm that they really do not wish to be informed of even a potentially serious finding. This approach adds unnecessary complexity to the management of incidental findings and places the investigator in a difficult position of judging what severity of findings would prompt the confirmatory contact and how hard to push the subject to reconsider their opt-out position. It is not clear that it is necessary or desirable to allow subjects to opt-out of notification. In the absence of evidence that subject recruitment for MRI research would be dramatically reduced unless subjects are offered this alternative, a more straightforward solution to this issue would to have such subjects opt-out of being research subjects for the study.

CONCLUSIONS

The potential for an incidental finding of definite or possible clinical significance is a predictable risk of participating in MRI research. Procedures for detecting and verifying incidental findings and disclosing them to research subjects must be included in the research protocol and consent process. Incidental findings can adversely impact subjects by causing anxiety or requiring additional evaluation and treatment with additional risk. Although there is some potential for health benefit if a life-threatening finding is observed and successfully treated, the clinical significance of most incidental findings is uncertain, so it is unclear that comprehensive screening for incidental findings is truly a benefit to participating in MRI research for the subject. Although the acquisition of extra clinically oriented images may facilitate screening for incidental findings, the acquisition of such data is ideally motivated by the research design rather than for the convenience of image interpretation or to protect against future litigation. For nonclinical studies in nonclinical populations it is reasonable to perform no specific screening for incidental findings as long as research subjects are clearly aware that the study has no diagnostic intent. However, even in this situation there is still a risk of an incidental finding and some mechanism for timely access to the clinical expertise to evaluate it must be in place. The most useful format for translating an incidental finding into clinical care is a written radiological report. If image quality is insufficient for formal interpretation, a recommendation to proceed to a clinical study to verify a questionable finding may be preferable to attempting to make clinical recommendations on incomplete data.

REFERENCES

Colletti, P. M. (2004). Size "H" oxygen cylinder: Accidental MR projectile at 1.5 tesla. *J Magn Reson Imaging, 19* (1): 141–143.

Emanuel, E. J., Wendler, D., et al. (2000). What makes clinical research ethical? *JAMA, 283* (20): 2701–2711.

Illes, J., Kirschen, M. P., et al. (2008). Practical approaches to incidental findings in brain imaging research. *Neurology, 70* (5): 384–390.

Illes, J., Rosen, A. C., et al. (2004). Ethical consideration of incidental findings on adult brain MRI in research. *Neurology, 62* (6): 888–890.

Katzman, G. L., Dagher, A. P., et al. (1999). Incidental findings on brain magnetic resonance imaging from 1000 asymptomatic volunteers. *JAMA, 282* (1): 36–39.

Kirschen, M. P., Jaworska, A., et al. (2006). Subjects' expectations in neuroimaging research. *J Magn Reson Imaging, 23* (2): 205–209.

Kumra, S., Ashtari, M., et al. (2006). Ethical and practical considerations in the management of incidental findings in pediatric MRI studies. *J Am Acad Child Adolesc Psychiatry, 45* (8): 1000–1006.

Vernooij, M. W., Ikram, M. A., et al. (2007). Incidental findings on brain MRI in the general population. *N Engl J Med, 357* (18): 1821–1828.

Weber, F., & Knopf, H. (2006). Incidental findings in magnetic resonance imaging of the brains of healthy young men. *J Neurol Sci, 240* (1–2): 81–84.

Wolf, S. M., Lawrenz, F. P., et al. (2008). Managing incidental findings in human subjects research: analysis and recommendations. *J Law Med Ethics, 36* (2): 219–248, 211.

Yue, N. C., Longstreth, W. T., Jr., et al. (1997). Clinically serious abnormalities found incidentally at MR imaging of the brain: Data from the cardiovascular health study. *Radiology, 202* (1): 41–46.

11

Neuroimaging in Clinical Psychiatry

MARTHA J. FARAH AND SETH J. GILLIHAN

"Psychiatrists remain the only medical specialists that never look at the organ they treat." This statement, from the website of psychiatrist Daniel Amen (www. amenclinics.com/amenclinics/clinics/information/about-us/; web pages cited in this chapter were accessed May 24th, 2012), highlights the puzzle with which this chapter is concerned. On the one hand, brain imaging technologies provide ever more sensitive measures of structure and function that are, in principle, relevant to cognitive and emotional functioning. Furthermore, the psychiatry research literature documents with image-based correlates for virtually every psychiatric disorder. On the other hand, notwithstanding a small number of practitioners (including Amen) who use functional brain imaging as a diagnostic tool, the established view in psychiatry is that brain imaging has no role to play in routine clinical care. Aside from its use to rule out potential medical causes of a patient's condition (for example, a brain tumor), neuroimaging is not used in the process of psychiatric diagnosis.

Diagnoses in psychiatry are based entirely on behavioral, not biological, criteria. We diagnose depression by asking the patient how he feels and whether his sleeping, eating, and other behaviors have changed. We diagnose attention deficit hyperactivity disorder (ADHD) by asking the patient, family members, and others about the patient's tendency to get distracted, act impulsively, and so on. For these and all other psychiatric illnesses described by the *Diagnostic and Statistical Manual* (DSM) of the American Psychiatric Association (APA), findings from imaging do not appear among the diagnostic criteria. In the words of Kim, Schulz, Wilde, and Yudofsky (2008) in *The American Psychiatric Publishing Textbook of Psychiatry*, "neuroimaging does not yet play a diagnostic role for any of the primary psychiatric disorders."

Several questions are raised by this eschewal of diagnostic brain imaging by most psychiatrists and its endorsement by a few. The first question is the puzzle referred to in the title of this chapter: Given that psychiatric disorders are brain disorders, and given the large literature on neuroimaging in psychiatry, why is imaging not useful for diagnosis? In addition, we can ask why some psychiatrists and other mental health professionals nevertheless maintain that neuroimaging does have a role to play in diagnosis. Indeed, for all the various stakeholders—practitioners, patients, and patient families—we can ask what motivates them to pursue neuroimaging in this context. Another set of questions concerns how patients, their families, and society stand to benefit or be harmed by the current use diagnostic neuroimaging in psychiatry. Finally, we can ask about the future prospects for neuroimaging in clinical psychiatry: How might diagnostic imaging eventually enter mainstream psychiatric practice? What is being done now to facilitate this transformation? And are there other more immediately promising applications of neuroimaging to clinical practice? Each of these questions is addressed below. We begin with a brief review of the role of neuroscience in contemporary psychiatry as a reminder of why imaging has prima facie relevance to diagnosis, and the some of the motivations for seeking imaging-based diagnosis.

DIAGNOSTIC NEUROIMAGING IN PSYCHIATRY: PLAUSIBILITY AND PROMISE

Most psychiatric treatment is "biological," in the sense that it operates directly on the brain. This includes medication for depression, anxiety, psychosis, and disorders of attention. It also includes such nonpharmacologic treatments as electroconvulsive therapy, neural stimulation, biofeedback, and surgery. Even talking psychotherapy, such as cognitive and behavioral therapy, is now understood to change the brain in ways that have been visualized by neuroimaging (DeRubeis, Siegle & Hollon, 2008). In the light of the biological nature of psychiatric treatments, one would expect psychiatric diagnosis to be biological as well.

The idea of diagnostic brain imaging is all the more plausible given that psychiatric illnesses have biological correlates that are apparent in both structural and functional brain imaging. Functional brain imaging, in particular, has been widely used in psychiatry research. For example, Medline returns hundreds of hits each for searches pairing diagnostic categories such as depression and schizophrenia with imaging methods such as PET and fMRI. Given the ability of brain imaging to reveal biological correlates of psychiatric disorders, it seems plausible that imaging would play some role in psychiatric diagnosis.

The idea of imaging-based diagnosis is not only plausible; it also promises to increase the validity of psychiatric diagnoses, as well as increasing the accuracy with which individual patients can be diagnosed. A valid category is one that, as Plato put it, "carves nature at its joints." According to this view, the task of science is to identify the ways in which potentially unique phenomena cluster into groups with underlying similarity in nature. Validity, in the context of psychiatric diagnosis, refers to many different ways in which a diagnostic category corresponds to the true clustering of psychiatric dysfunction in the world. Although debate continues as to whether the diagnostic categories of psychiatry can be drawn on the basis of purely biological factors or whether society's demands, beliefs, and values also play a role (see, e.g., Wakefield, 1992; Horwitz & Wakefield, 2007), all of the following are viewed as potential indicators of the validity of diagnostic categories: covariance among symptoms within a category and not between categories, the sharing of underlying etiology, similar courses of illness over time, and relations with genetic and other biological traits of a patient and with their treatment response (see Andreasen, 1995; Kendell & Jablensky, 2003; Kendler, 2006).

One of the earliest explicit calls for biological testing was made by Robins and Guze (1970) in their seminal paper on the validation of psychiatric diagnoses. These authors laid out a broad range of criteria by which diagnostic categories could be validated: clinical correlates, family history, treatment response, course, outcome, and what they termed "laboratory studies." Neuroimaging per se was not mentioned simply because it was so rudimentary in those days. But on the assumption that psychiatric disorders are brain disorders, one could not wish for a better indication of the validity of a diagnostic category than a measure of brain function found in all and only patients with that diagnosis.

Better diagnostic methods would of course also improve the accuracy with which individual patients can be diagnosed and thereby result in more patients receiving effective treatment. By capturing information about the underlying pathophysiology believed to cause the disorder, rather than behaviors that are one causal step removed from that pathophysiology, brain imaging promises to deliver a more direct and therefore potentially more accurate diagnosis. For similar reasons, image-based diagnosis could also increase the power of research to develop better treatments, by reducing the number of inappropriate research subjects included in study samples.

The related problems of validity and accuracy of current DSM categories are most pressing for patients whose history and behavior seem equally consistent with more than one diagnosis. Researchers have long been aware of the potential contribution of neuroimaging in such cases. For example, in an early and influential PET study of depression, Schwartz et al. (1987) wrote that such findings "may have value ... as a tool for the differential diagnosis" of bipolar and unipolar depression (p. 1370). In a survey of possible uses for SPECT in psychiatry, O'Connell et al. (1989) judged it to be "a promising technique that appears to have potential in differential diagnosis" (p. 152). The hope that neuroimaging can assist in differential diagnosis lives on, as expressed more recently by Brotman et al. (2010), who suggest that functional neuroimaging will help us to distinguish between different disorders with similar presentations: "Determining the neural circuitry engaged in processing neutral faces may assist in the differential diagnosis of disorders with overlapping clinical features" (pp. 61–62). As these quotes make clear, none of these authors viewed imaging as applicable to current practice, but were instead expressing great hope for its future potential.

SPECT CLINICS: PUTTING NEUROIMAGING TO WORK NOW

Some practitioners are already using brain imaging for psychiatric diagnosis. We view this practice as premature at best, but also as potentially informative concerning the forces acting to promote and impede the eventual incorporation of neuroimaging into diagnostic practice. In this real-world example we can see the intersecting motivations of clinicians, patients, and families. We can also see one nonhypothetical way in which diagnostic imaging affects nosology in practice,

The imaging method currently being used is single photon emission computed tomography (SPECT), a functional imaging method by which regional cerebral blood flow is measured by a gamma-emitting tracer in the blood. From these regional blood flow measures a three-dimensional, low-resolution image of brain activity is constructed.

The best known of the SPECT clinics are the Amen Clinics, founded by the psychiatrist and self-help author Daniel Amen, who was quoted at the outset of the article. There are now four Amen Clinics operating in the United States, the first of which opened in 1989, and plans for another two clinics in major US cities have been announced (http://70.32.73.82/blog/5534/changing-the-world-one-brain-at-a-time/). Other clinics offering SPECT-guided psychiatric diagnosis and treatment include Cerescan, Pathfinder Brain SPECT, Silicon Valley Brain SPECT Imaging Center, Dr. Spect Scan, and MindMatters of Texas. The use of brain imaging appears to be a selling point for these clinics; their websites all feature brain images prominently and the names of the first four leave no doubt about the emphasis they place on imaging for attracting patients (Chancellor & Chatterjee, 2011; Farah, 2009).

These clinics promise to diagnose and treat a wide range of psychiatric disorders in children and adults, and base their diagnoses on patient history and examination along with the results of SPECT scans. We will focus our discussion of SPECT-assisted psychiatric diagnosis on the Amen Clinics, because their website and publications offer much more information concerning their diagnostic procedures, diagnostic categories, and patient care philosophy than is publicly available from other clinics.

SPECT-Assisted Diagnosis

At the Amen Clinics, patients are typically scanned twice: once at rest and once performing cognitive tasks. The resulting scans are used in addition to more conventional diagnostic methods including clinical interview and diagnostic checklists.

The Amen Clinics use a system of diagnoses that does not correspond to the standard system defined by the DSM. For example, anxiety and depression are combined into a single superordinate category and seven subtypes, with names such as "temporal lobe anxiety and depression" and "overfocused anxiety and depression." Attention deficit hyperactivity disorder is also reconceptualized as having six subtypes, with names such as "limbic ADD" and "ring of fire ADD." The use of new diagnostic categories in conjunction with diagnostic brain imaging is not coincidental. The mutual influence of diagnostic tests and the categories to which patients are assigned by those tests is discussed later in this article.

The images are also used to identify certain dimensions of functioning that cut across diagnostic categories, associated with seven specific brain regions: prefrontal cortex, anterior cingulate gyrus, basal ganglia, deep limbic thalamus, temporal lobes, parietal lobes, and cerebellum. In a section of the website for professionals   (www.amenclinics.net/clinics/professionals/how-we-can-help/), Amen explains that "once we know the brain system or systems that are not functioning optimally, we can then target treatment to the system that needs help." The same section of the website includes the proposed diagnostic significance of these different systems and implications for therapy. Taking the basal ganglia system as an example, "increased basal ganglia activity is often associated with anxiety (left sided problems are often associated with irritability, right sided problems more often associated with inwardly directed anxiety). Often, we have seen increased activity in this part of the brain in our normal population as well. We have seen increased activity associated here with increased motivation. Clinical correlation is needed. We have seen relaxation therapies, such as bio-feedback and hypnosis, and cognitive therapies help calm this part of the brain. If clinically indicated, too much activity here may be helped by antianxiety medications, such as buspirone. Sometimes, if the finding is focal in nature (more one side than the other), anticonvulsant medications can also he helpful" (www.amenclinics.net/clinics/professionals/how-we-can-help/brain-science/basal-ganglia-system-bgs/).

Evidence of Usefulness

The Amen Clinics website states that they have performed almost 50,000 scans (www.amenclinics.com/clinics/patients/18-ways-spect-can-help-you/)—a   huge number which, with associated clinical data and analyzed appropriately, could provide important evidence on the value of SPECT scanning in diagnosis and the efficacy of Amen's approach to psychiatric care. Unfortunately, no such studies have been reported. The lack of empirical validation has led to widespread condemnation of diagnostic SPECT as premature and unproven.

In 2005 the American Psychiatric Association Council on Children, Adolescents, and Their Families issued a white paper that concluded, "at the present time, the available evidence does not support the use brain imaging for clinical diagnosis or treatment of psychiatric disorders in children and adolescents" (Flaherty et al., 2005). In a review of one of Amen's popular books, appearing in the *American Journal of Psychiatry*, Leuchter (2009) writes, "it is not clear how the SPECT image provides reliable information that informs clinical decisions ... There is also no evidence presented to justify exposing patients to the radiation of a SPECT scan and to support the considerable expense to patients, families and their insurers. " It was recently reported that the Brain Imaging Council of the Society of Nuclear Medicine proposed a test of Amen's methods by asking him to interpret a set of blinded SPECT scans, but the offer was declined (Adinoff & Devous, 2010). Amen (2010) has countered that the society did not "formally" approach him with this proposal.

In recent writings Amen and coauthors have argued for the value of SPECT in psychiatry by emphasizing its usefulness in complex and treatment-refractory cases (Amen, Trujillo, Newberg, Willeumier, Tarzwell, Wu, & Chaitin, 2011; Amen, Willeumier, & Johnson, 2012). They offer case studies ranging from failed marriage therapy to compulsive eating in which SPECT studies revealed evidence of previous head injury, toxin exposure, seizure disorder, or normal pressure hydrocephalus. Patients improved after appropriate treatment of these (neurological, by conventional terms) problems. Such peer-reviewed reports of individual cases do represent empirical evidence that is relevant to the usefulness of SPECT in psychiatry, but the role they support for neuroimaging is already recognized in mainstream psychiatry: the identification of medical causes for psychiatric symptoms. As case reports they are essentially existence proofs that SPECT can sometimes be useful, but do not tell us how often SPECT would be expected to yield diagnostically useful information for any given category of psychiatric patients. Furthermore, they do not address the more fundamental issue surrounding SPECT-assisted diagnosis: whether, and to what extent, SPECT can aid in the diagnosis of primary psychiatric disorders including DSM axis-I disorders such as depression, bipolar disorder, anxiety, and ADHD.

At present we have no evidence that would allow us to estimate the value added to psychiatric diagnosis by SPECT imaging. It is possible that SPECT scans are helpful in diagnosing some or most patients; alternatively, it is possible that the scans add nothing to the accuracy of diagnosis. It is even possible that that the scans add a red herring to the diagnostic process, leading physicians to less accurate diagnoses or less helpful treatment plans.

Although we lack information about the benefit side of the risk-benefit calculation for SPECT-aided diagnosis in psychiatry, we do know something about the risks, specifically the small but nonnegligible risks of radiation exposure from SPECT scanning (see, e.g., Amis et al., 2007). Incidental findings and false negatives could also be viewed as risks. An additional consideration for individuals seeking such a procedure is the cost: Depending on the location of the Amen Clinic, a complete work-up including the two SPECT scans costs \$3,575–\$4,125 (www.amenclinics.com/clinics/patients/). Patients and their families typically pay out-of-pocket for SPECT scans, as insurers will not pay for unproven methods of diagnostic testing.

Appeal of Diagnostic Neuroimaging to Practitioners, Patients, and Families

The vast majority of psychiatrists and psychologists do not use SPECT imaging when they diagnose patients. What might account for the small fraction who do? The list of possible answers includes a desire to practice a more biologically based form of psychiatry, despite the current prematurity of diagnostic imaging; the potential for brain scans to motivate patient compliance with treatment plans (www.amenclinics.net/clinics/professionals/how-we-can-help/direct-benefits-for-patients-and-families/), and the ability to attract patients who pay directly (as opposed to third-party payment) for the procedure and follow-up.

Why would patients and their families pay large sums of their own money for an unproven method? The answer is that most patients are unaware that diagnostic SPECT scanning in psychiatry lacks empirical support. In addition, Amen has become a familiar and trusted figure thanks to his numerous best-selling books and TV shows broadcast on public broadcasting service stations (which are not PBS productions, but self-produced shows that some have called infomercials for Amen's products and clinics; see Burton, 2008). Amen's positive image, coupled with the intuitively sensible notion that physicians should look at the organ they plan to treat, draws many patients.

If prospective patients come across criticisms of SPECT scanning for psychiatric diagnosis online or from a health care provider, they may be reassured by carefully composed statements from the Amen Clinics such as the following, from the Brain SPECT Informed Consent Form (available at www.markkosinsmd. com/PatientPortal/MyPractice.aspx?UCID={85CB5B0E-51C8-468B-AE34-174650EE240F}&TabID={4}) in answer to the question "Is the use of brain SPECT imaging accepted in the medical community?": "Brain SPECT studies are widely recognized as an effective tool for evaluating brain function in seizures, strokes, dementia and head trauma." In addition to sidestepping the question of SPECT for psychiatric diagnosis, the answer continues by discounting criticism as rooted in ignorance: "As with many new technologies or new applications of existing technologies, many physicians do not fully understand the application of SPECT imaging." The answer concludes with the statement that "psychiatric SPECT imaging is used in the academic setting in many centers in the US and abroad," which is only true if it refers to research uses of SPECT rather than the diagnostic purpose for which patients are giving consent. The Amen Clinics website also invokes the authority of Thomas Insel, director of the National Institute of Mental Health, quoting from a lecture in which he claimed that "brain imaging in clinical practice is the next major advance in psychiatry." Although Insel's statement was clearly about the future of imaging in clinical practice, and therefore might be taken as implying that imaging is not currently useful, it is quoted under the heading "The Future is Now" (www.amenclinics.com/clinics/patients/18-ways-spect-can-help-you/).

There are many reasons that patients and families may seek diagnostic SPECT scanning for psychiatric problems, beyond the claims just reviewed. Brain imaging has a high-tech allure that suggests advanced medical care. People may assume that the treatments available at these clinics, as well as the diagnostic methods, are cutting edge. In addition, there is a strong allure in imaging's visual proof that psychological problems have a physical cause. The Amen Clinics cite several ways in which patients and their families may find this helpful.

First, the images can reduce feelings of stigma and guilt. Brain imaging provides a concrete reminder that psychiatric disorders are disorders of brain function. As the clinic website states, "SPECT scans help patients better understand their problems, decreasing shame, guilt, stigma and self-loathing." By demonstrating visually that patients' psychological problems are associated with brain dysfunction, imaging may help them feel less responsible for their illness or less personally stigmatized by it; they can believe that "it's not me, it's my brain" (see Dumit, 2003).

For similar reasons, the families of patients may also feel relieved by brain imaging. This is especially true for parents, who may worry that their child's illness was caused by their own behavior toward the child—their absence from the home, the allowance of too much TV, or inadequate discipline, for example. An abnormal brain scan can be seen as proof that the child's brain, and not his or her upbringing, is responsible for the psychiatric disorder. Of course, the alternatives of brain and upbringing are not mutually exclusive. Any behavioral trait must have a basis in the brain whether its causes are genetic or environmental. Nevertheless, insofar as images depict the biological basis of mental illness, they shift the attention toward a more physical, deterministic understanding of the disorders and away from responsibility, blame, and other moral concepts that add to the burden on patients and families. Amen points this out as a benefit when he states that "a SPECT scan can help families understand the underlying medical reasons for a problem, which helps decrease shame, self blame and conflict" (www.amenclinics.com/clinics/patients/18-ways-spect-can-help-you/).

Patients and families may also appreciate the similar role played by SPECT scans in legal contexts, "helping judges and juries understand difficult behavior" (www.amenclinics.com/clinics/patients/18-ways-spect-can-help-you/). Functional neuroimaging has been introduced as evidence in a variety of roles in the United States and other legal systems—most often as evidence of brain injury in tort cases, but also by the defense in criminal trials (Patel et al., 2007). In the latter case it is most often used for mitigation in the sentencing phase of criminal trials, where it can provide concrete, visual evidence of a person's abnormal or diminished faculties (Hughes, 2010).

PROSPECTS FOR DIAGNOSING PSYCHIATRIC ILLNESS WITH NEUROIMAGING

Current and Foreseeable Diagnostic Practices

At the time of writing, psychiatric illnesses are currently classified according to one of two similar systems: the fourth revised edition of the Diagnostic and Statistical Manual of Mental Disorders (DSM-IV-TR), and the tenth edition of the International Classification of Diseases (ICD-10). Psychiatric diagnosis is poised for change, however, with the revision of the DSM. New disorders defined by new constellations of signs and symptoms are being considered for addition to the DSM-5, scheduled for release in May 2013. Some disorders already covered by previous editions may be subdivided differently or merged in the DSM-5. For each diagnostic category, criteria for inclusion and exclusion are being reviewed and updated in light of current knowledge. In relation to this last change, the incorporation of genetic and neurobiological measures has been considered (Regier, Narrow, Kuhl, & Kupfer, 2009).

Why will brain imaging not figure in these new diagnostic criteria? The consensus answer is that, despite the value of brain imaging in understanding mental disorders, it would be premature to include brain imaging among diagnostic criteria for the next DSM (Agarwal et al., 2010; Hyman, 2007; Miller, 2010; Miller & Holden, 2010). Whereas biological evidence will figure more prominently in the DSM-5 than in any previous edition, its role is expected to be in the validation of the categories themselves rather than in the criteria for diagnosing an individual patient (Hyman, 2010).

Current Obstacles to Diagnostic Neuroimaging

Why has diagnostic neuroimaging not yet found a place in psychiatric practice? Generic answers such as "psychiatric neuroimaging is in an early stage of development" or "medicine is a conservative field" contain kernels of truth but do not adequately address the question. After all, psychiatry research has made use of neuroimaging for three decades, and psychiatry has eagerly pursued new treatment technologies, including transcranial, deep brain, and vagal nerve stimulation, as well as drugs borrowed from specialties as diverse as epilepsy (Yatham, 2004) and sleep medicine (Ballon & Feifel, 2006). How, then, can we explain the lack of a role for neuroimaging in psychiatric diagnosis? What obstacles currently lie on the path to the use of such methods? The answer involves the nature of imaging as well as the nature of psychiatric diagnosis itself.

Limitations Related to Imaging

The vast majority of current neuroimaging research in psychiatry compares two groups of subjects: those with an illness and healthy control subjects. For functional neuroimaging studies the subjects may be resting or performing a task involving cognitive or emotional processing. The words "two," "groups," and "task," above, each represent important limitations on the ability to translate such research into diagnostic tests, related to *specificity*, *sensitivity*, and *standardization*, respectively.

We will start by examining the *sensitivity* of imaging studies of psychiatric patients. Although the accuracy and reliability of individual subject scans has increased since the early days of brain imaging with PET, SPECT, and fMRI due to improvements in image acquisition methods as well as data analysis, the vast majority of psychiatric neuroimaging studies aggregate data from groups of subjects for analysis. In contrast, diagnosis must be applied to individuals, not groups. When structural and functional findings from individual subjects are examined, a high degree of variability is observed, even within groups of healthy and ill subjects. More problematic for diagnostic purposes, the distributions of healthy and ill subjects generally overlap (see Gillihan & Parens, 2011). In the language of diagnostic tests, imaging studies are generally not highly sensitive to the difference between illness and health.

Standardization might appear to be a premature concern for an approach to diagnosis that is nowhere near ready for widespread clinical use. In the context of diagnostic testing, the term "standardization" often refers to the specification of all details of the protocol that might vary from lab to lab and could influence the results. Here we refer to a broader but related issue, namely, the many obvious and fundamental ways in which protocols differ between imaging studies, in particular functional imaging studies. The patterns of activation obtained in studies of psychiatric patients depend strongly on the tasks performed by the subjects and the statistical comparisons examined by the researchers afterwards. Although this seems obvious when applied to cognitive neuroscience studies of normal subjects, it is easier to lose sight of when considering studies of psychiatric patients, where results may be summarized by stating that certain regions are under- or overactive, or more or less functionally connected, in particular patient groups. Of course such summaries are fundamentally incomplete unless they include information about what task evoked the activation in question: Were the patients resting, processing emotional stimuli (e.g., fearful faces), trying not to process emotional stimuli (e.g., emotional Stroop task), or engaged in effortful cognition (e.g., task switching)? The fact that any imaging study's conclusions are relative to the tasks performed adds further complexity to the problem of seeking consistently discriminating patterns of activation for control subjects and patients with different disorders.

Another limitation imposed by imaging concerns *specificity*. When researchers compare subjects from only two categories—patients from a single diagnostic category to healthy subjects—the most that they can learn is how brain activation in a single illness differs from healthy brain activation. Of course, the dilemma faced by a diagnosing clinician is rarely "Does this person have disorder X or is he healthy?" Rather, it is typically "Does this person have disorder X, Y or Z?" For all we know, the pattern that distinguishes people with disorder X from healthy people is not unique to X, but is shared with a whole alphabet of other disorders. It might be nothing more than a sign of psychopathology per se, and thus provide less specificity for diagnosis than a brief clinical exam. Because few imaging studies directly compare brain activation across multiple disorders, or use sufficiently standardized methods that their results can be directly compared with the results of other studies, we lack good evidence on the likely specificity of brain imaging for diagnosis.

The best we can do at present is to compare the results of brain imaging across studies, with admittedly different tasks and methods of analysis, in order to assess specificity. On the face of things, there is considerable similarity of imaging results across different diagnoses. For example, a meta-analysis of neuroimaging studies of anxiety disorders reported common areas of activation (amygdala, insula) across PTSD, social phobia, and specific phobia—suggesting that neuroimaging has yet to reveal patterns of neural activity that are unique to specific anxiety disorders (Etkin & Wager, 2007). Abnormalities of amygdala activation also have been reported consistently in neuroimaging studies of depression. For example, Gotlib and Hamilton (2008) reviewed the literature on neuroimaging of depression and concluded that "most consistently, the amygdala and subgenual [anterior cingulate cortex] appear to be overactive in [major depressive disorder], and the [dorsolateral prefrontal cortex] underactive" (p. 160). A similar pattern of results has been reported in bipolar disorder; Keener and Phillips (2007) summarized the relevant neuroimaging results as showing increased activity in emotion processing regions (including the amygdala) and decreased activity in executive regions (e.g., dorsolateral prefrontal cortex). In schizophrenia, a disorder primarily of thought rather than of mood, amygdala hyperactivity is again often observed, along with lower dorsolateral prefrontal activity (Berman & Meyer-Lindenberg, 2004). Psychopathy (which shares features with the *DSM* diagnosis of antisocial personality disorder) has been associated with similar neural patterns; in their recent review, Wahlund and Kristiansson (2009) stated that "a dysfunctional amygdala has been suggested as one of the core neural correlates of psychopathy.... Aside from the amygdala, frontal lobe dysfunction has been suggested in psychopaths" (p. 267).

More sophisticated methods of image analysis may hold promise for discerning the underlying differences among the many disorders that feature similar regional abnormalities. By taking into account the nature of the task used to evoke brain activity and functional relationships among different activated or resting brain areas, we may be able to revise the initial impression that all imaging of psychopathology involves the usual suspects such as limbic hyperactivity and prefrontal hypoactivity. In addition, new multivariate statistical approaches to image analysis enable the discovery of spatial and temporal patterns within brain images that distinguish between task conditions or types of subject more effectively than traditional statistical methods (Haynes & Rees, 2006). These methods have only begun to be applied to clinical disorders but show promise for increasing the specificity of brain imaging markers for psychiatric illness (Bray et al., 2009; Calhoun et al., 2008).

Finally, as methods of acquiring and analyzing brain images continue to develop, it bears remembering that imaging will never measure all aspects of brain function. There is no guarantee that it will be able to capture those aspects most characteristic or defining of the psychiatric disorders. Regional differences in brain activity measured on a spatial scale discernible through current functional imaging methods, or neurochemical differences discernible through PET or SPECT, are not the only ways in which disordered brains can differ from healthy brains. Although neuroimaging research has demonstrated differences among brain activity in different psychiatric disorders, it is an open empirical question whether current or future imaging methods will reveal sufficiently sensitive and specific features of brain function to ever serve as diagnostic tests.

Limitations Related to Current Diagnostic Categories

Another set of reasons why progress toward diagnostic imaging in psychiatry has been slow concerns the nature of the diagnostic categories themselves. The categories of DSM are intended to be both valid and reliable. As discussed earlier, validity refers to the correspondence between diagnostic categories and the ways in which psychiatric disorders are truly structured in nature. Reliability refers to the degree to which the categories' criteria can be used consistently by any appropriately trained clinician, so that different diagnosticians will arrive at the same diagnosis for each patient.

Good, or at least improved, reliability was one of the signal achievements of the DSM-III, and has carried over to DSM-IV. Unfortunately, validity continues to be more difficult to achieve. This is not surprising given how closely validity is related to scientific understanding, and how complex and poorly understood psychiatric illness continues to be (Robert, 2007). To the extent that our psychiatric categories do not correspond to natural kinds (Quine, 1969), we should probably not expect perfect correspondence with brain physiology as revealed by imaging.

As an illustration of how far from being natural kinds our current diagnostic categories are, consider the diagnostic criteria for one of the more common serious disorders, major depressive disorder. According to the DSM-IV-TR, patients must report at least one of the two symptoms of depressed mood or anhedonia and at least four of an additional eight symptoms. It is therefore possible for two patients who do not share a single symptom to both receive a diagnosis of major depressive disorder. In addition to heterogeneity within the diagnostic categories of psychiatry, there are also commonalities of symptoms between categories. For example, impulsivity, emotional lability and difficulty with concentration each occur in multiple disorders.

THE PRESENT AND FUTURE OF BRAIN IMAGING IN PSYCHIATRY

Coevolution of Science, Diagnostic Tests, and Diagnostic Categories

We are currently far from being able to use brain imaging for psychiatric diagnosis. Yet all of the limitations of imaging and diagnosis just reviewed may eventually be overcome. By what path might this occur?

Imaging markers of diagnostic categories may emerge from basic research on psychopathology and prove to be highly diagnostic. Alternatively, it is possible that the relatively atheoretical multivariate statistical approach mentioned earlier could provide the first candidate neural signatures of psychiatric disorders. By whatever method the candidate neural signatures are identified, large-scale validation trials will be needed before they can enter routine clinical use. This promises to be a lengthy and expensive process, which could easily fill the interval between two or more editions of the DSM.

Whether the path to imaging-based diagnosis involves translation of newly discovered mechanisms of pathophysiology, brute force number crunching, or both, we cannot assume that it will preserve current nosology. Brain imaging may succeed in delineating categories of patients based on abnormalities in brain function, but these categories may not be the same as the categories of the DSM. Indeed, given the heterogeneity within diagnostic categories and the overlap between categories just noted, it seems likely that our nosology will be forced to change. If the mismatch between imaging markers and diagnostic categories is not drastic, the

DSM categories may change incrementally—for example, by revisions of individual diagnostic criteria for specific disorders.

More revolutionary change is also possible. Psychiatrists do not view DSM categories as ground truth, and the validity of the current system of categories has been widely questioned (e.g., Radden, 1994). The potential for imaging research to disrupt the gradual, iterative approach to nosological change was anticipated by Kendler and First (2010), who wrote "the iterative model assumes continuity over time in ... the methods used to determine validity.... What happens if dramatic technical breakthroughs in genetics, imaging or neuroscience cast the problems of psychiatric nosology in an entirely new light? The application of such new methods to our nosology would likely disrupt the smooth evolutionary approach of the iterative model" (p. 263). If a new nosology based on imaging is proven to have clinical utility—for example, enabling better treatment decisions—then imaging may prompt a radical reconceptualization of psychiatric diagnosis and entirely new diagnostic categories may emerge. Indeed, the existence of categories per se has been questioned, with some experts proposing to characterize patients in terms of where they fall on different dimensions of psychological functioning, which may be more or less severely impaired, rather than assigning them to discrete categories. This system may better capture the ways in which patients differ from one another and from healthy people (Krueger, Watson & Barlow, 2005).

It is interesting to note that both the emergence of new diagnostic categories and the use of dimensional classification schemes are presaged by the uses of SPECT just reviewed. Although SPECT-aided psychiatric diagnosis has no basis in evidence and is regarded with extreme skepticism by most experts, it nevertheless reveals the tension that can be expected between the DSM's system of categories, on the one hand, and the kinds of diagnoses that are more naturally built on functional brain imaging results, on the other.

Recall that the Amen Clinics' diagnostic system includes entirely new types of diagnoses such as overfocused anxiety and depression and ring of fire ADD, which do not correspond to anything in the DSM. They are instead based on a combination of behavioral observations and SPECT findings. Examples of the latter include "increased anterior cingulate gyrus activity and increased basal ganglia and/or deep limbic activity at rest and during concentration" for overfocused anxiety and depression (www.amenclinics.net/conditions/Anxiety_Issues/) and "marked overall increased activity across the cortex, may or may not have low prefrontal cortex activity" for ring of fire ADD (www.amenclinics.net/conditions/ADHD/). Although the validity and clinical utility of these categories is far from clear, given the absence of any peer-reviewed evidence supporting them, they demonstrate in a concrete way how the inclusion of neuroimaging data in the diagnostic process can change not only that process but the diagnoses themselves. Imagine that patients sharing some features of a conventional diagnosis without meeting all criteria are found to group into distinct categories according to their patterns of brain activation. It would be reasonable to consider these new groupings good candidates for new and more valid diagnostic categories.

Similarly, in the Amen Clinics' approach we also see the use of dimensions of functioning that cut across diagnostic categories, in the form of the seven different anatomically defined systems described earlier (prefrontal cortex, anterior cingulate cortex, basal ganglia system, deep limbic system thalamus, temporal lobes, parietal lobes, and cerebellum; see www.amenclinics.net/clinics/professionals/how-we-can-help/). Setting aside the question of the validity of these systems, it is apparent that the organization of the brain into functionally and anatomically distinct systems, combined with the graded nature of activation in those systems, fits naturally with a dimensional rather than categorical system for characterizing patients.

In sum, there are a priori reasons to expect our diagnostic system to change as imaging data are incorporated. There is also the illustrative (if not adequately evaluated) example of SPECT-aided psychiatric diagnosis, where the use of imaging has led to redrawn categories and cross-cutting dimensional classifications of patients. Although changes to our diagnostic systems may well be inevitable as neuroimaging provides new information linking brain function with psychiatric symptoms, and those changes can be expected to improve validity and clinical utility, these changes will likely come slowly. Where diagnoses are concerned there are strong arguments for conservatism.

The current system of categories is valuable in part simply because we have used it for so long and therefore much of our clinical knowledge is relative to this system (e.g., First & Kendler, 2010). As Hyman (2002) put it, "we should not tinker with existing diagnoses without a very high threshold because even small changes in diagnostic criteria may have negative consequences. They may alter the apparent prevalence of disorders, confound family and longitudinal studies, alter treatment development by affecting regulatory agencies … " (p. 6). For these reasons it is appropriate for the influence of brain imaging on psychiatric diagnosis to be more evolutionary than revolutionary. In keeping with this approach, DSM diagnoses have so far changed in a gradual and piecemeal manner through multiple editions of the manual, with most disorders retaining their defining criteria and a minority being subdivided, merged, added, and eliminated in the light of new research findings.

An attempt to reconcile the need for consistency with the promise of more neurobiologically based classifications can be found in the research domain criteria (RDoC) for psychiatry research, proposed by the US National Institute of Mental Health. This is "a long-term framework for research … [with] classifications based on genomics and neuroscience as well as clinical observation, with the goal of improving treatment outcomes" (Insel et al., 2010). The RDoC system, still under construction at the time of writing (see www.nimh.nih.gov/research-funding/rdoc/nimh-research-domain-criteria-rdoc.shtml), is organized into five domains: negative valence, positive valence, cognitive processes, social processes, and arousal/regulatory processes. Within the domains are more specific functions related to known neural circuits, which vary dimensionally from normal levels of function to abnormal. Examples include fear, in the negative valence domain, associated with "amygdala, hippocampus, interactions with ventromedial PFC," and working memory, in the cognitive processes domain, associated with "dorso-lateral PFC, other areas in PFC." The use of RDoC across research labs, in parallel with DSM categories, may ultimately lead to the development of a new diagnostic system that would both be more valid and also possibly more consistent with the use of imaging as a diagnostic test.

NONDIAGNOSTIC USES OF IMAGING IN CLINICAL PSYCHIATRY

We have seen that, for reasons to do with both the nature of neuroimaging and the nature of psychiatric diagnosis, imaging is far from providing useful diag-nostic information in psychiatry. However, neither is it without immediate clini-cal promise. Here we summarize several promising roles for imaging other than diagnosis.

New treatments can be suggested by imaging research, exemplified by the use of deep brain stimulation (DBS) in area 25 for the treatment of depression. Based on findings from functional and structural neuroimaging studies, Mayberg and colleagues developed a neural model of major depressive disorder that included the influence of hyperactivity in the subgenual cingulate cortex on other regions important for mood. The initial test of this model in a treatment setting used electrodes implanted in this region of the brain in six individuals with treatment refractory depression. Four of the six patients in this group achieved remission of their depression (Mayberg et al., 2005). Subsequent work with a larger group of patients confirmed the efficacy of DBS for treatment resistant depression (Lozano et al., 2008). Imaging is also being used to assess the effectiveness of this technique and personalize the placement of electrodes (Hamani et al., 2009).

Another potential application of imaging to clinical care involves the predic-tion of treatment response. Treatments for some psychiatric disorders take weeks or months to produce a therapeutic effect, and not all treatments are equally effec-tive for all patients. The ability to predict a patient's response to a given treatment can therefore save considerable time and suffering. Although not currently part of clinical care, there is reason for optimism concerning its feasibility in several different disorders (Evans, Dougherty, Pollack & Rauch, 2006). In addition to the prediction of treatment response in those already diagnosed with an illness, imag-ing can aid prediction of disease onset in asymptomatic individuals. For example, research has shown that structural MRI can predict first episodes of schizophre-nia in individuals who are at genetically increased risk (McIntosh et al., 2011), enabling early or even preventive treatment to be offered to those most likely to benefit from it.

Finally, functional neuroimaging can be used as a treatment itself, by providing patients with a real time measure of regional brain activity to use in biofeedback training (deCharms, 2008). This technique has been used to enhance pain control in chronic pain patients by deCharms and colleagues (2005). DeCharms (2008) has also noted the potential of the method for treating depression and addiction.

Neuroimaging will be likely to enter clinical use with the applications just reviewed before it finds a general role in diagnosis. Nevertheless, attempts to diagnose with the help of imaging will undoubtedly continue. Practitioners have a financial incentive to offer this service and patients are attracted by the promise of more scientific diagnosis and treatment, as well as relief from blame and stigma. Neuroimaging has strong prima facie relevance to psychiatric diagnosis that can only be dispelled by careful reflection on the technical limitations of imaging and the historical and pragmatic nature of psychiatric nosology.

ACKNOWLEDGMENTS

Editors' note: This chapter is based on an article that appeared in October 2012 in the *American Journal of Bioethics—Neuroscience* and is used with permission here.

REFERENCES

Adinoff, B., & Devous, M. (2010). Scientifically unfounded claims in diagnosing and treating patients. *American Journal of Psychiatry, 167*(5), 598.

Agarwal, N., Port, J. D. Bazzocchi & Renshaw, P. F. (2010). Update on the use of MR for assessment and diagnosis of psychiatric disease. *Radiology, 255*, 23–41.

Amen, D. (2010). Brain SPECT imaging in clinical practice. *American Journal of Psychiatry, 167*(9), 1125.

Amen, D. G., Trujillo, M., Newberg, A., Willeumier, K., Tarzwell, R., Wu, J. C. & Chaitin, B. 2011. Brain SPECT imaging in complexpsychiatric cases: An evidence-based, underutilized tool. *Open Neuroimaging Journal 5*, 40–48.

Amen, D., Highum, D., Licata, R. et a;. (2012).Specific ways brain SPECT imaging enhances psychiatric practice. *Journal of Psychoactive Drugs, 44*(2), 96–106.

Amis, E. S., Butler, P. F., Applegate, K. E., et al. (2007). American College of Radiology white paper on radiation dose in medicine. *Journal of the American College of Radiology, 4*(5), 272–284.

Ballon, J. S., & Feifel, D. (2006). A systematic review of modafinil: potential clinical uses and mechanisms of action. *The Journal of Clinical Psychiatry, 67*(4), 554.

Berman, K., & Meyer-Lindenberg, A. (2004). Functional brain imaging studies in schizophrenia. In D. Charney & E. Nestler (Eds.), *Neurobiology of Mental Illness*. Oxford New York: Oxford University Press

Bray, S., Chang, C., & Hoeft, F. (2009). Applications of multivariate pattern classification analyses in developmental neuroimaging of healthy and clinical populations. *Frontiers in Human Neuroscience, 3*, 32.

Brotman M. A., Rich, B. A., Guyer, A. E., Lunsford, J. R., Horsey, S. E., Reising, M. M., et al. (2010). Amygdala activation during emotion processing of neutral faces in children with severe mood dysregulation versus ADHD or bipolar disorder. *American Journal of Psychiatry 167*:61–69.

Burton, R. (2008). Brain Scam. *Salon.* www.salon.com/2008/05/12/daniel_amen.

Chancellor, B. & Chatterjee, A. (2011): Brain Branding: When Neuroscience and Commerce Collide, *American Journal of Bioethics Neuroscience, 2*(4), 18–27

deCharms, R. C. (2008). Applications of real-time fMRI. *Nature Reviews Neuroscience, 9*, 720–729.

deCharms, R. C., Maeda, F., Glover, G. H., Ludlow, D., Pauly, J. M., Soneji, D., et al. (2005). Control over brain activation and pain learned by using real-time functional MRI. *Poceedings of the National Academy of Sciences, 102*(51), 18626–18631.

Derubeis, R. J., Siegle, G. J., & Hollon, S. D. (2008). Cognitive therapy versus medication for depression: treatment outcomes and neural mechanisms. *Nature Reviews Neuroscience, 9,* 788–796.

Dumit, J. (2003). Is it me or my brain? Depression and neuroscientific facts. *Journal of Medical Humanities, 24,* 35–47.

Etkin, A., & Wager, T. D. (2007). Functional neuroimaging of anxiety: A meta-analysis of emotional processing in PTSD, social anxiety disorder, and specific phobia. *American Journal of Psychiatry 164,* 1476–1488.

Evans, K. C., Dougherty, D. D., Pollack, M. H., & Rauch, S. L. (2006). Using neuroimaging to predict treatment response in mood and anxiety disorders. *Annals of Clinical Psychiatry, 18*(1), 33–42.

Farah, M. J. (2009). A picture is worth a thousand dollars (Editorial), *Journal of Cognitive Neuroscience, 21,* 623–624.

Flaherty, L. T., Arroyo, W., Chatoor, I., Edwards, R. D., Ferguson, Y. B., Kaplan, S., et al. (2005). Brain imaging and child and adolescent psychiatry with special emphasis on SPECT. American Psychiatric Association, Council on Children, Adolescents and Their Families.

Gillihan, S. J., & Parens, E. (2011). Should we expect "neural signatures" for DSM diagnoses?. *Journal of Clinical Psychiatry, 72*(10), 1383–1389.

Hamani, C., Mayberg, H. S., Snyder, B., Giacobbe, P., Kennedy, S., & Lozano, A. (2009). Deep brain stimulation of the subcallosal cingulate gyrus for depression: Anatomical location of active contacts in clinical responders and a suggested guideline for targeting. *Journal of Neurosurgery, 111,* 1209–1215.

Haynes, J. D., & Rees, G. (2006). Decoding mental states from brain activity in humans. *Nature Reviews Neuroscience, 7*(7), 523–534.

Horwitz, A. V., & Wakefield, J. C. (2007). *The Loss of Sadness: How Psychiatry Transformed Normal Sorrow Into Depressive Disorder.* New York: Oxford University Press.

Hughes, V. (2010). Science in court: Head case. *Nature, 464,* 340–342.

Hyman, S. E. (2002). Neuroscience, genetics, and the future of psychiatric diagnosis. *Psychopathology. 35*(2–3), 139–144.

Hyman, S. E. (2007). Can neuroscience be integrated into DSM-V? *Nature Reviews Neuroscience 8,* 725–732.

Hyman, S. E. (2010). The diagnosis of mental disorders: the problem of reification. *Annual Review of Clinical Psychology, 27*(6), 155–179.

Insel, T., Cuthbert, B., Garvey, M., Heinssen, R., Pine, D. S., Quinn, K., et al. (2010). Research domain criteria (RDoC): toward a new classification framework for research on mental disorders. *American Journal of Psychiatry, 167*(7), 748–751.

Keener, M. T., & Phillips, M. L. (2007). Neuroimaging in bipolar disorder: A critical review of current findings. *Current Psychiatry Report, 9,* 512–520.

Kendell, R ., & Jablensky, A. (2003). Distinguishing between the validity and utility of psychiatric diagnoses. *The American Journal of Psychiatry, 160* (1): 4–12.

Kendler, K. S. (2006). Reflections on the relationship between psychiatric genetics and psychiatric nosology. *The American Journal of Psychiatry, 163* (7): 1138–1146.

Kendler K. S., & First, M. B . (2010). Alternative futures for the DSM revision process: iteration v. paradigm shift. *British Journal of Psychiatry, 197,* 263–265.

Kim, H. F., Schulz, P. E., Wilde, E. A., & Yudofsky, S. C. (2008). Laboratory Testing and Imaging Studies in Psychiatry. In R. E. Hales, S. C. Yudofsky, & G. O., Gabbard (Eds.),

The American Psychiatric Publishing Textbook of Psychiatry 5th Edition (pp. 19–72). Arlington, VA: American Psychiatric Publishing.

Krueger, R. F., Watson, D., & Barlow, D. H. (2005). Introduction to the special section: Toward a dimensionally based taxonomy of psychopathology. *Journal of Abnormal Psychology, 114*(4), 491–493.

Leuchter, A.F. (2009). Healing the hardware of the soul by Daniel Amen. Book review. *American Journal of Psychiatry, 166*, 625.

Lozano, A. M., Mayberg, H. S., Giacobbe, P., Clement, H., Craddock, R. C., & Kennedy, S. H. (2008). Subcallosal cingulate gyrus deep brain stimulation for treatment-resistant depression. *Biological Psychiatry, 64*, 461–467.

Mayberg, H. S., Lozano, A., Voon, V., McNeely, H., Seminowicz, D., Hamani, C., et al. (2005). Deep brain stimulation for treatment-resistant depression. *Neuron, 45*, 651–660.

McIntosh, A. M., Owens, D. C., Moorhead, W. J., Whalley, H. C., Stanfield, A. C., Hall, J., et al. (2011) Longitudinal volume reductions in people at high genetic risk of schizophrenia as they develop psychosis. *Biological Psychiatry, 69* (10), 953–958.

Miller, G. (2010). Beyond DSM: Seeking a brain-based classification of mental illness. *Science, 327*, 1437.

Miller, G., & Holden, C. (2010). Proposed revisions to psychiatry's canon unveiled. *Science, 327*, 770–771.

O'Connell, R. A., Van Heertum, R. L., Billick, S. B., Holt, A. R., Gonzalez, A., Notardonato, H., et al. (1989). Single photon emission computed tomography (SPECT) with [123] IMP in the differential diagnosis of psychiatric disorders. *Journal of Neuropsychiatry, 1*, 145–153.

Patel, P., Meltzer, C. M., Mayberg, H. S., & Levine, K. (2007). The role of imaging in United States Courtrooms, *Neuroimaging Clinics of North America, 17* (4), 557–567.

Quine, W.V.O . 1969. Natural Kinds. In *Ontological Relativity and Other Essays*. New York: Columbia University Press.

Radden, J. (1994). Recent criticism of psychiatric nosology: A review. *Philosophy, Psychiatry & Psychology, 1*(3), 193–200.

Regier, D. A., Narrow, W. E., Kuhl, E. A., & Kupfer, D. J. (2009). The conceptual development of DSM-V. *American Journal of Psychiatry, 166*(6), 645–650.

Robert, J. S. (2007). Gene maps, brain scans, and psychiatric nosology. *Cambridge Quarterly of Healthcare Ethics, 15*, 209–218.

Robins, E., & Guze, S. B. (1970). Establishment of diagnostic validity in psychiatric illness: Its application to schizophrenia. *American Journal of Psychiatry, 126*, 983–987.

Schwartz, J. M., Baxter, L. R., Mazziotta, J. C., Gerner, R. H., & Phelps, M. E. (1987). The differential diagnosis of depression: Relevance of positron emission tomography studies of cerebral glucose metabolism to the bipolar-unipolar dichotomy. *Journal of the American Medical Association, 258*, 1368–1374.

Tan, H. Y., Callicott, J. H., & Weinberger, D. R. (2007). Dysfunctional and compensatory prefrontal cortical systems, genes and the pathogenesis of schizophrenia. *Cerebral Cortex, 17*(1), 171–181.

Wahlund, K., & Kristiansson, M. (2009). Aggression, psychopathy and brain imaging: Review and future recommendations. *International Journal of Law and Psychiatry, 32*, 266–271.

Wakefield, J. C . (1992). The concept of mental disorder: on the boundary between biological facts and social values. *American Psychologist, 47,* 73–88.

Weisberg, D. S., Keil, F. C., Goodstein, J., Rawson, E. & Gray, J. R. (2008). The seductive
allure of neuroscience explanations. *Journal of Cognitive Neuroscience, 20,* 470–477.

Yatham, L. N. (2004). New anticonvulsants in the treatment of bipolar disorder. *Journal
of Clinical Psychiatry, 65,* 28–35.

PART FOUR

Severe Brain Damage

12

Brain Death

STEVEN LAUREYS

What is death, and how do medical practitioners determine whether a human being is alive or dead? The answers to these questions have varied over history, but at present both answers typically involve brain function. Given the central importance of death for a wide array of societal concerns, from estate law and religious practices to patient care and organ donation, brain death raises a host of neuroethical issues.

BRAIN DEATH: HISTORY AND CONCEPTUAL FOUNDATIONS

In the most general terms, death can be defined as the "permanent cessation of the critical functions of the organism as a whole" (Bernat, 1998; Loeb, 1916). The organism as a whole is an old concept in theoretical biology (1916) that refers to its unity and functional integrity—not to the simple sum of its parts—and encompasses the concept of an organism's critical system (Korein & Machado, 2004). Critical functions are those without which the organism as a whole cannot function: control of respiration and circulation, neuroendocrine and homeostatic regulation, and consciousness. Death is defined by the irreversible loss of all these functions.

The relevance of the brain to this conception of death lies in its essential role in regulating the functions of the organism as a whole. Two formulations of brain death are relevant in this regard, known as "whole brain" and "brainstem" formulations. A third formulation, sometimes referred to as "neocortical death," does not entail the cessation of whole organism functioning and will be dealt with next, in a separate section.

Whole brain and brainstem death are both defined as the irreversible cessation of the organism as a whole, but differ in their anatomical interpretation. Because many areas of the brain above the brainstem (including the neocortex, thalami, and basal ganglia) cannot be accurately tested for clinical function in a comatose patient, most bedside tests for brain death (such as cranial nerve reflexes and apnoea testing) directly measure function of the brainstem alone.

The idea that death is the irreversible absence of brain function has roots extending back as far as medieval times. Moses Maimonides (1135–1204) argued that the spasmodic jerking observed in decapitated humans did not represent evidence of life as their muscle movements were not indicative of presence of central control (Bernat, 2002). However, it was not until the invention of the positive pressure mechanical ventilator by Bjorn Ibsen in the 1950s and the widespread use of high-tech intensive care in the 1960s that cardiac, respiratory, and brain function could be truly dissociated. Patients with severe brain damage could now have their heartbeat and systemic circulation provisionally sustained by artificial respiratory support. Such profound unconscious states had never been encountered before, as, until that time, all such patients had died instantly due to cessation of respiration.

In the mid-twentieth century there were several discussions of a neurocentric definition of death among European authors (Wertheimer, Jouvet & Descotes, 1959; Lofstedt & von Reis, 1956). French neurologists Mollaret and Goulon first discussed the clinical, electrophysiological and ethical issues of what is now known as brain death, using the term "coma dépassé" (irretrievable coma) (Mollaret & Goulon, 1959). Unfortunately, their paper was written in French and remained largely unnoticed by the international community. In 1968, the Ad Hoc Committee of Harvard Medical School, which included ten physicians, a theologian, a lawyer, and a historian of science, published a milestone paper defining death as irreversible coma. The Harvard report represents a sea change in medicolegal conceptions of death. In the words of Joynt (1984), it "opened new areas of law, and posed new and different problems for theologist and ethicist... it has made physicians into lawyers, lawyers into physicians, and both into philosophers."

The concept of brain death as cessation of brainstem functioning (Pallis & Harley, 1996) came some years later, when neuropathological studies showed that damage to the brainstem was critical for brain death (Mohandas & Chou, 1971). On the basis of these findings, brain death came to be regarded in the United Kingdom as complete, irreversible loss of brainstem function ("Criteria for the diagnosis," 1995; "Diagnosis of brain death," 1976); "If the brainstem is dead, the brain is dead, and if the brain is dead, the person is dead" (Pallis & Harley, 1996).

Brain death means human death determined by neurological criteria. It is an unfortunate term, as it misleadingly suggests that there are two types of death: brain death and "regular" death (Bernat, 2002). There is, however, only one type of death, which can be measured in two ways—by cardiorespiratory or neurological criteria. This misapprehension might explain much of the public and professional confusion about brain death. Bernat and colleagues have distinguished three levels of discussion: the definition or concept of death (a philosophical matter); the anatomical criteria of death (a philosophical/medical matter); and the practical testing, by way of clinical or complementary examinations, that death has occurred (a medical matter) (Bernat, Culver & Gert, 1981).

DISTINGUISHING BRAIN DEATH FROM THE VEGETATIVE STATE

The term "neocortical brain death" is an unfortunate phrase that continues to be responsible for much confusion. It was initially proposed as an alternative formulation of brain death in the early days of the brain death concept (Veatch, 1975). According to this formulation, death is the irreversible loss of the capacity for consciousness and social interaction. By application of this consciousness- or personhood-centred definition of death, its proponents classify patients in a permanent vegetative state and anencephalic infants as dead. Although the vegetative state is now understood to be a distinct clinical entity, and is described more fully in the following chapter, there is considerable need to clarify the distinction, even among specialists. Slightly less than half of neurologists and nursing home directors surveyed in the United States believed that patients in a vegetative state could be declared dead (Payne, Taylor, Stocking, & Sachs, 1996).

Based on the neocortical definition of death, patients in a vegetative state following an acute injury or chronic degenerative disease and anencephalic infants are considered dead. Depending on how "irreversible loss of capacity for social interaction" (Veatch, 1976) is interpreted, even patients in a permanent "minimally conscious state" (Giacino et al., 2002), who, by definition, are unable to functionally communicate, could be regarded as dead. I argue that despite its theoretical attractiveness to some, this concept of death cannot be reliably implemented using anatomical criteria nor in reliable clinical testing.

First, our current scientific understanding of the necessary and sufficient neural correlates of consciousness is incomplete at best (Laureys, 2005; Baars, Ramsoy, & Laureys, 2003). In contrast to brain death, for which the neuroanatomy and neurophysiology are both well-established, anatomopathology, neuroimaging, and electrophysiology cannot, at present, determine human consciousness. Therefore, no accurate anatomical criteria can be defined for a higher brain formulation of death.

Second, clinical tests would require the provision of bedside behavioral evidence showing that consciousness has been irreversibly lost. There is an irreducible philosophical limitation in knowing for certain whether any other being possesses a conscious life (Chalmers, 1998). Consciousness is a multifaceted subjective first-person experience and clinical evaluation is limited to evaluating patients' responsiveness to the environment (Majerus, Gill-Thwaites, Andrews, & Laureys, 2005). As previously discussed, patients in a vegetative state, unlike patients with brain death, can move extensively, and clinical studies have shown how difficult it is to differentiate automatic from willed movements (Rochazka, Clarac, Loeb, Rothwell, & Wolpaw, 2000). This results in an underestimation of behavioral signs of consciousness and, therefore, a misdiagnosis, which is estimated to occur in about one third of patients in a chronic vegetative state (Childs, Mercer, & Childs, 1993; Andrews, Murphy, Munday, & Littlewood, 1996). In addition, physicians frequently erroneously diagnose the vegetative state in elderly residents with dementia in nursing homes (Volicer, Berman, Cipolloni, & Mandell, 1997). Clinical testing for absence of consciousness is much more problematic than testing for absence of wakefulness, brainstem reflexes, and apnoea in whole brain or brainstem death. The vegetative state is one end of a spectrum of awareness, and the subtle differential diagnosis between this and the minimally conscious state necessitates repeated evaluations by experienced examinors. Practically, the neocortical death concept also implies the burial of breathing corpses.

Third, complimentary tests for neocortical death would require provision of confirmation that all cortical function has been irreversibly lost. Patients in a vegetative state are not apallic, as previously thought (Ore, Gerstenbrand, & Lucking, 1977; Ingvar, Brun, Johansson, & Samuelsson, 1978), and may show preserved islands of functional pallium or cortex. Recent functional neuroimaging studies have shown limited, but undeniable, neocortical activation in patients in a vegetative state, disproving the idea that there is complete neocortical death in the vegetative state. However, as previously stated, results from these studies should be interpreted cautiously for as long as we do not fully understand the neuronal basis of consciousness. Again, complimentary tests for proving the absence of the neocortical integration that is necessary for consciousness are, at present, not feasible and unvalidated.

The absence of whole brain function in brain death can be confirmed by means of cerebral angiography (nonfilling of the intracranial arteries), transcranial Doppler ultrasonography (absent diastolic or reverberating flow), brain imaging (absence of cerebral blood flow: hollow-skull sign), or EEG (absent electrical activity). In contrast to brain death, in which prolonged absent intracranial blood flow proves irreversibility (Bernat, 2004), the massively reduced—but not absent—cortical metabolism observed in the vegetative state (Schiff et al., 2002; Levy et al., 1987; De Volder et al., 1990; Tommasino, Grana, Lucignani, Torri, & Vazio, 1995; Laureys et al., 1999; Boly et al., 2004) cannot be regarded as evidence for irreversibility. Indeed, fully reversible causes of altered consciousness, such as deep sleep (Maquet et al., 1997) and general anaesthesia (Alkire et al., 1995; Alkire, Haier, Shah, & Anderson, 1997; Alkire et al., 1999), have shown similar decreases in brain function, and the rare patients who have recovered from a vegetative state have been shown to resume near-normal activity in previously dysfunctional associative neocortex (Laureys, Lemaire, Maquet, Phillips, & Franck, 1999; Laureys, Faymonville, Moonen, Luxen, & Maquet, 2000).

However, proponents of the neocortical death formulation might counter-argue that because all definitions of death and vegetative state are clinical, finding some metabolic activity in functional neuroimaging studies does not disprove the concept (as these studies are measuring nonclinical activities), although this does contrast with the validated nonclinical laboratory tests used to confirm whole brain death.

Finally, proving irreversibility is key to any concept of death. The clinical testing of irreversibility has stood the test of time only in the framework of whole brain or brainstem formulations of death. Indeed, since Mollaret and Goulon first defined their neurological criteria of death more than 45 years ago (Mollaret & Goulon, 1959), no patient in apnoeic coma who was properly declared brain (or brainstem) dead has ever regained consciousness (Pallis & Harley, 1996; Bernat, 2005; Wijdicks, 2001). This cannot been said for the vegetative state, in which permanent is probabilistic—the chances of recovery depend on a patient's age, aetiology, and time spent in the vegetative state ("Persistent vegetative state," 1994). Unlike brain death, for which the diagnosis can be made in the acute setting, the vegetative state can only be regarded as statistically permanent after long observation periods, and even then there is a chance that some patients might recover. However, it should be stressed that many anecdotes of late recovery are difficult to substantiate and it is often difficult to know how certain the original diagnosis was.

BRAIN DEATH ACROSS LEGAL SYSTEMS AND RELIGIONS

Under the US Uniform Determination of Death Act ("Uniform Determination," 1997), a person is dead when physicians determine, by applying prevailing clinical criteria, that cardiorespiratory or brain functions are absent and cannot be retrieved (Beresford, 2001). The neurocentric definition is purposefully redundant, requiring a determination that "all functions of the entire brain, including the brain stem" have irreversibly ceased ("Uniform determination," 1997). In 1971, Finland was the first European country to accept brain death criteria. Since then, all EU countries have accepted the concept of brain death. However, although the required clinical signs are uniform, less than half the European countries that have accepted brain death criteria require technical confirmatory tests, and approximately half require more than one physician to be involved (Haupt & Rudolf, 1999). Confirmatory tests are not mandatory in many third-world countries because they are simply not available. In Asia, death based on neurological criteria has not been uniformly accepted and there are major differences in regulation. India follows the UK criteria of brainstem death ("Transplantation of Human Organs Bill," 1992). China has no legal criteria and there seems to be some hesitation among physicians to disconnect the ventilator in patients with irreversible coma (Diringer & Wijdicks, 2001). Japan now officially recognizes brain death, although the public remains reluctant—possibly as a result of the heart surgeon Sura Wada, who was charged with murder in 1968 after removing a heart from a patient who was allegedly not brain dead ("Death: Contemporary Controversies," 1999). Australia and New Zealand have accepted whole brain death criteria (Pearson, 1995).

Both Judaism and Islam have a tradition of defining death on the basis of absence of respiration, but brain death has now become an accepted definition of death for these religions (Beresford, 2001). The Catholic Church has stated that the moment of death is not a matter for the church to resolve. More than 10 years before the Harvard criteria were established, anaesthesiologists who were concerned that new resuscitation and intensive care technologies designed to save lives sometimes appeared to only extend the dying process sought advice from Pope Pius XII. The Pope, up-to-date with (even, surprisingly, in advance of) modern day medicine, ruled that there was no obligation to use extraordinary means to prolong life in critically ill patients (Pius XII, 1957). Therefore, withholding or withdrawing life-sustaining treatment from patients with acute irreversible severe brain damage became morally accepted.

Many prominent progressive Catholic theologists have accepted the idea of therapeutic futility in patients in an irreversible vegetative state, and have defended the decision to withdraw nutrition and hydration in well-documented cases (Schotsman, 1993). However, the Catholic Church continues to maintain a clear line between brain death and the vegetative state. Pope John Paul II stated that the cessation of artificial life-sustenance to patients in a permanent vegetative state could never be morally accepted, whatever the situation (Pope John Paul II, 2004). The official Catholic position de-emphasizes the reality of irreversibility in long-standing vegetative state and does not consider artificial nutrition and hydration to be treatments. So far, it has not changed practices in the United States, where withdrawal of life-sustaining treatment from patients in an irreversible vegetative state remains a settled view; a view that was endorsed by the US Supreme Court in the case of Nancy Cruzan, and that is held by many other medical, ethical and legal authorities (Gostin, 2004).

ETHICAL ISSUES CONCERNING BRAIN DEATH

Since the concept of brain death was first introduced into medical and legal practice, there have been calls to expand the category to include vegetative patients, which is the neocortical conception of death. In my view, this is conceptually inadequate and practically unfeasible. Clinical, electrophysiological, neuroimaging, and postmortem studies now provide clear and convincing neurophysiological and behavioral distinctions between brain death and the vegetative state. Similar lines of evidence also provide compelling data that neocortical death cannot be reliably demonstrated and is an insufficient criterion for establishing death.

Death is a biological phenomenon for which we have constructed pragmatic medical, moral and legal policies on the basis of their social acceptance (Bernat, 2001). The decision of whether a patient should live or die is a value judgment over which physicians can exert no specialized professional claim. The democratic traditions of our pluralistic society should permit personal freedom in patients' decisions to choose to continue or terminate life-sustaining therapy in cases of severe brain damage. Like most ethical issues, there are plausible arguments supporting both sides of the debate. However, these issues can and should be tackled without changes being made to the current neurocentric definition of death. The benefits of using living humans in a vegetative state as organ donors do not justify the harm to society that could ensue from sacrificing the dead donor principle (Bernat, 2001).

Many of the controversial issues relating to the death and end of life in patients with brain damage who have no hope of recovery result from confusion or ignorance on the part of the public or policy makers about the medical reality of brain death and the vegetative state. Therefore, the medical community should improve educational and public awareness programmes on the neurocentric criteria and testing of death, stimulate the creation of advance directives as a form of advance medical care planning, continue to develop clinical practice guidelines, and more actively encourage research on physiological effects and therapeutic benefit of treatment options in patients with severe brain damage.

What is the future of death? Improving technologies for brain repair and prosthetic support for brain functions (for example, stem cells, neurogenesis, neural computer prostheses, cryonic suspension and nano-neurological repair) might one day change our current ideas of irreversibility and force medicine and society to once again revise its definition of death.

ACKNOWLEDGMENTS

This chapter is adapted from an article published in the *Nature Reviews Neuroscience* in 2004, volume 6, pages 899–909, and is used with permission of publisher and author.

REFERENCES

A definition of irreversible coma. (1968). Report of the Ad Hoc Committee of the Harvard Medical School to Examine the Definition of Brain Death. *JAMA, 205,* 337–340.

Agardh, C. D., Rosen, I., & Ryding, E. (1983). Persistent vegetative state with high cerebral blood flow following profound hypoglycemia. *Ann. Neurol., 14,* 482–486.

Alkire, M. T., et al. (1995). Cerebral metabolism during propofol anesthesia in humans studied with positron emission tomography. *Anesthesiology, 82,* 393–403.

Alkire, M. T., et al. (1999). Functional brain imaging during anesthesia in humans: effects of halothane on global and regional cerebral glucose metabolism. *Anesthesiology, 90,* 701–709.

Alkire, M. T., Haier, R. J., Shah, N. K. & Anderson, C. T. (1997). Positron emission tomography study of regional cerebral metabolism in humans during isoflurane anesthesia. *Anesthesiology, 86,* 549–557.

American Academy of Neurology. (1989). Position of the American Academy of Neurology on certain aspects of the care and management of the persistent vegetative state patient. *Neurology, 39,* 125–126.

American Congress of Rehabilitation Medicine. (1995). Recommendations for use of uniform nomenclature pertinent to patients with severe alterations of consciousness. *Arch. Phys. Med. Rehabil, 76,* 205–209.

An appraisal of the criteria of cerebral death. (1977). A summary statement. A collaborative study. *JAMA, 237,* 982–986.

ANA Committee on Ethical Affairs. (1993). Persistent vegetative state: report of the American Neurological Association Committee on Ethical Affairs. *Ann. Neurol, 33,* 386–390.

Andrews, K. (2004). Medical decision making in the vegetative state: withdrawal of nutrition and hydration. *NeuroRehabilitation, 19,* 299–304.

Andrews, K., Murphy, L., Munday, R., & Littlewood, C. (1996). Misdiagnosis of the vegetative state: retrospective study in a rehabilitation unit. *BMJ, 313,* 13–16.

Annas, G. J. "Culture of life" politics at the bedside—The case of Terri Schiavo. (2005). *N. Engl. J. Med, 352,* 1710–1715.

Arnold, R. M., & Youngner, S. J. (1993). The dead donor rule: Should we stretch it, bend it, or abandon it? *Kennedy Inst. Ethics J, 3,* 263–278.

Asai, A., et al. (1999.) Survey of Japanese physicians' attitudes towards the care of adult patients in persistent vegetative state. *J. Med. Ethics 25,* 302–308.

Baars, B., Ramsoy, T., & Laureys, S. (2003). Brain, conscious experience and the observing self. *Trends Neurosci, 26,* 671–675.

Bakran, A. (1998). Organ donation and permanent vegetative state. *Lancet, 351,* 211–212; discussion 212–213.

Beauchamp, T. L., & Childress, J. F. (1979). *Principles of biomedical ethics* (New York: Oxford University Press.

Beecher, H. K. (1970). Definitions of "life" and "death" for medical science and practice. *Ann. NY Acad. Sci, 169,* 471–474.

Beresford, H. R. (2001). Legal aspects of brain death. In E.F.M. Wijdicks (Ed.), *Brain death* (pp. 151–169). Philadelphia: Lippincott Williams & Wilkins.

Bernat, J. L. (1998). A defense of the whole-brain concept of death. *Hastings Cent. Rep, 28,* 14–23.

Bernat, J. L. (2001). Philosophical and ethical aspects of brain death. In E.F.M. Wijdicks (Ed.), *Brain death* (pp. 171–187). Philadelphia: Lippincott Williams & Wilkins.

Bernat, J. L. (2002a). The biophilosophical basis of whole-brain death. *Soc. Philos. Policy, 19,* 324–342.

Bernat, J. L. (2002b) Brain death. In J. L. Bernat (Ed.), *Ethical issues in neurology* (pp. 243–281. Boston: Butterworth Heinemann.

Bernat, J. L. (2002c). Medical futility. In J. L. Bernat (Ed.), *Ethical issues in neurology* (pp. 215–239. Boston: Butterworth Heinemann).

Bernat, J. L. (2002d). The persistent vegetative state and related states. In J. L. Bernat (Ed.), *Ethical issues in neurology (pp.* 283–305. Boston: Butterworth Heinemann).

Bernat, J. L. (2005). The concept and practice of brain death. In S. Laureys (Ed.), *The Boundaries of consciousness: Neurobiology and neuropathology* (pp. 369–379). Amsterdam: Elsevier).

Bernat, J. L., Culver, C. M. & Gert, B. (1981). On the definition and criterion of death. *Arch. Intern. Med, 94,* 389–394.

Bernstein, I. M., et al. (1989). Maternal brain death and prolonged fetal survival. *Obstet. Gynecol, 74,* 434–437.

Boly, M., et al. (2004). Auditory processing in severely brain injured patients: differences between the minimally conscious state and the persistent vegetative state. *Arch. Neurol, 61,* 233–238.

Brierley, J. B., Graham, D. I., Adams, J. H., & Simpsom, J. A. (1971). Neocortical death after cardiac arrest. A clinical, neurophysiological, and neuropathological report of two cases. *Lancet, 2,* 560–565.

British Medical Association. (2001). *Withholding or withdrawing life-prolonging medical treatment: Guidance for decision making* 2nd ed. (London: BMJ Books).

Buchner, H., & Schuchardt, V. (1990). Reliability of electroencephalogram in the diagnosis of brain death. *Eur. Neurol, 30,* 138–141.

Cabeza, R., et al. (1997). Age-related differences in neural activity during memory encoding and retrieval: a positron emission tomography study. *J. Neurosci. 17,* 391–400.

California Court of Appeal, Second District, Division 2. (1983). Barber v. Superior Court of State of California; Nejdl v. Superior Court of State of California. *Wests Calif. Report, 195,* 484–494.

Canadian Neurocritical Care Group. (1999). Guidelines for the diagnosis of brain death. *Can. J. Neurol. Sci. 26,* 64–66.

Cassem, N. H. (1974). Confronting the decision to let death come. (1974). *Crit. Care Med, 2,* 113–117.

Chalmers, D. J. (1998). The problems of consciousness. *Adv. Neurol, 77*, 7–16; discussion 16–18.

Childs, N. L., Mercer, W. N., & Childs, H. W. (1993). Accuracy of diagnosis of persistent vegetative state. *Neurology, 43*, 1465–1467.

Conrad, G. R., & Sinha, P. (2003). Scintigraphy as a confirmatory test of brain death. *Semin. Nucl. Med, 33*, 312–323.

Council on Ethical and Judicial Affairs, American Medical Association. (1992). Decisions near the end of life. *JAMA, 267*, 2229–2233.

Council on Ethical and Judicial Affairs. A. M. A. (1995). The use of anencephalic neonates as organ donors. *JAMA, 273*, 1614–1618.

Council on Scientific Affairs and Council on Ethical and Judicial Affairs. (1990). Persistent vegetative state and the decision to withdraw or withhold life support. *JAMA, 263*, 426–430.

Cranford, R. E. (1984). Termination of treatment in the persistent vegetative state. *Semin. Neurol, 4*, 36–44.

Cranford, R. (2005). Facts, lies, and videotapes: the permanent vegetative state and the sad case of Terri Schiavo. *J. Law Med. Ethics, 33*, 363–371.

Crisci, C. (1999). Chronic "brain death": meta-analysis and conceptual consequences. *Neurology, 53*, 1370; author reply 1371–1372.

Criteria for the diagnosis of brain stem death. (1995). Review by a working group convened by the Royal College of Physicians and endorsed by the Conference of Medical Royal Colleges and their Faculties in the United Kingdom. *J. R. Coll. Physicians Lond, 29*, 381–382.

Danze, F., Brule, J. F., & Haddad, K. (1989). Chronic vegetative state after severe head injury: clinical study; electrophysiological investigations and CT scan in 15 cases. *Neurosurg. Rev, 12* (Suppl. 1), 477–499.

De Volder, A. G., et al. (1990). Brain glucose metabolism in postanoxic syndrome. Positron emission tomographic study. *Arch. Neurol. 47*, 197–204.

Diagnosis of brain death. (1976). Statement issued by the honorary secretary of the Conference of Medical Royal Colleges and their Faculties in the United Kingdom on 11 October 1976. *Br. Med. J, 2*, 1187–1188.

Diringer, M. N., & Wijdicks, E. F. M. (2001). Brain death in historical perspective. In E.F.M. Wijdicks (Ed.), *Brain death* (pp. 5–27). Philadelphia: Lippincott Williams & Wilkins).

Ducrocq, X., et al. (1998). Consensus opinion on diagnosis of cerebral circulatory arrest using Doppler-sonography: Task Force Group on cerebral death of the Neurosonology Research Group of the World Federation of Neurology. *J. Neurol. Sci. 159*, 145–150.

Engelhardt, K. (1998). Organ donation and permanent vegetative state. *Lancet, 351*, 211; author reply 212–213.

Feldman, D. M., Borgida, A. F., Rodis, J. F., & Campbell, W. A. (2000). Irreversible maternal brain injury during pregnancy: a case report and review of the literature. *Obstet. Gynecol. Surv, 55*, 708–714.

Fost, N. (2004). Reconsidering the dead donor rule: is it important that organ donors be dead? *Kennedy Inst. Ethics J, 14*, 249–260.

Frankl, D., Oye, R. K., & Bellamy, P. E. (1989). Attitudes of hospitalized patients toward life support: a survey of 200 medical inpatients. *Am. J. Med, 86*, 645–648.

Gervais, K. G. (1986). *Redefining death* (New Haven: Yale University Press).

Giacino, J. T., et al. (2002). The minimally conscious state: definition and diagnostic criteria. *Neurology, 58,* 349–353.

Gillon, R. (1998). Persistent vegetative state, withdrawal of artificial nutrition and hydration, and the patient's "best interests." *J. Med. Ethics, 24,* 75–76.

Gostin, L. O. (1997). Deciding life and death in the courtroom. From Quinlan to Cruzan, Glucksberg, and Vacco—a brief history and analysis of constitutional protection of the "right to die." *JAMA, 278,* 1523–1528.

Gostin, L. O. (2005). Ethics, the constitution, and the dying process: the case of Theresa Marie Schiavo. *JAMA, 293,* 2403–2407.

Haupt, W. F., & Rudolf, J. (1999). European brain death codes: a comparison of national guidelines. *J. Neurol, 246,* 432–437.

Hoffenberg, R., et al. (1997). Should organs from patients in permanent vegetative state be used for transplantation? International Forum for Transplant Ethics. *Lancet, 350,* 1320–1321.

Homer-Ward, M. D., Bell, G., Dodd, S., & Wood, S. (2000). The use of structured questionnaires in facilitating ethical decision-making in a patient with low communicative ability. *Clin. Rehabil, 14,* 220.

Horsley, V. (1894). On the mode of death in cerebral compression and its prevention. *O. Med. J,* 306–309.

Ingvar, D. H., Brun, A., Johansson, L., & Samuelsson, S. M. (1978). Survival after severe cerebral anoxia with destruction of the cerebral cortex: the apallic syndrome. *Ann. NY Acad. Sci, 315,* 184–214.

Jennett, B. (2002). *The vegetative state. Medical facts, ethical and legal dilemmas* (Cambridge: Cambridge University Press).

Jennett, B. (2005). The assessment and rehabilitation of vegetative and minimally conscious patients: definitions, diagnosis, prevalence and ethics. *Neuropsychological Rehabilitation, 15,* 163–165.

Jennett, B., & Plum, F. (1972). Persistent vegetative state after brain damage. A syndrome in search of a name. *Lancet, 1,* 734–737.

Joynt, R. J. (1984). Landmark perspective: a new look at death. *JAMA, 252,* 680–682.

Kantor, J. E., & Hoskins, I. A. (1993). Brain death in pregnant women. *J. Clin. Ethics, 4,* 308–314.

King, T. T. (1998). Organ donation and permanent vegetative state. *Lancet, 351,* 211; discussion 212–213.

Korein, J., & Machado, C. (2004). Brain death: updating a valid concept for 2004. In C. Machado & D. A. Shewmon (Eds.), *Brain death and disorders of consciousness* (pp. 1–21). New York: Kluwer Academic/Plenum.

Lang, C. J. (1999). Chronic "brain death" meta analysis and conceptual consequences. *Neurology, 53,* 1370–1371; author reply 1371–1372.

Laureys, S. (2005). The functional neuroanatomy of (un)awareness: lessons from the vegetative state. *Trends in Cognitive Sciences, 9,* 556–559.

Laureys, S., et al. (1999). Impaired effective cortical connectivity in vegetative state: preliminary investigation using PET. *Neuroimage, 9,* 377–382.

Laureys, S., et al. (2004). Cerebral processing in the minimally conscious state. *Neurology, 63,* 916–918.

Laureys, S., Faymonville, M. E., & Berre, J. (2000). Permanent vegetative state and persistent vegetative state are not interchangeable terms, [online], <http://bmj.com/cgi/eletters/321/7266/916#10276> *British Medical Journal.*

Laureys, S., Faymonville, M. E., Moonen, G., Luxen, A., & Maquet, P. (2000). PET scanning and neuronal loss in acute vegetative state. *Lancet, 355,* 1825–1826.

Laureys, S., Lemaire, C., Maquet, P., Phillips, C., & Franck, G. (1999). Cerebral metabolism during vegetative state and after recovery to consciousness. *J. Neurol. Neurosurg. Psychiatry, 67,* 121–122.

Layon, A. J., D'Amico, R., Caton, D., & Mollet, C. J. (1990). And the patient chose: medical ethics and the case of the Jehovah's Witness. *Anesthesiology, 73,* 1258–1262.

Levy, D. E., et al. (1987). Differences in cerebral blood flow and glucose utilization in vegetative versus locked-in patients. *Ann. Neurol. 22,* 673–682.

Lock, M. (1999). The problem of brain death: Japanese disputes about bodies and modernity. In S. J. Younger, R. M. Arnold, & R. Schapiro (Eds.), *The definition of death: Contemporary controversies* (pp. 239–256. Baltimore: John Hopkins University Press).

Loeb, J. (1916). *The organism as a whole* (New York: G. P. Putnam's Sons).

Loewy, E. H. (1987). The pregnant brain dead and the fetus: must we always try to wrest life from death? *Am. J. Obstet. Gynecol, 157,* 1097–1101.

Lofstedt, S., & von Reis, G. (1956). Intracranial lesions with abolished passage of X-ray contrast throughout the internal carotid arteries. *Pacing Clin. Electrophysiol, 8,* 199–202.

Majerus, S., Gill-Thwaites, H., Andrews, K., & Laureys, S. (2005). Behavioral evaluation of consciousness in severe brain damage. In S. Laureys (Ed.), *The boundaries of consciousness: Neurobiology and neuropathology* (pp. 397–413). Amsterdam: Elsevier).

Maquet, P., et al. (1997). Functional neuroanatomy of human slow wave sleep. *J. Neurosci. 17,* 2807–2812.

McMillan, T. M. (1997). Neuropsychological assessment after extremely severe head injury in a case of life or death. *Brain Inj, 11,* 483–490. Erratum *Brain Inj. 11,* 775 (1997).

McMillan, T. M., & Herbert, C. M. (2004). Further recovery in a potential treatment withdrawal case 10 years after brain injury. *Brain Inj, 18,* 935–940.

Mohandas, A., & Chou, S. N. (1971). Brain death. A clinical and pathological study. *J. Neurosurg, 35,* 211–218.

Mollaret, P., & Goulon, M. (1959). Le coma dépassé. *Rev. Neurol, 101,* 3–15.

The Multi-Society Task Force on PVS. (1994). Medical aspects of the persistent vegetative state (1). *N. Engl. J. Med, 330,* 1499–1508.

The Multi-Society Task Force on PVS. (1994). Medical aspects of the persistent vegetative state (2). *N. Engl. J. Med, 330,* 1572–1579.

O'Malley, C. D. (1951). The life and time of Andreas Vesalius. *Ann. West Med. Surg, 5,* 191–198.

Ore, G. D., Gerstenbrand, F., & Lucking, C. H. (1977). *The Apallic Syndrome* (Springer, Berlin).

Pallis, C., & Harley, D. H. (1996). *ABC of brainstem death* (London: BMJ).

Payne, K., Taylor, R. M., Stocking, C., & Sachs, G. A. (1996). Physicians' attitudes about the care of patients in the persistent vegetative state: a national survey. *Ann. Intern. Med, 125,* 104–110.

Payne, S. K., & Taylor, R. M. (1997). The persistent vegetative state and anencephaly: problematic paradigms for discussing futility and rationing. *Semin. Neurol, 17,* 257–263.

Pearson, I. Y. (1995). Australia and New Zealand Intensive Care Society Statement and Guidelines on Brain Death and Model Policy on Organ Donation. *Anaesth. Intensive Care, 23,* 104–108.

Pernick, M. S. (1988). Back from the grave: recurring controversies over defining and diagnosing death in history. In R. M. Zaner (Ed.), *Death: Beyond whole-brain criteria* (pp. 17–74). Dordrecht: Kluwer Academic.

Pius XII. (1957). Pope speaks on prolongation of life. *Osservatore Romano, 4,* 393–398.

Plows, C. W. (1996). Reconsideration of AMA opinion on anencephalic neonates as organ donors. *JAMA, 275,* 443–444.

Plum, F., & Posner, J. B. (1966). *The diagnosis of stupor and coma* (F. A. Davis, Philadelphia).

Poe, E. A. (1981). *The Complete Edgar Allan Roe Tales* (pp. 432–441). New York: Avenel Books.

Pope John Paul II. (2004). Address of Pope John Paul II to the Participants in the International Congress on "Life-Sustaining Treatments and Vegetative State: Scientific Advances and Ethical Dilemmas," Saturday, 20 March 2004. *NeuroRehabilitation, 19,* 273–275.

President's Commission for the Study of Ethical Problems in Medicine and Biomedical and Behavioral Research. (1981). *Defining death: A report on the medical, legal and ethical issues in the determination of death.* Washington, DC: U S Government Printing Office.

President's Commission for the Study of Ethical Problems in Medicine and Biomedical and Behavioral Research. (1983). *Deciding to forego life-sustaining treatment: a report on the ethical, medical, and legal issues in treatment decisions* (pp. 171–192). Washington, DC: US Government Printing Office).

Prochazka, A., Clarac, F., Loeb, G. E., Rothwell, J. C., & Wolpaw, J. R. (2000). What do reflex and voluntary mean? Modern views on an ancient debate. *Exp. Brain Res, 130,* 417–432.

The Quality Standards Subcommittee of the American Academy of Neurology. (1995). Practice parameters for determining brain death in adults (summary statement). *Neurology, 45,* 1012–1014.

Quill, T. E. (2005). Terri Schiavo—a tragedy compounded. *N. Engl. J. Med. 352,* 1630–1633.

Saposnik, G., Bueri, J. A., Maurino, J., Saizar, R., & Garretto, N. S. (2000). Spontaneous and reflex movements in brain death. *Neurology, 54,* 221–223.

Saposnik, G., Maurino, J., Saizar, R., & Bueri, J. A. (2005). Spontaneous and reflex movements in 107 patients with brain death. *Am. J. Med, 118,* 311–314.

Schiff, N. D., et al. (2005). fMRI reveals large-scale network activation in minimally conscious patients. *Neurology, 64,* 514–523.

Schiff, N. D., et al. (2002). Residual cerebral activity and behavioural fragments can remain in the persistently vegetative brain. *Brain, 125,* 1210–1234.

Schotsmans, P. (1993). The patient in a persistent vegetative state: an ethical re-appraisal. *Int. J. Phil. Theol, 54,* 2–18.

Seifert, J. (1993). Is "brain death" actually death? *Monist, 76,* 175–202.

Shewmon, A. D. (2001). The brain and somatic integration: insights into the standard biological rationale for equating "brain death" with death. *J. Med. Philos, 26,* 457–478.

Shewmon, D. A. (1998). Chronic "brain death": meta-analysis and conceptual consequences. *Neurology, 51,* 1538–1545.

Shewmon, D. A. (1999). Spinal shock and "brain death": somatic pathophysiological equivalence and implications for the integrative-unity rationale. *Spinal Cord, 37,* 313–324.

Shewmon, D. A., Holmes, G. L., & Byrne, P. A. (1999). Consciousness in congenitally decorticate children: developmental vegetative state as self-fulfilling prophecy. *Dev. Med. Child Neurol, 41,* 364–374.

Shiel, A., & Wilson, B. A. (1998). Assessment after extremely severe head injury in a case of life or death: further support for McMillan. *Brain Inj, 12,* 809–816.

Smedira, N. G., et al. (1990). Withholding and withdrawal of life support from the critically ill. *N. Engl. J. Med. 322,* 309–315.

Smith, D. R. (1986). Legal recognition of neocortical death. *Cornell Law Rev, 71,* 850–888.

Stacy, T. (1992). Death, privacy, and the free exercise of religion. *Cornell Law Rev, 77,* 490–595.

Starr, T. J., Pearlman, R. A., & Uhlmann, R. F. (1986). Quality of life and resuscitation decisions in elderly patients. *J. Gen. Intern. Med, 1,* 373–379.

Tommasino, C., Grana, C., Lucignani, G., Torri, G., & Fazio, F. (1995). Regional cerebral metabolism of glucose in comatose and vegetative state patients. *J. Neurosurg. Anesthesiol, 7,* 109–116.

The transplantation of human organs bill (1992). Republic of India, Bill No. LIX-C.

Truog, R. D., & Robinson, W. M. (2003). Role of brain death and the dead-donor rule in the ethics of organ transplantation. *Crit. Care Med, 31,* 2391–2396.

Truog, R. D. (1997). Is it time to abandon brain death? *Hastings Cent. Rep, 27,* 29–37.

Truog, R. D. (2000). Organ transplantation without brain death. *Ann. NY Acad. Sci, 913,* 229–239.

Uhlmann, R. F., Pearlman, R. A., & Cain, K. C. (1988). Physicians' and spouses' predictions of elderly patients' resuscitation preferences. *J. Gerontol, 43,* M115–M121.

Uniform Determination of Death Act. (1997). *598 (West 1993 and West Supp. 1997)* (Uniform Laws Annotated (U. L. A.)).

Veatch, R. M. (2004). Abandon the dead donor rule or change the definition of death? *Kennedy Inst. Ethics J, 14,* 261–276.

Veatch, R. M. (1976). *Death, dying, and the biological revolution. Our last quest for responsibility* (New Haven: Yale University Press).

Veatch, R. M. (1975). The whole-brain-oriented concept of death: an outmoded philosophical formulation. *J. Thanatol, 3,* 13–30.

Volicer, L., Berman, S. A., Cipolloni, P. B., & Mandell, A. (1997). Persistent vegetative state in Alzheimer disease. Does it exist? *Arch. Neurol. 54,* 1382–1384.

Walters, J., Ashwal, S., & Masek, T. (1997). Anencephaly: where do we now stand? *Semin. Neurol, 17,* 249–255.

Wertheimer, P., Jouvet, M., & Descotes, J. (1959.) A propos du diagnostic de la mort du sysème nerveux dans les comas avec arrêt respiratoire traités par respiration artificielle. *Presse Med, 67,* 87–88.

Wijdicks, E. F. (2001.) The diagnosis of brain death. *N. Engl. J. Med, 344,* 1215–1221.

Wijdicks, E. F., & Bernat, J. L. (1999). Chronic "brain death": meta-analysis and conceptual consequences. *Neurology, 53,* 1369–1370; author reply 1371–1372.

Wijdicks, E. F. M. (2001a). Title. In E.F.M. Wijdicks (Ed.), *Brain death* (Philadelphia: Lippincott Williams & Wilkins).

Wijdicks, E. F. M. (2001b). Confirmatory testing of brain death in adults. In E.F.M. Wijdicks (Ed.), *Brain death* (pp. 61–90). Philadelphia: Lippincott Williams & Wilkins.

World Medical Association. (1989). *Statement on persistent vegetative state. Adopted by the 41st World Medical Assembly Hong Kong, September 1989.* www.wma.net/e/policy/p11.htm.

13

Disorders of Consciousness Following Severe Brain Injury

JOSEPH J. FINS AND NICHOLAS D. SCHIFF

CONSCIOUSNESS IN THE SEVERELY DAMAGED BRAIN: AN OVERVIEW OF THE CLINICAL DISORDERS

Physicians and neuroscientists have distinguished among a number of different disorders of consciousness associated with severe brain injuries (Fins, 2007a). Figure 13.1 illustrates the relations among these disorders in terms of their clinical course. To inform clinical practice and elevate public policy discussions about these patients, we will review the clinical features and aspects of the pathophysiology of these disorders, beginning with a brief review of brain death, and then considering in turn coma, the vegetative state (VS), and the minimally conscious state (MCS). We will highlight new insights into these conditions that have come from neuroimaging studies and argue for caution in the translation of these research findings into clinical practice.

Brain Death

As the previous chapter describes in greater detail, brain death is the irreversible cessation of whole brain function including both the cortex and brain stem.

The assessment of a patient for brain death is clinical, based on history and physical examination. It is neither a diagnosis made by electrophysiologic criteria (EEG) nor imaging criteria. Despite Hollywood's depiction of flat-line EEGs being associated with brain death, these tests are often not done.

Coma

Coma is an eyes-closed state of unresponsiveness associated with severe brain trauma, loss of blood flow or oxygenation, and a variety of other causes (see

Posner et al., 2007 for review). Most comas are self-limited, usually lasting two to three weeks, after which changes in state are observed. Coma requires the presence of bihemispheric brain dysfunction at the cortical level, thalamic level or brainstem level. In all cases, coma reflects a functional disturbance of brain stem and basal forebrain arousal systems. Only large rostral dorsal-medial pontine, mesencephalic and paramedian thalamic lesions, or global damage to the cerebral hemispheres can produce sustained coma (see Plum, 1991).

The most common causes of coma include traumatic brain injuries, cerebral vascular accidents (large cerebral and brain stem ischemic strokes, acute ruptured cerebral aneurysms, and large cerebral or subtentorial hemorrhages), severe anoxic injury due to cardiac arrest, drowning, and more diffuse processes often associated with systemic or cerebral disorders.

Although functional imaging like functional magnetic resonance imaging (fMRI) or positron emission tomography (PET) can help to assess functional status, conventional neuroimaging studies like a CT scan, which can show structural evidence of injury, play a larger role in the assessment of coma—for example, to identify fractures, bleeding, mass occupying collections, and midline shift and herniation (the displacement of the brain down toward the brain stem in the setting of increased cranial pressure). Functional imaging may eventually help elucidate mechanisms of injury, although few studies have addressed acute coma as the movement of patients with ventilators outside of intensive care units is potentially risky and typically done only for diagnostic purposes.

Vegetative State

Recovery from coma may lead to the "locked-in" state, whereby higher brain functions are preserved but are disconnected from motor outflow (generally with the exception of preserved eye movement, which can be used for communication). More commonly, recovery from coma results in the vegetative state, whereby brain stem function is preserved but higher cortical functions are not, as shown in Figure 13.1. The vegetative state was first described by Jennett and Plum in 1972 as a "state of wakeful unresponsiveness" in which the eyes are open but there is no awareness of self, others, or the environment. Clinically, vegetative patients demonstrate sleep/wake cycles, blinking, eye movements, and even the startle reflex, though these actions are not purposeful or indicative of intent (Jennett and Plum, 1972). Instead, these are autonomic behaviors comparable to the maintenance of homeostasis via the brain stem's control of heart rate and respiration.

Understanding whether or not a patient in vegetative state may recover further requires consideration of the time from onset of the vegetative state and the etiology of injury. Although the term "persistent vegetative state" was used in the original paper by Jennet and Plum, it is now generally accepted practice to speak of the vegetative state without a modifying adjective. When vegetative state has lasted three months after anoxic injury and 12 months after traumatic injury it is considered permanent by existing guidelines, though individual exceptions have been noted (Multi-Society Task Force 1994; Fins, 2005).

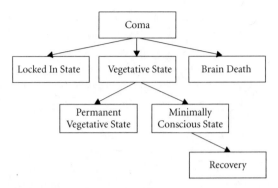

Figure 13.1 Disorders of consciousness that follow severe brain injury, showing pathways between different disorders.

Generally, neuroimaging of the vegetative state has revealed functional evidence of low metabolic activity comparable to general anesthesia as measured by FDG-PET. Functional studies of vegetative state patients using fPET and fMRI methods have established clear loss of functional integration in the vegetative state. Using evoked potentials and concurrent PET imaging, Laureys and colleagues demonstrated that responses to noxious somatosensory stimuli in vegetative state patients were limited to the primary sensory areas, contralateral thalamus, and the midbrain (Laureys et al., 2002). These regions of the brain were "functionally disconnected" from secondary somatosensory areas and higher-level associative areas. Similar findings were obtained by this investigative group using auditory stimuli (Fins, 2008a).

Minimally Conscious State

MCS patients show clear but episodic behavioral evidence of awareness of self, others, or the environment. Brief behavioral events exhibited by MCS patients reveal elements of intention, attention, and memory and are thus considered to reflect a minimum level of consciousness. Such behaviors may be restricted to tracking a visitor coming into the room, following a command, or grasping for a ball (Giacino et al., 2002).

Although these intermittent behavioral displays can make it hard to distinguish the minimally conscious patient from a patient in the vegetative state, the presence of consciousness makes this distinction a critical one. On a neurophysiological level, studies of MCS patients have demonstrated functionally integrated network responses (Boly et al., 2004; Schiff et al., 2005; Coleman et al., 2007) in contrast to the functional disconnection of the vegetative state described by Laureys and colleagues (Laureys et al., 2002; Fins, 2008a).

Evidence of selective responses to spoken language in MCS patients using neuroimaging methods was demonstrated by our team in 2005 (Schiff et al., 2005). When presented with narratives spoken by family members, MCS patients activated language networks involving the superior and middle temporal gyrus. When the same digitized recordings were played in reverse, with the same frequency spectrum but absent normal grammar and syntax, these language networks did not activate. This contrasts with the response of normal controls who showed similar patterns of activation to both the forward and time-reversed recordings. Based on the self-report of naïve normal subjects who suggest an effort to decode the time-reversed stimuli, one inference is that the MCS patient's lack of activation to the time-reversed stimuli reflects a lack of effort to find identifiable language, much as the tourist traveling abroad would listen for word fragments.

For our purposes, the key issue is not the differential response of MCS to forward or time-reversed speech stimuli but the capability of MCS patients to demonstrate widespread network activations.

Emergence from MCS

A longitudinal perspective is also necessary to capture any structural changes that may occur over time and which may correlate with functional improvements. Such appear to play a role in the case of Terry Wallis, a 39-year-old nursing home patient who had carried the vegetative diagnosis for 19 years following traumatic brain injury in 1984. He gained international headlines when he spontaneously began to speak in the summer of 2003, emerging from the minimally conscious state (Schiff and Fins, 2003).[1]

Henning Voss and colleagues at Weill Cornell Medical College used diffusion tensor imaging (DTI)—a method of mathematical analysis of MRI data that measures the restriction of the motion of water molecules—to assess the structural integrity of Wallis's brain. While an overall severe reduction in white matter connectivity in Wallis's brain was identified by DTI consistent with global diffuse axonal injury (DAI) or shearing of axons, a large area of increased fractional anisotropy, a measure of restricted water diffusion, was present in the parietal-occipital regions of the brain. A follow-up study 18 months later showed a marked increased in fractional anisotropy within the cerebellar white matter, findings that correlated with improved motor function across the time interval between the two studies (Voss et al, 2006).

Based on these quantitative data and recent studies in the experimental literature, Voss et al. proposed that sprouting of new axonal connections may have played a part in the subject's recovery. The findings suggest a potential biological mechanism for late recovery in patients with diffuse axonal injury (but may not play a role in other patients with different underlying etiologies of brain damage). This hypothesis will require further testing in many MCS patients. Of note is a recent prospective cohort study of patients recovering after severe traumatic brain injuries that showed similar DTI findings in examination roughly one year past, with initial assessment near the time of injury (Sidaros et al., 2008). It is unlikely that slow axonal regrowth is the sole explanation for the changes observed in the subject, and additional unidentified factors that must have influenced the patient's recovery remain unknown. A possibility is the introduction of pharmacologic neuromodulation (paroxetine) approximately two years prior to his recovery of speech, as the patient showed persistent behavioral dose-related dependence on this drug for more than a year after recovery of speech (unpublished observations).

NEW INSIGHTS AND ETHICAL CHALLENGES FROM NEUROIMAGING

Of late much attention has been paid to neuroimaging and disorders of consciousness. With these nascent technologies, investigators have been able to better delineate the pathophysiology of the vegetative and minimally conscious states and mechanisms of recovery as some patients transgress these diagnostic categories en route to greater functional restoration or full recovery.

As already mentioned, recent diffusion tensor imaging studies have suggested that structural changes in some cases may reflect axonal sprouting and correlate with behavioral improvements seen in following severe brain injury (Sidaros et al., 2008). Similar changes have been observed in an instance of very late recovery from the minimally conscious state (Voss et al., 2006). Other neuroimaging techniques suggest that slow changes in brain structure may play an important role in recovery of brain function (Voss et al., 2006; Danielsen et al., 2003). There is no doubt that this investigative work is rich and needs to be bolstered and sustained.

Functional imaging studies have also demonstrated surprising evidence in some individual MCS patients of large-scale cerebral network activity, suggesting a degree of functionality at odds with behavior evident on clinical exam (Boly et al., 2004; Schiff et al., 2005). Indeed, one particularly striking contrast between imaging and behavior was reported for a vegetative patient by Owen and colleagues (2006).

In their landmark paper in *Science,* Owen and colleagues described a patient who remained in vegetative state at five months by behavioral criteria, with neuroimaging evidence of mental command following in response to requests to perform imagined actions (Owen et al., 2006). Specifically, when their patient was asked to imagine walking through her home, fMRI studies showed network activation of the parahippocampal gyrus, posterior parietal cortex, and the lateral premotor cortex—areas associated with task performance in normal subjects. When asked to imagine playing tennis, there was activity in the supplementary motor areas. Finally, when presented with linguistically ambiguous words, she activated the middle and superior temporal gyrus bilaterally and the left inferior frontal region. All of these responses were characterized as being indistinguishable from normals.

These findings are striking because they demonstrate that this patient was in fact responding to her environment and doing so via intact, integrated cerebral networks that are inconsistent with the vegetative state diagnosis. This presents a bit of a quandary. If this patient was vegetative her response raises a host of questions, both ethical and political, especially one year after the death of Terri Schiavo (Fins, 2006, 2008b). If Owen's subject could respond in this way, might not the same have been possible for Terri Schiavo? Would these data make decisions made during the Schiavo case seem precipitous in retrospect and remove the certainty with which experts asserted that she was permanently vegetative?

Although the Owen paper was heralded as a case of awareness in the vegetative state, such a statement is incompatible with the definition of the vegetative state offered by Jennett and Plum. Rather, we believe that the activations she demonstrated in response to external stimuli represent the first example of how neuroimaging data might lead to diagnostic refinement of diagnostic criteria that have been based on behavioral criteria (Fins & Schiff, 2006). However, without attention to clinical nuance and deference to investigative rigor, neuroimaging studies may lead to more confusion than clarity in disorders of consciousness, especially in the setting of diagnostic confusion or conflation.

Recent extensions of the approach take in the Monti et al. (2010) study illustrate the challenges in this area of research. Bardin et al. (2011) studied a group of severely brain-injured subjects (one locked-in, one emerged from MCS, and four MCS) and found several interesting discordances of behavioral and neuroimaging assessments of command following and communication. One patient judged in vegetative state by clinical exam for two years but identified on some (but not all) examinations by the research team as fulfilling MCS criteria showed evidence of both command following and attempted communication using an fMRI based motor imagery signal. This patient's highest level of behavior was restricted to inconsistent demonstration of command following and communication using a single direction of eye-movement of only the left eye. Conversely, a patient nearly completely locked-in after severe traumatic brain injury who could reliably communicate with small head movements demonstrated fMRI-based command following with motor imagery. Importantly, that patient could not demonstrate communication using the same signal despite accurate communication before and after the fMRI scanning session using head movements. These findings may reflect variation in blood flow signals in the injured brain or a true dissociation of the capacity to engage in motor imagery while holding an independent task instruction—that is, an interference effect of the experimental conditions. Either explanation indicates that variations in performance of such novel behaviors may be seen in brain-injured subjects compared with normals.

Another more recent study demonstrates the caution needed in these early days of employing such diagnostic technologies. A study by Cruse et al. (2011) using a pattern classification of EEG signals suggests that three subjects judged as in vegetative state by behavioral criteria, out of a cohort of 16, could successfully carry out a very high-level cognitive task for which only seven in 10 could generate a detectable EEG signal. Approaches to EEG signal analysis in such patient populations should require an assessment of the *physiological plausibility* of the signals through demonstration of the actual EEG data. That is, the data should make *biological sense*. Goldfine et al. (2011), using quantitative EEG assessments of motor imagery in an earlier study of MCS and locked-in patients—but examining the spectral content of the raw EEG signal—demonstrated that brain-injured subjects may perform the task using very different frequency bands and spatial organization of brain activity. Moreover, given that such EEG signals are typically much weaker than those measured in normal subjects, EEG based methods must have higher standards of demonstration than those currently applied in fMRI research. Because of this high bar, EEG findings should be considered insufficiently demonstrated without such basic cross-checks to validate methodology.

Although we believe that neuroimaging will, over time, lead to refinements of nosological frameworks, this nascent technology, viewed in isolation from clinical data, can inadvertently promote category errors (Fins, et al., 2008). Just consider how misleading images of resting metabolism of brains in the vegetative state would be when viewed in isolation, divorced from the clinical facts: both the patients history and the clincial exam. (Schiff, et a.,l 2002). In the Schiff et al paper, five images of vegetative patients are presented, which appear so very different. Out of context, one might reasonably—and falsely presume—that there are varying degrees of clinically meaningful functional variations and anticipated level of consciousness due to differential metabolic activity.

We recall the biostatistical theorem that the incidence and prevalence of a condition is what most determines the specificity and sensitivity of any test. This Bayesian concept becomes problematic when those who utilize these tests do not have robust nosological, demographic, and pathophysiological frameworks within which to interpret results. To utilize these tests, clinically and in the research context, one needs to begin with an informed and well-justified pretest probability of the outcome. This overarching point leads to our argument developed below that neuroimaging and clinical neurological assessment can and must move forward in tandem, and never in isolation. One without the other is perilous, especially if we view images in isolation of the clinical exam.

So, instead of viewing neuroimaging and bedside assessments in isolation, let us suggest that we see these two realms of assessment as inextricably linked and analogous to the genetic relationship of genotype and phenotype before Avery, Macleoud, and McCarty discovered (Lederberg, 1994), and Watson and Crick described the genetic material and parsed the genetic code (Watson, 1968). How do we create this amalgam in the absence of a comparable mechanistic framework to understand structure-function relationships? At this juncture in the evolution of this field, where there is not direct linear relationship between what is seen on neuroimaging findings (here akin to genotype) and the behavioral manifestations seen on clinical exam (here akin to phenotype) (Fins, 2007a), how is clinical and imaging data integrated responsibly and prudentially?

This brings us back to Owen's case, and the seemingly sharp divergence between the behavioral and imaging findings. Previously, we have argued that this vegetative patient was actually in a state of nonbehavioral MCS at the boundary zone between VS and MCS by virtue of the fact that she responded in an integrated fashion, albeit solely by imaging findings and not by behavioral criteria (Fins & Schiff, 2006). Patients recovering from traumatic brain injury are allowed one year in vegetative state before further improvement is considered statistically highly unlikely (Jennett, 2002) and she remained within this window when recovery into the minimally conscious state remained possible.

To continue our analogy with genotype and phenotype, this heterozygosity of discordant imaging and behavioral findings would be akin to the white pea plant in Mendel's garden, which was white in one generation but would become red over time. This change in phenotype becomes comprehensible once the nature of inheritance is understood fully and specific genetic sequences for red or white colored leaves are identified. In neuroimaging we are embarking on the analogous mapping effort. Here we are trying to correlate observable phenomena on neuroimaging—some might even view this as a type of phenotype—with what can be observed at the bedside (Fins et al., 2008).

As such, the Owen study is a critical landmark in the field of disorders of consciousness. It provides an unambiguous demonstration of how neuroimaging can provide, in principle, an operational alternative to bedside assessment of a patient's level of function. Because command following is a cardinal feature of MCS and the behavioral findings up to evidence of command following are unambiguous, the possibility of neuroimaging assessment represents a major advance.

The other issue raised by the Owen case is the importance of the temporal grid and of time (Fins 2009). We cannot stress enough the importance of moving beyond reports of snapshots of brain states and attempting to capture functional status against a timeline. Patients recovering from a disorder of consciousness need a movie—not a snapshot—to be properly assessed. This is especially critical from a functional point of view, before the vegetative state becomes permanent. Again returning to the Owen et al. paper, their dramatic findings make much more clinical sense when the image is not viewed a single point in time but rather a point on a prognostic trajectory. By 11 months the subject as we unequivocally minimally conscious, fixating on a mirror held at a 45-degree angle to her face. She made this transition to overt MCS within in the time limits laid out by the MSTF report for traumatic brain injury, and at 22 months she was verbalizing and smiling, raising the chance that she was in the process of transition when the scan was performed at five months.

These fMRI findings (and any other neuroimaging findings) in vegetative state patients must be considered within the context of the natural history of recovery from vegetative state following different types of severe brain injuries. There are important differences in levels of uncertainty of outcome for individual patients based on both clinical findings obtained from bedside examinations and varying etiologies of injury. To properly interpret neuroimaging studies it must be recognized that the vegetative state is very often a transitional one when no strong negative clinical predictors are identified (e.g., loss of pupillary and corneal responses). As measurements are applied to recovery over time we should expect that a significant percentage of patients (for example, 20 to 40% of vegetative state patients studied at the same time after traumatic brain injury as Owen's patient) may show evidence of some unsuspected residual cerebral processing capacity that could reflect the known time frames of expected further functional recovery. It is also import that studies are clear about the context of time of injury and whether or not the observations might cast doubt on known outcome probabilities for the permanence of the vegetative state following cardiac arrest or traumatic brain injury. These have been established in large prospective studies (see Posner et al., 2007 for review).

But even if the patient had not progressed from nonbehavioral MCS at five months to classic MCS by behavioral criteria at 11 months, the larger point about these data suggest that it is very important to plot a timeline. That is, a determination of a brain state should not be made based on a single point in time. One needs to know where a patient or subject is on a timeline to delimit what brain states remain possible and which recoveries might be precluded or achievable. Having a perspective on the Owen study helps to contextualize the rather sensational, albeit seminal, observation that a patient who was vegetative was responding in an integrative fashion heretofore unexpected. Given a longer view, we now appreciate that she was likely following a natural transition into the minimally conscious state, where if such a response had already been demonstrated it would have been expected.

PROFOUND CHALLENGES

At this juncture, neuroimaging is yielding more questions than answers concerning disorders of consciousness, and to some this may come as a disappointment. We would prefer to see the questions posed by neuroimaging as opportunities to deepen our quest for understanding, to better inform our modes of inquiry, and to generate the next round of questions and challenges. The novel yet lingering questions that follow upon the unexpected findings of the Owen study is but one example of the sorts of questions we should anticipate in the coming years. While we could lament the definitiveness of neuroimaging technologies, we prefer to revel in how these modalities properly complicate rather than falsely simplify the problem space of the mind and brain.

To be sure, neuroimaging will help clarify many clinical situations. The refinement of methods capable of eliminating the probability of consciousness will be helpful to families facing tragic decisions to withhold or withdraw life-sustaining therapies. Ironically, it will be more hopeful findings—like evidence of consciousness in the absence of behavioral manifestations of the same—that will lead to ambiguity and raise ethical questions about goals of care and proportionality.

Neuroimaging cannot possibly provide answers to these values questions, but it can point to situations where we have too easily dismissed the presence of ethical choices. Studies like Adrian Owen's and those that will follow give credence to the possibility of consciousness in the absence of behavioral manifestations and compels us to be ethically more attuned to our obligations and responsibilities to individuals who might otherwise be misunderstood and ignored. And they force us to ponder ways to remediate the burdens potentially experienced by these patients and their families—that is, how to confirm and respond to the potential isolation experienced by patients who appear at some level conscious but remain behaviorally dormant on examination.

These responses will lead the field forward toward efforts to restore functional communication for these isolated individuals. We suspect that this will emerge as a primary goal of care in order to provide these patients with an opportunity to communicate about their current state and to defend their interests. Knowledge of these preferences will help us meet a strong affirmative ethical obligation to patients who retain that ability to communicate and harbor residual cognitive capacity.

Neuroimaging may provide a mechanism for communication in the long term, thereby decreasing their perceptions of isolation. But in the short term this technology will help to legitimate the concerns of family members and validate the possibility that patients who are minimally conscious have the neural substrate that might enable them to communicate.

The mother of a subject we have studied at Weill Cornell said it best in an interview conducted as part of an IRB-approved study (Fins & Hersh, 2011). She was speaking of her 24-year-old daughter, who had sustained a brain-stem stroke and was determined to be at a minimum, minimally conscious. In response to a question about neuroimaging research and the establishment of functional communication, she said:

> But I think it [the research] still has implications for treatment. It should. It would be important if you had information, at least don't share it with me but share it with Dr. [X, her referring doctor] so he knew everything so that he could decide treatment and he and I could use that speculative information how we want to treat [the patient] because we need it in order to decide on treatment and plans. I want to be as aggressive as I can to help her communicate. But how can I, so I have to convince people that she could communicate. And the only, if I could convince people that she would be able to if we could just find a pathway and here's a speculative idea, so let's try it.[2]

This brief quote shows the import of neuroimaging as an evolving technology and its ability to accomplish two critical tasks. First, dire findings may further clarify the futility of the situation and make decisions to withhold or withdraw care less painful for family members. On a more positive note, evidence of integrative networks may help validate the observations of family members and indicate that a loved one may be aware and conscious, even in the face of absent behavioral manifestations.

But if we were asked by a colleague in the neuro-ICU about the role of neuroimaging, we would still be still be cautious (Fins, 2011). These assessments remain investigational and need to be validated for widespread clinical utility. At the very least they should give us pause and force us to reevaluate prospects of higher recovery in patients who have a discordance between their exam and neuroimaging studies.

Such data will help counter nihilistic tendencies that lead to the neglect of these patients (Fins, 2003), and hopefully prompt more attentive care, which seeks to optimize residual cognitive function and provide these isolated and sequestered patients with the right to consciousness (Fins, 2010).

Central to that effort will be efforts to restore functional communication. This is an objective shared by most respondents. It is also a goal that neuroimaging and EEG techniques may eventually in the aggregate help to achieve through the development of imaging-based/EEG mind-brain interfaces that might provide some patients a vector out: a way to express their feelings, report their pain, share their thoughts (Fins & Suppes, 2011).

We are still a long way from this objective. But this aspirational therapeutic goal—an almost expressive application of the technology—should be seen as the natural consequence of our increased ability to assess cognitive function in these patients.

1. Name used with permission of Wallis family.
2. Transcription: IN316W (NY Visit) February 26, 2008.

REFERENCES

Ad Hoc Committee of the Harvard Medical School to Examine the Definition of Brain Death. (1968). A definition of irreversible coma. *The Journal of the American Medical Association, 205,* 337–340.

Bardin, J. C., Fins, J. J., Katz, D. I., Hersh, J., Heier, L. A., Tabelow, K., et al. (2011). Dissociations between behavioural and functional magnetic resonance imaging-based evaluations of cognitive function after brain injury. *Brain, 134* (Pt3), 769–782.

Boly, M., Faymonville, M. E., Peigneux, P., Lambermont, B., Damas, P., Del Fiore, G., et al. (2004). Auditory processing in severely brain injured patients: differences between the minimally conscious state and the persistent vegetative state. *Archives of Neurology, 61,* 233–238.

Cruse, D., Chennu, S., Chatelle, C., Bekinschtein, T. A., Fernandez-Espejo, D., Pickard, J. D., et al. (2011). Bedside detection of awareness in the vegetative state: a cohort study. *Lancet, 6736*(11), 61224–61225. DOI:10.1016/S0140

Danielsen, E. R., Christensen, P. B., Arlien-Soborg, P., & Thomsen, C. (2003). Axonal recovery after severe traumaticbrain injury demonstrated in vivo by 1H MR spectroscopy. *Neuroradiology, 45,* 722–724.

Fins, J. J. (1995) Across the divide: Religious objections to brain death. *Journal of Religion and Health, 34* (1), 33–39.

Fins, J. J. (2003) Constructing an ethical stereotaxy for severe brain injury: Balancing risks, benefits and access. *Nature Reviews Neuroscience, 4,* 323–327.

Fins, J. J. (2005) Rethinking disorders of consciousness: New research and its implications. *The Hastings Center Report, 35,* 22–24.

Fins, J. J. (2006) Affirming the right to care, preserving the right to die: Disorders of consciousness and neuroethics after Schiavo. *Supportive & Palliative Care, 4,* 169–178.

Fins, J. J. (2007a). Border zones of consciousness: Another immigration debate? *American Journal of Bioethics-Neuroethics, 7,* 51–54.

Fins, J. J. (2007b). Ethics of clinical decision making and communication with surrogates. In J. Posner, C. Saper, N. D. Schiff, & F. Plum F. (Eds.) *Plum and Posner's diagnosis of stupor and coma,* 4th ed. New York: Oxford University Press.

Fins, J. J. (2008a). Neuroethics & neuroimaging: Moving towards transparency. *American Journal of Bioethics, 8,* 46–52.

Fins, J. J. (2008b). Brain injury: The vegetative and minimally conscious states. In M. Crowley (Ed.), *From Birth to Death and Bench to Clinic: The Hastings center briefing book for journalists, policymakers, and campaigns* (pp. 15–19). Garrison, NY: The Hastings Center.

Fins, J. J (2009). The ethics of measuring and modulating consciousness: The imperative of minding time. *Progress in Brain Research, 177C,* 371–382.

Fins, J. J. (2010). Minds apart: Severe brain injury. In M. Freeman (Ed.), *Law and Neuroscience, Current Legal Issues* (pp. 367–384). New York: Oxford University Press.

Fins, J. J. (2011). Neuroethics, neuroimaging & disorders of consciousness: Promise or peril? *Transactions of the American Clinical and Climatological Association, 122:* 336–346.

Fins, J. J., & Hersh, J. (2011). Solitary advocates: The severely brain injured and their surrogates. In B. Hoffman, N. Tomes, M. Schlessinger, & R. Grob (Eds.), *Transforming Health Care from Below: Patients as Actors in U.S. Health Policy* (pp. 21–42). New Brunswick, NJ: Rutgers University Press.

Fins, J. J., & Suppes, A. (2011). Brain injury and the culture of neglect: Musings on an uncertain future. The body and the state. *Social Research: An International Quarterly, 78* (3), 731–746.

Fins, J. J., Illes, J., Bernat, J. L., Hirsch, J., Laureys, S., Murphy, E., and participants of the Working Meeting on Ethics, Neuroimaging and Limited States of Consciousness. (2008). Neuroimaging and disorders of consciousness: Envisioning an ethical research agenda. *American Journal of Bioethics, 8*(9), 3–12.

Fins, J. J., & Schiff, N. D. (2006). Shades of gray: New insights from the vegetative state. *The Hastings Center Report, 36,* 8.

Giacino, J. T., Ashwal, S., Childs, N., et al. (2002). The minimally conscious state: definition and diagnostic criteria. *Neurology, 58,* 349–353.

Gitierrez E. (1997). Japan's house of representatives passes brain-death bill. *Lancet, 349,* 1304.

Goldfine, A. M., Victor, J. D., Conte, M. M., Bardin, J. C., & Schiff, N. D. (2011). Determination of awareness in patients with severe brain injury using EEG power spectral analysis. *Clinical Neurophysiology 122* (11), 2157–2168.

Jennett, B., & Plum, F. (1972). Persistent vegetative state after brain damage. A syndrome in search of a name. *Lancet, 1,* 734–737.

Kimura, R. (1991). Japan's dilemma with the definition of death. *Kennedy Institute of Ethics Journal, 1,* 123–131.

Laureys, S., Faymonville, M. E., Degueldre, C., Fiore, G. D., Damas, P., Lambermont, B., et al. (2000). Auditory processing in the vegetative state. *Brain, 123,* 1589–1601.

Laureys, S., Faymonville, M. E., Peigneux, P., et al. (2002). Cortical processing of noxious somatosensory stimuli in the persistent vegetative state. *Neuroimage, 17,* 732–741.

Lederberg, J. (1994). The transformation of genetics by DNA: An anniversary celebration of Avery, MacLeod and McCarty (1944). *Genetics, 136* (2), 423–426.

The Multi-Society Task Force on PVS. (1994). Medical aspects of the persistent vegetative state (1 and 2). *New England Journal of Medicine, 330* (21),1499–1508 and (22),1572–1579.

Monastersky, R. (2008). When the brain breaks down: Rapid discoveries in genomics and neuroscience stop short of the doctor's office. *The Chronicle of Higher Education.* http://chronicle.com/cgi-bin/printable.cgi?article=http://chronicle.com/weekly/v55/i15/15b00701.htm.

Monti, M. M., Vanhaudenhuyse, A., Coleman, M. R., Boly, M., Pickard, J. D., Tshibanda, L., et al. (2010). Willful modulation of brain activity in disorders of consciousness. *New England Journal of Medicine, 362* (7), 579–589.

New Jersey Declaration of Death Act. P.L. (1991). Chapter 90 (to be codified as chapter 6A of title 26 of the revised statutes.) Section 5: exemption to accomodate personal religious beliefs. *Kennedy Institute of Ethics Journal, 1,* 289–292.

New York State Department of Health. The Determination of Death. Adopted Regulation 10 N.Y.C.R.R. 400.16.

New York State Task Force on Life and the Law. (1989). The Determination of Death. Minority report by Rabbi J. David Bleich. 2nd ed.

Owen, A. M., Coleman, M. R., Boly, M., Davis, M. H., Laureys, S., & Pickard, J. D. (2006). Detecting awareness in the vegetative state. *Science, 313,* 1402.

Plum, F. (1991). Coma and Related Global Disturbances of the Human Conscious State. In E. Jones & P. Peters (Eds.), *Cerebral Cortex, Vol. 9.* New York: Plenum Press.

Posner, J., Saper, C., Schiff, N., & Plum, F. (2007). *Plum and Posner's Diagnosis of Stupor and Coma,* 4th ed. New York: Oxford University Press.

President's Commission for the Study of Ethical Problems in Medicine and Biomedical and Behavioral Research. (1981). *Defining death: Medical, legal and ethical issues in the determination of death.* Washington, DC: US Government Printing Office.

Schiff, N. D., & Fins, J. J. (2003). Hope for "comatose" patients. *Cerebrum, 5,* 7–24.

Schiff, N. D., Ribary, U., Moreno, D, R., Beattie, B., Kronberg, E., Blasberg, R., Giacino, J., McCagg, C., Fins, J. J., Llinas, R., & Plum, F. (2002) Residual Cerebral Activity and Behavioural Fragments Can Remain in the Persistently Vegetative Brain. *Brain,* 125, 1210–1234.

Schiff, N. D., Rodriguez-Moreno, D., Kamal, A., Kim, K. H., Giacino, J., Plum, F., et al . (2005). fMRI reveals large-scale network activation in minimally conscious patients. *Neurology, 64,* 514–523.

Sidaros, A., Engberg, A. W., Sidaros, K., Liptrot, M. G., Herning, M., Petersen, P., et al. (2008). Diffusion tensor imaging during recovery from severe traumatic brain injury and relation to clinical outcome: a longitudinal study. *Brain, 131,* 559–572.

Uniform Determination of Death Act. (1990). 12 Uniform Laws Annotated 320 (Supp.).

Voss, H. U., et al. (2006). Possible axonal regrowth in late recovery from minimally conscious state. *Journal of Clinical Investigation, 116,* 2005–2011.

Watson, J. D. (1968). *The Double Helix. New York:* Doubleday.

14

Personhood, Consciousness, and Severe Brain Damage

MARTHA J. FARAH

The concept of personhood is closely intertwined with concepts of morality in most people's minds, and bioethical arguments sometimes turn on the question of whether someone or something is a person. For example, if some animals are persons, as has been argued by animal rights advocates (e.g., Francione, 1993), then many of the ways in which these animals are routinely treated in agriculture, research labs, and zoos are morally impermissible. The question of the personhood or potential personhood of fetuses figures prominently in debates about abortion (Dunaway, 2010).

Although normal adult humans are prototypical persons, severe brain damage can bring an adult human's personhood into question. We are often left wondering whether a person remains in the living but unresponsive and uncommunicative body of a human being. Such uncertainty brings with it questions about our moral obligations toward such patients. Can we withhold nutrition and hydration? Must we administer life-saving treatments for medical illnesses such as infections? Should we provide analgesia for conditions or procedures that persons such as ourselves would find painful?

This chapter addresses three main goals. The first is to review the concept of personhood, a familiar and intuitive concept that plays a central role in ethics and law but is nevertheless difficult to define precisely (Farah & Heberlein, 2007). After surveying the literature on criteria for personhood, three general, interrelated aspects of personhood will be taken to be a working definition with which to address the second goal: relating the concept of personhood to our clinical knowledge about severely brain-damaged humans. The limitations of traditional clinical methods will be reviewed, motivating a new approach using functional neuroimaging. The third goal of this chapter is to review the ways in which imaging has been used for assessing the mental life of severely brain-damaged patients, distinguishing among different forms of inference using functional imaging data and differing strengths and weaknesses of these approaches.

PERSONHOOD IN ETHICS AND LAW

Persons have a special moral status distinct from all other objects in most systems of ethics and law. Persons, and not other things, are moral agents who can be held responsible for their actions, and can thus deserve credit or blame. Indeed, even when the injury caused by a person was not intended, but merely the unfortunate consequence of intentional negligence, common law and the model penal code hold the person responsible. In contrast, we do not assign blame to nonpersons. For example, if a falling tree branch kills someone, we do not regard the branch or its behavior as morally wrong.

More relevant to the neuroethics of severe brain damage, persons are moral patients as well as moral agents. Injuring or failing to help a person is morally wrong in a way that similar actions toward other kinds of object are not. Bioethical discussions of rights generally pertain to the rights of persons (e.g., Universal Declaration of Human Rights, 1948). The vague but frequently invoked bioethical concept of dignity also seems closely related to personhood and has been defined as "the presumption that one is a person whose actions, thoughts and concerns are worthy of intrinsic respect" (Nuffield Council on Bioethics, 2002, cited by Macklin, 2003). The four principles of bioethics, autonomy, nonmaleficence, beneficence, and justice (Beauchamp & Childress, 2001) apply specifically to persons. For example, in their book *Principles of Biomedical Ethics*, Beauchamp and Childress refer to the first three principles thus: "Morality requires not only that we treat persons autonomously and refrain from harming them, but also that we contribute to their welfare" (p. 165), and frame the need for the fourth principle thus: "Standards of justice are needed whenever persons are due benefits or burdens because of their particular properties or circumstances" (p. 226).

ELEMENTS OF PERSONHOOD

The attempt to define personhood has occupied philosophers for centuries. The definitions proposed differ in their specific details and in the degree of emphasis placed on different criteria. It is safe to say that no definition succeeds in capturing the concept to everyone's satisfaction, and the most precisely crafted definitions invariably involve drawing arbitrary lines between persons and nonpersons. In the words of Dennett (1978, p. 285), "there can be no way to set a "passing grade" [for an entity to be considered a person] that is not arbitrary."

Nevertheless, commonalities can be found among both historical and contemporary philosophical approaches to personhood. It is these commonalities, which accord well with most people's intuition as well as philosophical theories of personhood, that will be the focus of this section.

The three most frequently invoked elements for defining personhood are cognition, conscious self-awareness, and relationships with others. Although these elements are conceptually fairly distinct and can in principle be separated from one another, in reality they are quite closely associated with one another. For example, most high-level cognition takes place consciously. Rational decision making, involving the weighing of reasons for different courses of actions, is normally carried out with conscious awareness. Similarly, relationships with others are normally based on the sharing of experiences and exchange of ideas by verbal and nonverbal communication. These processes require conscious cognition. In the attempt to capture the meaning of the term "person" and distinguish between persons and nonpersons, different theorists have emphasized one or more of the three criteria, namely cognition, conscious self-awareness, and relationships with others.

Almost all definitions of the concept "person" feature cognitive abilities, such as rationality and a conception of past and future. Indeed the earliest explicit definition, offered by Boethius in the sixth century, equated a person with "an individual substance of a rational nature" (Singer, 1994). For Locke (1997), a person was "an intelligent being that has reason and reflection, and can consider itself the same thinking being in different times and places." Kant's formulation also includes intelligence, but mainly for its role in enabling one to act morally. At the heart of moral action, for Kant, was the ability to distinguish between persons and things and treat them accordingly (Kant, 1948).

In addition to cognitive capacity, two other essential ingredients associated with personhood include self-awareness (also mentioned by Locke, above) and the capacity for relationships with other persons. These are explicit in many modern definitions of personhood. For example, Dennett (1978) incorporates criteria of intelligence and cognitive capacities essential for relationships, such as the ability to understand other persons and to use language, as well as the capacity to be "conscious in some special way" not shared by other animals (p. 270). Joseph Fletcher (1979) proposes 15 criteria for personhood, beginning with intelligence. He suggests that "below IQ 40 individuals might not be persons; below IQ 20 they are definitely not persons." Among his other "marks of personhood" are the ability to relate to others, including to feel concern for others, and self-awareness. Tooley (1972) proposes that something is a person "if it possesses the concept of a self as a continuing subject of experiences and other mental states, and believes that it is itself such a continuing entity." Feinberg (1980, p. 189) suggests that "persons are those beings who are conscious, have a concept and awareness of themselves, are capable of experiencing emotions, can reason and acquire understanding, can plan ahead, can act on their plans, and can feel pleasure and pain." According to Englehardt (1986, p. 107), "what distinguishes persons is their capacity to be self-conscious, rational, and concerned with worthiness of blame or praise." And from Rorty (1988, p. 43): "A person is . . . (a) capable of being directed by its conception of its own identity and what is important to that identity, and (b) capable of interacting with others, in a common world. A person is that interactive member of a community, reflexively sensitive to the contexts of her activity, a critically reflective inventor of the story of her life."

In sum, there are many different perspectives on what it means to be a person, but most feature human-like intelligence or rationality; conscious awareness, including awareness of the self; and the ability to understand, care for, and relate to other persons. These definitions correspond reasonably well with the commonsensical notions we use when we ask whether the person is still "there" after severe brain damage, or whether they are, in layman's terms, a "vegetable."

PERSONHOOD AFTER SEVERE BRAIN DAMAGE: OVER- AND UNDERATTRIBUTION

The elements of personhood that emerge from the foregoing brief review are among the human traits we most value in one another. However tragic a loss of mobility or perception following neurological damage can be, the person him or herself is not lost. As the title of Christopher Reeves's (1999) autobiography expressed it, after his devastating spinal cord injury he was "Still Me." In contrast, with unresponsive patients of the kind discussed by Fins and Schiff and Laureys, it is unclear whether or not the patient continues to be a person in the sense defined earlier. This is because the usual method of ascertaining whether someone is capable of cognition, awareness, or relating to others involves communicating verbally or behaviorally, and this is unreliable with these patients.

Consider how we normally assess cognition, state of conscious awareness, or the ability to relate to others: We pose questions, or when patients are incapable of speaking, we determine whether they can understand what is being said by giving commands like "squeeze my hand" or "close your eyes." We look for signs of recognition when a loved one enters the room or speaks. We tap or pinch the body or make a noise to engage attention. Unfortunately, these methods can both overestimate and underestimate a patient's mental status. Let us begin with the phenomenon of overestimation.

The risk of overattributing personhood to a patient lacking cognitive ability, awareness, and the capacity to relate to others was painfully evident in the case of Terri Schiavo. This severely brain-damaged woman, who died in 2005 and whose autopsy was consistent with her having been in a persistent vegetative state for many years, was the subject of a bitter family battle. Her husband requested the withdrawal of artificial nutrition and hydration, based on the belief that she would not have wanted to linger in a vegetative state. Her parents believed their daughter was still a person, a thinking and aware individual with whom they had a loving relationship, and therefore fought to sustain her life. The conflict eventually became public and highly politicized.

Although the legal issues were complex and extended well beyond the question of Schaivo's mental status, they would probably not have been so difficult to resolve if all parties had held the same view of her cognitive ability, awareness, and ability to relate to others. However, different observers came to very different conclusions about Schaivo's personhood on the basis of her behavior. For example, a lawyer who visited her wrote: "I was pleasantly surprised to observe Terri's purposeful and varied behaviors…I never imagined Terri would be so active, curious, and purposeful. She watched people intently, obviously was attempting to communicate with each one in various ways and with various facial expressions and sounds" (www.rense.com/general63/skdpe.htm). Even her court-appointed advocate, who believed that she was in a persistent vegetative state, reported sensing a "presence:" "Her eyes are not shut, she's breathing on her own and she makes noises," he said. "You want so much to say, 'Terri, give me a sign!' It's not a cucumber lying in a bed" (www.theledger.com/apps/pbcs.dll/article?AID=/20050126/

NEWS/501260353/1004). Watching Terri Schiavo in the videos available on YouTube, it is difficult not feel you are seeing a person interacting with others and aware of her surroundings.

However, clinical and experimental neuroscience have taught us some surprising things about the range of behaviors that can emerge from a decorticate brain without cognition or consciousness. Such behaviors include orienting with eye and head movements toward sights and sounds, generating facial expressions, and producing nonverbal vocalizations that have meaning for us, if not the person producing them, such as cries and laughter. In light of this, we must interpret the behavior of such patients cautiously and approach our own responses to them with a measure of skepticism. The most natural interpretations for certain behaviors are not the only ones. Following a mother's face with one's eyes is not necessarily a sign of love or even recognition.

In addition to the surprising behavioral repertoire possible without conscious thought, there is a second reason that we may attribute the elements of personhood to vegetative human beings: We are hardwired to interpret behavior in terms of mental states. In the psychology literature this tendency is part of a suite of abilities termed "theory of mind," and in most situations we apply our theory of mind automatically, without weighing alternative reasons for the behavior. For a particularly striking demonstration of this fact, consider the typical response to the robot Kismet (Brooks, 2002). Kismet was part of a research effort at the MIT artificial intelligence lab to design machines that interact socially with humans. Kismet was been programmed to gaze at humans who approach it, orient to salient objects moving within its field of view, pull back avoidantly if an object is thrust forward at it, and so on. Visitors to the AI lab attributed all manner of cognitive and emotional states to this robot on the basis of a fairly small set of simple behaviors, and even some lab members became attached to it. If a contraption made of metal and plastic can evoke such feelings, imagine how strong an illusion of personhood could be projected by a living human being emitting these same kinds of simple behaviors (see Farah & Heberlein, 2007, for other examples of illusory personhood).

Behavioral assessments of nonresponsive patients have the potential to mislead us in the opposite direction as well, resulting in individuals with the traits of personhood being wrongly classified as nonpersons or former persons. One reason for this is that cognition and consciousness can fluctuate over time, such that clinical evaluations may not capture a patient what a patient is capable of. The discovery that more than 40% of patients with disorders of consciousness are misdiagnosed, and that patients diagnosed as vegetative may be intermittently capable of following commands and responding to questions (Andrews, 1996; Schnakers, 2009), led to the creation of a new diagnostic category: the minimally conscious state. For this reason it is now recommended that multiple assessments of mental state be performed before a patient is classified as vegetative, and that staff and family remain vigilant for signs of awareness (Giacino et al., 2002).

Above and beyond problems of fluctuating mental status, there are patients in whom the main problem is an inability to communicate by any behavioral method despite relatively intact mental status. Such patients are said to be "locked-in," a depressingly apt phrase that describes near-complete or complete paralysis, the result of interruption of outgoing (efferent) motor connections, most often by stroke. These patients find themselves treated as vegetative, and may try for months or even years to signal their awareness to medical staff and family members (Laureys et al., 2005). In its most classic form, a degree of preserved voluntary eye movement allows communication, for example answering questions with an upward gaze for "yes" or spelling words by selecting one letter at a time with eye movements. For other patients, the de-efferentation is more complete and no voluntary behavior is possible (Bauer, Gerstenbrand, & Rumpl, 1979). The personhood of such individuals would be unlikely ever to be recognized.

BRAIN IMAGING AS A WINDOW INTO PERSONHOOD: A THREE-PART TAXONOMY

Families and medical personnel alike have wondered what, if anything, goes on in the mind of someone as noncommunicative as a vegetative or minimally conscious patient. Is the patient still a person—someone with the capacity for cognition, consciousness, and relationships with others who is in effect locked in, unable to demonstrate these capacities behaviorally? The advent of functional brain imaging has given us a new and qualitatively different way of assessing mental life from the traditional behavioral methods by which disorders of consciousness are still classified and diagnosed.

Brain imaging can play many different roles in advancing our understanding the effects of severe brain damage and the process of recovery from it (Fins & Schiff, chapter 13, this volume). Among these roles, Fins and Schiff highlight several: Diagnostic categories may become better justified or need to be redrawn as imaging provides a better understanding of underlying pathophysiology (Fins & Schiff, 2006). Prognosis, a vexing challenge with coma and its aftermath, may be significantly improved once sufficiently large and well designed studies have correlated early structural and functional imaging results with later functional outcomes. Imaging will undoubtedly also teach us much about the mechanisms of recovery (e.g., Voss et al., 2006) and may suggest targets for therapeutic intervention (Schiff et al., 2007).

In addition to these uses of imaging, there is another that is particularly relevant to the question of personhood after severe brain damage: the assessment of cognition and conscious awareness. To the extent that brain activity is a reliable correlate of the elements of personhood reviewed earlier—cognitive ability, conscious awareness, and relating to others—then it provides new a way of assessing them (Farah, 2008). This is a potentially important advance because it allows us to bypass the behavioral systems that may lead us to overestimate or underestimate patients' mental life.

Functional neuroimaging has been used in many ingenious ways to assess the cognitive capacities of severely brain-damaged patients. The role of imaging, and the inferential logic linking images with conclusions about mental status, differ across studies. In what follows I offer a taxonomy of the ways in which functional brain imaging has been used to infer the mental status of severely brain-damaged patients. The logic of the experimental designs and their consequent strengths and weaknesses differ across the three cases. It is therefore important to bear these differences in mind and avoid treating them as equivalent when interpreting the results of studies and attempting to integrate the results of multiple studies. The ideal next steps, to strengthen the conclusions of the research and advance knowledge, are also different for the different groups of studies.

Preserved Functional Networks

One approach is to show preserved high-level cognitive processing by functional regions or networks in patients' brains, of the kind normally performed by persons in a conscious state. This approach is exemplified by the pioneering study of Schiff et al. (2005). As described by Fins and Schiff in their chapter, they scanned patients in MCS while playing them recordings of either meaningful speech or backward speech. The difference in brain activation in response to forward and backward speech was used as a measure of their brains' processing of meaningful speech per se, that is, without auditory processing that is common to both meaningful and meaningless speech sounds. Surprisingly, the MCS patients showed patterns of brain activity that were qualitatively similar to those evoked in normal healthy subjects, suggesting that large-scale networks underlying language comprehension were to some degree preserved in these severely brain-damaged individuals. Although the authors of the study were careful to interpret these results as indicating preserved language function but not necessarily consciousness, the results have been described elsewhere as indicative of consciousness (Carey, 2005).

Imaging studies of vegetative, as opposed to minimally conscious, patients have yielded mixed evidence of the preserved functional systems of processing. In the largest and most rigorous study of its kind, fifteen carefully evaluated patients meeting criteria for persistent vegetative state were subjected to painful stimulation while being scanned. Like normal subjects, they showed activation of midbrain, thalamic, and primary sensory cortical areas. However, unlike normal subjects, higher cortical areas normally involved in responding to painful stimuli, such as the anterior cingulate cortex (ACC) were not activated (Laureys et al., 2002). This is consistent with the idea that higher level, integrative processing is absent in such patients.

However, single case studies of vegetative patients have occasionally shown preserved brain responses to meaningful stimuli. For example, a patient whose face recognition system responded to photographs of faces (Menon et al., 1988) was described as either "upper boundary vegetative state or lower boundary minimally conscious state" (Laureys, Owen, & Schiff, 2004). The vegetative patient of Owen et al. (2006), most often discussed in relation to the third type of study reviewed below, also demonstrated brain activation in language areas corresponding to high-level linguistic distinctions. Specifically, she showed increased brain activity when presented with sentences containing ambiguous words, in the same region as for normal subjects, consistent with the additional cognitive processing required for resolving the ambiguity of such sentences. This patient later recovered consciousness.

Findings of preserved neural information processing are consistent with some cognitive capacity in these patients, although not the kinds of cognitive capacity featured in the definitions of personhood reviewed earlier, such as rationality, moral conscience, a sense of past, present, and future, or a sense of self. Nor do such findings inform us about patients' capacities for conscious experience. When used for this purpose, it is with the implicit assumption that higher cognitive processes, such as those involved in speech and face recognition, cannot be carried out unconsciously. This assumption is supported by everyday experience, but it is not true under all circumstances. As pointed out by Levy (2008), the cognitive psychology literature contains many examples of dissociated cognition and conscious awareness. More to the point, brain damage can lead to just this type of dissociation between preserved cognitive and neural information processing, on the one hand, and conscious awareness, on the other (Farah, O'Reilly, & Vecera, 1993).

In short, even high-level forms of neural information processing not shared with animals, such as speech recognition, do not tell us about the capacity of a brain for rationality, sense of self, or any of the other cognitive capacities frequently named in connection with personhood. In addition, these findings do not tell us about a patient's state of conscious awareness. To address these questions, different approaches are needed.

Neural Correlates of Consciousness

A second approach to assessing the mental life of severely brain-damaged patients addresses the presence of consciousness directly, by building on previous cognitive neuroscience research on the neural correlates of consciousness. It involves showing preservation of patterns of activity that have been demonstrated, in previous research, to distinguish conscious from unconscious processing.

Laureys, Owen, and Schiff (2004) found that earlier published studies of global brain metabolism at rest show that the brains of locked in patients are almost as active as those of healthy and awake individuals, consistent with the known mental status of such patients, while the activity measured in comatose, vegetative, and minimally conscious patients' brains is more like that of a sleeping or anesthetized person. Of course, global brain activity is not the most specific correlate of consciousness. These authors went on to review activity in specific brain areas associated with awareness of self and environment. These include primarily the prefrontal and medial parietal cortices, based on measurements of activity in these areas in the normal conscious state and across a variety of states in which conscious awareness is diminished, including general anesthesia, sleep, and absence seizures. When resting medial parietal activity is compared across the diagnostic categories discussed here, it is highest for normal control subjects, next highest in locked in patients, lower in minimally conscious patients, and lowest in vegetative patients. These results are reassuringly consistent with current clinical beliefs about these diagnostic categories.

Another example of this approach was provided by Boly et al. (2004). Rather than limit the neural correlates of consciousness to regional activations during rest, they considered also the relationships in activation across regions during stimulation. They used simple clicks as probes to activate the brain and looked to see what parts of the brain became active and whether the activity in these different parts was intercorrelated. The implicit assumption here is that consciousness arises with certain patterns of brain activity, specifically activity in medial frontal and parietal areas and activity that is correlated across these and other brain areas. Boly et al. sought patterns of activation assumed to mark conscious auditory perception and found them in minimally conscious but not vegetative patients.

Again, an assumption is implicit in the interpretation of these data: The assumption that the patterns of brain activation distinguishing the minimally conscious and vegetative patients are the patterns that would distinguish conscious and unconscious responses to stimuli in the same experimental task when given to conscious and unconscious healthy humans. Just as the studies reviewed in the previous section compared patients' preserved neural processing to the normal processing observed in normal subjects (e.g., Owen et al., 2006; Schiff et al., 2005), studies using the neural correlates of consciousness approach must demonstrate the validity of their purported correlates. Unfortunately, this is a more challenging research task, as it requires experimental manipulations of consciousness in normal brains in the context of the same task or stimulus activation being used with the patients.

To summarize, brain activity can be used as an indicator of consciousness by identifying those patterns that characterize different states of consciousness in normal healthy subjects and relating patients' patterns to these states in the normal brain. We have ample evidence concerning the differences between resting brain activity in normal humans whose level of consciousness has been manipulated by anesthesia, sleep, or seizures. Studies of resting activity in groups of severely brain-damaged patients have yielded results that are consistent with clinical inferences about their states of consciousness. Studies of brain activity evoked in response to stimulation promise to deliver additional insight into the functioning of brain systems required for consciousness, but are currently limited by the absence of validation for candidate neural correlates of consciousness in stimulus or task activation contexts. Currently there is a paucity of nonresting imaging studies in which conscious and unconscious processing have been directly compared. To advance research using this approach we need more such studies, using different subject populations and different methods of manipulating conscious awareness of stimuli, to provide a more general and reliable "brain signature" of consciousness.

Brain Activity as Response Surrogate

The third and final way in which functional brain imaging has been used is as a surrogate for overt behavior in examining patients. This was first accomplished by Owen et al. (2006) in a groundbreaking study that used brain activation rather than behavior to assess a patient's ability to follow verbal commands. At the time of testing this patient met criteria for being in a vegetative state. The commands were to imagine playing tennis or taking a walk through the rooms of one's home.

When normal subjects performed these mental imagery tasks, they activated the supplementary motor area when imagining playing tennis and the parahippocampal place area while imagining the walk. Stunningly, the patient activated the same areas in response to the same spoken instructions. In other words, a behaviorally uncommunicative patient was able to comply with verbal commands.

The promise of this new approach and its potential to reveal that some vegetative patients are conscious persons has begun to be explored more systematically. Monti et al. (2010) reported the results of their attempt to use functional brain imaging as a surrogate for behavior in a series of 54 patients with severe brain injury. Twenty-three of these patients were in a vegetative state by clinical criteria and 31 in a minimally conscious state. As in the earlier study, patients were asked to imagine playing tennis and walking through their homes. Of the 54 patients, five showed the appropriate pattern of imagery-related activity following instructions to imagine tennis and walking. For two of these patients, traditional clinical examinations before and after the scans were consistent with the vegetative state; only brain imaging showed there to be a thinking person in the scanner.

Of course, this approach also requires certain assumptions to proceed from data to interpretation, most notably the assumption that command-following cannot be accomplished without conscious awareness. If it were the case that hearing the request to imagine playing tennis could automatically and unconsciously trigger motor imagery, then these results would not be evidence of consciousness. However, this seems implausible. After all, when a patient squeezes the examiner's hand on request, we take that as evidence of consciousness, and don't ask whether the squeeze could have been triggered unconsciously.

Could brain activity be used by patients to communicate? The researchers attempted to harness the patients' imagery command-following ability to enable the patients to answer yes-no questions, and succeeded with one patient. Their protocol involved asking the patient a question (e.g., "Is your father's name Alexander?") and instructing them to respond during the imaging session by using one type of mental imagery (either tennis imagery or house imagery) for "yes" and the other for "no." Each question has a true and false version (e.g., the patient's father was indeed named Alexander; he was also asked "Is your father's name Thomas?"), and during different scanning sessions he was sometimes instructed to use tennis imagery for "yes" or house imagery for "yes." For five our of the six questions asked, the patient reliably answered correctly, and for the sixth question the response was not incorrect but merely absent.

Returning to the elements of personhood discussed earlier, this performance does not necessarily demonstrate fully intact rationality, sense of self, and so on. Nevertheless, it does indicate the ability to comply with a request, follow changing rules concerning the mapping between imagery task and response, and recall facts about the self. As with imagery command-following, behavioral responses in tasks such as these are also normally taken as evidence of consciousness in the clinical assessment of patients. Finally, by demonstrating a cooperative interaction with the experimenters, these brain-imaging results evince a kind of rudimentary interpersonal process. These are capacities which, when signaled though spoken language or nonverbal behavior rather than brain activation, are routinely considered sufficient for considering a patient to be a person.

In sum, functional brain imaging has been used in a variety of ways to better assess the personhood of severely brain-damaged patients. The use of brain activity as a surrogate for behavioral responses has been particularly effective in demonstrating preserved capacity for cognition, consciousness and interpersonal communication in such patients. This has led to the important discovery that some patients who appear to be definitively vegetative on rigorous clinical examination may in fact be persons by the criteria of cognition, awareness, and capacity for relating to others.

REFERENCES

Andrews, K., Murphy, L., Munday, R., & Littlewood, C. (1996). Misdiagnosis of the vegetative state: retrospective study in a rehabilitation unit. *British Medical Journal, 313,* 13–16.

Bauer, G., Gerstenbrand, F., & Rumpl, E. (1979). Varieties of the locked-in syndrome. *Journal of Neurology, 221* (2), 77–91.

Beauchamp, T. L., & Childress, J. F. (2001). *Principles of biomedical ethics.* New York: Oxford University Press.

Boly, M., Faymonville, M. E., Peigneux, P., Lambermont, B., Damas, P., et al. (2004). Auditory processing in severly brain injured patients: differences between the minimally conscious state and the persistent vegetative state. *Achives of Neurology, 61* (2), 233–238.

Brooks, R. A. (2002). *Flesh and machines.* New York: Pantheon.

Dennett, D. (1978). *Brainstorms: Philosophical essays on wind and psychology.* Cambridge, MA: The MIT Press.

Dunaway, R. M. (2010). The Personhood strategy: The state's prerogative to take back abortion law. *Willamette Law Review, 327.*

Engelhardt, H.T.J. (1986). *The foundations of bioethics.* New York: Oxford University Press.

Farah, M. J. (2008). Neuroethics and the problem of other minds: Implications of neuroscience evidence for the moral status of brain-damaged patients and nonhuman animals. *Neuroethics, 1,* 9–18.

Farah, M. J., & Heberlein, A. S. (2007). Personhood and neuroscience: naturalizing or nihilating? *The American Journal of Bioethics, 7* (1), 37–48.

Farah, M. J., O'Reilly, R. C., & Vecera, S. P. (1993). Dissociated overt and covert recognition as an emergent property of a lesioned neuro network. *Psychological Review, 100* (4), 571–588.

Feinberg, J. (1980). Abortion. In T. Regan (Ed.), *Matters of life and death* (pp. 188–189). Philadelphia: Temple University Press.

Fins, J. J., & Schiff, N. D. (2006). Shades of gray: New insights into the vegetative state. *The Hastings Center Report, 36* (6), 8.

Fletcher, J. 1979. *Humanhood: Essays in biomedical ethics.* Buffalo, NY: Prometheus Books.

Francione, G. L. (1993). Animals, properties and legal welfarism: "Unnecessary" suffering and the "humane" treatment of animals. *Rutgers Law Review, 46,* 721.

Giacino, J. T., Ashwal, S., Childs, N., Cranford, R., Jennett, B., et al. (2002). The minimally conscious state: definitions and diagnostic criteria. *Neurology, 58* (3), 349–353.

Kant, I. (1948). *Groundwork of the metaphysics of morals*. In J. J. Paton (Ed.), *The moral law: Kant's groundwork of the metaphysics of morals*. London: Hutchinson.

Laureys, S., Faymonville, M. E., Peigneux, P., Damas, P., Lambermont, B., et al. (2002). Cortical processing of noxious somatosensory stimuli in the persistent vegetative state. *Neuroimage, 17* (2), 732–741.

Laureys, S., Owen, A. M., & Schiff, N. D. (2004). Brain function in coma, vegetative state and related disorders. *Lancet Neurology, 3* (9): 537–546.

Locke, J. (1997). *An essay concerning human understanding*. Harmondsworth: Penguin Books.

Macklin, R. (2003). Dignity is a useless concept. *BMJ, 327*, 1419.

Monti, M. M., Vanhaudenhyse, A., Coleman, M. R., Boly, M., Pickard, J. D., et al. (2010). Willful modulation of brain activity in disorders of consciousness. *New England Journal of Medicine, 362* (7), 579–589.

Nuffield Council on Bioethics. (2002). *The ethics of patenting DNA*. London: Nuffield Council on Bioethics.

Reeves, C. (1999). *Still Me*. New York: Ballentine Books.

Rorty, A. O. (1988). *Mind in action: Essays in the philosophy of mind*. Boston: Beacon Press.

Schiff, J. T., Giacino, J. T., Kalmar, K., Victor, J. D., Baker, K., et al. (2007). Behavioral improvements with thalamic stimulation after severe traumatic brain injury. *Nature, 448* (7153), 600–603.

Schiff, N. D., et al. (2007).Behavioural improvements with thalamic stimulation after severe traumatic brain injury. *Nature, 448*, 600–603.

Schiff, N. D., Rodriguez-Moreno, D., Kamal, A., Kim, K. H., Giacino, J. T., et al. (2005). fMRI reveals large-scale network activation in minimally conscious patients. *Neurology, 64* (3), 514–523.

Schnakers, C., Vanhaudenhuyse, A., Giacino, J., et al. (2009). Diagnostic accuracy of the vegetative and minimally conscious state: Clinical consensus versus standardized neurobehavioral assessment. *BMC Neurology, 9*, 35.

Singer, P. (1994). *Rethinking life and death: The collapse of our traditional ethics*. Oxford: Oxford University Press.

Tooley, M. (1972). Abortion and infanticide. *Philosophy and Public Affairs, 2*, 37–65.

Voss, H. U., Uluc, A. M., Dyke, J. P., Watts, R., Kobylarz, E. J., et al. (2006). Possible axonal regrowth in late recovery from the minimally conscious state. *J Clin Invest, 116* (7), 2005–2011.

PART FIVE

New Treatments, New Challenges

15

Functional Neurosurgery and Deep Brain Stimulation

MATTHIS SYNOFZIK

The use of stereotactic surgery had already started over a century ago, in 1908, when Sir Victor Horsley and Robert Clarke introduced a new apparatus to insert a probe or needle under accurate control into subcortical stuctures in an experimental animal model (Schurr & Merrington, 1978). These first experimental attempts were further elaborated throughout the twentieth century, carried on by landmark technical developments in the field of neurosurgery such as minimal-invasive surgery, image-guided surgery, neuronavigation, and radiosurgery. Today, stereotactic and functional neurosurgery forms a distinct branch of neurosurgery which, according to the World Society of Stereotactic and Functional Neurosurgery, "utilizes dedicated structural and functional neuroimaging to identify and target discrete areas of the brain and to perform specific interventions (for example ablation, neurostimulation, neuromodulation, neurotransplantation, and others) using dedicated instruments and machinery in order to relieve a variety of symptoms of neurological and other disorders and to improve function of both the structurally normal and abnormal nervous system."

This innovative field, albeit highly successful in the last decade, encounters several ethical issues when applied to clinical practice (Ford & Henderson, 2006; Ford & Kubu, 2006). First, unlike many other surgical practices, functional neurosurgery has the risk of modifying essential features of a patient's personhood, including mood, personality, and cognitive abilities. Second, it can result in nonreversible brain destruction and uses brain devices in which the precise neurophysiological mechanisms and long-term consequences are often not fully understood. Third, it presents unique challenges concerning perioperative management (e.g., surgery in awake patients in case of electrode placements or epilepsy resections) and postoperative care (e.g., long-term patient follow-up and continuous parameter optimization following DBS surgery). Fourth, it relies on often difficult to interpret and complex neuroimaging results for generating hypotheses, justifying interventions, and guiding clinical decision making (Ford & Kubu, 2005). Fifth, it inherits the historical legacy of abuses in the early use of neurosurgical procedures, such as frontal lobotomy during early psychosurgery (Fins, 2003).

Some of these challenges are not unique to functional neurosurgery, but apply to neuropharmacological drug therapy and psychotherapy as well (e.g., risk of modyfing personality traits, lack of knowledge about the exact mechanism of action and the long-term consequences, need of continuous follow-up and treatment optimization). Other challenges might loose large parts of their difficulty once they are conceptually scrutinized and clarified (e.g., the references to risks to personhood or to the legacy of psychosurgery). Thus, detailed ethical analyses for each functional neurosurgical method and its applications are warranted to determine more closely the specific ethical challenges and the validity of particular normative arguments.

THE RAPID EXPANSION OF DBS INDICATIONS AND APPLICATIONS

One of the most rapidly growing interventions that raises several of the ethical challenges in functional neurosurgery is electrical deep brain stimulation (DBS) of specific brain circuits (see Figure 15.1). It has gained widespread acceptance in the symptomatic treatment of various neurological disorders including Parkinson's disease, essential tremor, and dystonia, where its benefit has already

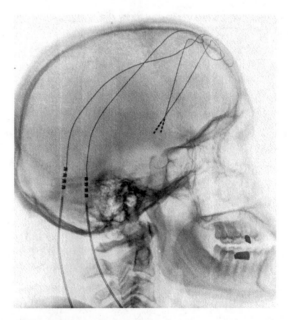

Figure 15.1 **Lateral X-ray image of two DBS leads** in an exemplary patient with DBS to the nucleus accumbens for treatment of major depression. (Courtesy of Thomas Schlaepfer, University of Bonn, Germany.)

been demonstrated by large scale randomized controlled trials (Deuschl et al., 2006; Kupsch et al., 2006). Correspondingly, indications of DBS are rapidly growing: First piloting trials have demonstrated preliminary efficacy in improving some features of minimally conscious state (Schiff et al., 2007; Yamamoto et al., 2005) or Alzheimer's disease (Laxton et al., 2010). Moreover, DBS is now being extended to treatment of other neuropsychiatric conditions such as Gilles de la Tourette syndrome (Porta et al., 2009), obsessive-compulsive disorder (OCD) (Abelson et al., 2005; Greenberg et al., 2006a; Mallet et al., 2008; Nuttin et al., 2003), major depression disorder (MD) (Lozano et al., 2008; Malone et al., 2008; Schlaepfer et al., 2008), or substance addiction (Kuhn et al., 2007; Mantione, van de Brink, Schuurman, & Denys, 2010; Muller et al., 2009). In fact, following the success of DBS for Parkinson's disease and essential tremor, DBS for psychiatric disorders is also already moving from experimental to clinical use. This is evidenced by the growing number of worldwide DBS implantations in neuropsychiatric cases, the increasing amount of publications presenting outcome data (Goodman & Insel, 2009), and the fact that two large pivotal trials of DBS for major depression have been launched (ANS, 2007; Medtronic, 2006).

So how should we weight the abovementioned ethical challenges with respect to neuropsychiatric DBS? What other ethical concerns come to the fore in the light of this rapid expansion of indications and increasingly widespread application? And, from a clinical point of view, at what point should we recommend DBS to an individual patient with a psychiatric disorder—and, in turn, when should we discourage operation? In the following discussion, some approaches to these questions will be outlined. Rather than giving a comprehensive overview over all ethical issues involved in DBS, these approaches aim to provide some practical guidance on clinical decision making with particular emphasis on psychiatric DBS.

BEYOND CONCERNS OVER DBS-INDUCED PERSONALITY CHANGES AND EARLY PSYCHOSURGERY

It is important to acknowledge at the outset the limitations of some commonly raised concerns in connection with the neuroethics of DBS and other functional neurosurgery procedures. Personality change is the focus of one of these concerns. For example, several authors view change of personality as a negative criterion for psychiatric DBS, claiming that DBS "should not be used to modify a person's individual character traits" (Hildt, 2006), that "changing the patient's identity is ethically problematic" (Witt, Kuhn, Timmermann, Zurowski, & Woopen, 2011), or by assessing DBS-induced changes of personality variables only under the category of risks (Ford & Kubu, 2006), but not also under the category of benefits. Based on a naturalistic, nonnormative concept of personality (Synofzik & Schlaepfer, 2008; for criticism see Witt et al., 2011), this association seems utterly misleading.

First of all, the potential of DBS to change aspects of mood and cognitive behavior—and thus important elements of personality—cannot per se be taken as an ethical criterion to mark ethical limitations of psychiatric DBS, as modification of these features is not an unwanted, coincidental side effect, but rather the main intended outcome of this intervention (Synofzik & Schlaepfer, 2008). If, for example, mood and cognitive behavior did not change in a patient with OCD or MD, DBS could not be considered an effective treatment. In fact, even DBS in movement disorder patients changes some basic aspects of one's personality, if it is effective—and so does psychotherapy and psychopharmacology (Synofzik & Schlaepfer, 2008). Second, there are no convincing a priori reasons why it should not be legitimate to modify some features of a patient's personality, as long as the patient benefits from this modification and consents to it. For example, if a certain DBS parameter setting would allow to improve not only motor behavior but also mood in a PD patient who was dysphoric throughout his whole life (unrelated to the neurodegenerative disorder) (Müller & Christen, 2011), and if this mood improvement would yield an affective level that is normal for most people without reaching the thresholds of hypomania, there might be good ethical reasons to justify and even welcome this DBS-induced change in personality (Synofzik, Schlaepfer, & Fins, 2012). Similarly, there is evidence that DBS to the nucleus accumbens might improve cognition in patients with depression independently from improving the anhedonic component of depression (Schlaepfer et al., unpublished data). Even if such an improvement in cognition would yield higher cognitive levels in a patient than during his premorbid state, this change might well be justified, as long as the patient benefits from this change and agrees to it. Thus, the ethically decisive question is not whether DBS alters a patient's personality or not, but whether it does so *in a good or bad way from the patient's own perspective* (Synofzik & Schlaepfer, 2008).

The ethically problematic history of early psychosurgery is another common source of criticism concerning the ethics of DBS (Kringelbach & Aziz, 2009; Miller, 2009). Although today's DBS endeavor can certainly learn some general lessons from the fallacies of early psychosurgery (e.g., the need of regulated, systematic research following transparent commonly shared standards [Fins, 2003]), the allusion to early psychosurgery is of only limited value for a detailed systematic discussion of the ethical issues in current neuropsychiatric DBS. This is simply due to the fact that neuropsychiatric DBS does not stand in a continuous line of tradition with early psychosurgery, but was rather adopted to psychiatry from the fields of pain therapy and movement disorders treatment (Hariz, Blomstedt, & Zrinzo, 2010). Correspondingly, it differs from early psychosurgery in almost all important ethical variables (see Table 15.1) (Synofzik & Schlaepfer, 2008).

ETHICAL CRITERIA FOR ASSESSING NEUROPSYCHIATRIC DBS

DBS for psychiatric disorders does not require any specific bioethical criteria. This is important to acknowledge not only since allusions to personality concepts or to the history of psychosurgery might be conceptually misleading. It is also important in order to avoid a possible tendency of "benevolent stigmatization" concerning psychiatric patients (Synofzik & Clausen, 2011): Several claims on the alleged vulnerability of psychiatric patients in the DBS decision-making process are

Table 15.1 DIFFERENCES IN ETHICAL VARIABLES BETWEEN EARLY PSYCHOSURGERY AND CURRENT ABLATIVE SURGERY AND DEEP-BRAIN STIMULATION. FOR CASE OF ILLUSTRATION, DESCRIPTION OF VARIABLES IS PARTLY SIMPLIFIED (ADAPTED FROM SYNOFZIK & SCHLAEPFER, 2008)

Ethical variable	Early psychosurgery	Currently used ablative surgery	Deep brain stimulation
Medical indication	Often disturbing, intolerable patients with unclear, vague indication	Patients with reduced health-related quality of life	Patients with reduced health-related quality of life
Surgical method	Imprecise open surgery	Precisely planned, ablative sterotactic intervention (thermocoagulation)	Precisely planned, stereotactic insertion of electrodes
Selection of targeted brain area	Crude prior clinical observation	Hypothesis-driven; based on previous animal lesion experiments and patient studies	Hypothesis-driven based on lesional surgery (initially) and extensive prior functional neuroimaging data (currently)
Invasiveness	Irreversible lesion of larger brain areas	Irreversible lesion of specific small brain area	Largely reversible damage; removable hardware
Adaptation to effects and side effects	Impossible	Impossible	Constant parameter optimization in order to maximize effect and reduce side effects
Patient information	None	Informed consent after counseling	Informed consent after counseling
Decision-making process	Treating physician only	Interdisciplinary conference and patient's own informed preferences	Interdisciplinary conference and patient's own informed preferences
Primary goal	Altering wholesale personality structures	Improving specific aspects of respective psychiatric disorder	Improving specific movement parameters or specific aspects of respective psychiatric disorder

(continued)

Table 15.1 (*Continued*)

Ethical variable	Early psychosurgery	Currently used ablative surgery	Deep brain stimulation
Follow-up care	No specific	Interdisciplinary team	Interdisciplinary team, continuous follow-up dates
Treatment standards	None	Clearly defined inclusion and exclusion criteria; intervention decided by committee on a case by case basis	Clearly defined inclusion and exclusion criteria; randomized controlled trials using sham-stimulation (recently)

either not evidence-based (Dunn et al., 2011) or apply to nonpsychiatric patients as well, and thus seem to result from an unduly overprotective attitude (Synofzik & Clausen, 2011). In fact, the same bioethical criteria can be applied for DBS in psychiatric disorders as for DBS in movement disorders or for any other biomedical intervention. According to these commonly shared criteria (Beauchamp & Childress, 2008), DBS has to (1) benefit the patient (principle of *beneficence*), (2) do no harm to the patient (principle of *nonmaleficence*), and (3) reflect individual patient preferences (principle of *respect for autonomy*). Consequently, there is a certain ethical obligation to *offer* DBS to a psychiatric patient when presumed benefits outweigh harms and when it corresponds to his or her true preferences and goals. In turn, there is an ethical obligation to *discourage* DBS when presumed harms outweigh benefits and/or when it does not correspond to his or her true preferences and goals. But how exactly can each of these criteria be contextualized in the domain of neuropsychatric DBS?

DETERMINING BENEFIT IN DBS

The assessment of benefit in DBS has to account for several heterogeneous aspects. First of all, there has to be sufficient empirical evidence that DBS is, at least in principle, effective in improving the condition at stake—and, preferably, that it is more effective than both nonsurgical measures and ablative surgery. Recent studies in OCD or MD have demonstrated first evidence of DBS effectiveness at least in some patients (Greenberg et al., 2006b; Lozano et al., 2008; Malone et al., 2008; Mayberg et al., 2005; Schlaepfer et al., 2008). Given the fact that patients studied were selected on the basis of being treatment-refractory to pharmacotherapy and psychotherapy, these findings are certainly significant. However, the ability to adequately determine true therapeutic effectiveness from these data is limited by small sample sizes (less than one hundred patients total) and the open-label, uncontrolled nature of most study designs (in particular single case studies or retrospective case studies without any control groups, for example, in DBS for addiction (Kuhn et al., 2009; Kuhn et al., 2007; Muller et al., 2009) or for aggressive behavior (Franzini, Marras, Ferroli, Bugiani, & Broggi, 2005; Kuhn et al., 2008)). Moreover, currently available data indicate that a significant share of patients (e.g., 50 to 75% of OCD (Abelson et al., 2005; Greenberg et al., 2006b) and 25 to 50% of MD patients (Lozano et al., 2008; Malone et al., 2008; Mayberg et al., 2005; Schlaepfer et al., 2008) fail to show long-term response to DBS-treatment. As individual prognostic predictors for lasting treatment effects still remain unclear, any assumption about an individual's potential benefit from DBS is still stochastic in nature.

Contrary to the currently prevailing perspective (Bell, Mathieu, & Racine, 2009; Glannon, 2009), however, *treatment refractoriness* does not necessarily have to be a mandatory criterion for likely benefit in the future: DBS might prove to be so superior in OCD and MD that especially patients who have *not* been on medication for longer time and whose social and physical life is *not yet* devastated by disease might benefit more from DBS and thus present the best candidates for DBS. This scenario seems provocative at first glance, but parallels a lesson recently learned in DBS for PD (Mesnage et al., 2002; W. M. Schupbach et al., 2007; Welter et al., 2002) and primary dystonia (Isaias, Alterman, & Tagliati, 2008). Correspondingly, several studies are underway assessing benefits of DBS in early stages of PD in relatively young PD patients (ClinicalTrials.gov Identifier: NCT00354133 [Kiel, Germany], and NCT00282152 [Vanderbilt, USA]).

In order to provide an actual *benefit* to the very individual patient, DBS not only has to be effective, that is, improve scores in movement disorder or psychopathology rating scales, but also has to demonstrate that these quantitative improvements are indeed associated with *an actual improvement in the individual patient's life* (Synofzik & Schlaepfer, 2008). DBS-induced restoration of some physiological parameters and/or cognitive and physical functions does not necessarily mean that the patient is better off—that is, receives a net benefit from DBS (Glannon, 2008; Synofzik & Schlaepfer, 2008). For true benefit, statistically significant effectiveness is only a necessary, not a sufficient, condition and it might, in certain cases, only present an invalid surrogate parameter. Or, in more vivid words: Patients undergo DBS surgery not to stop their tremor or improve on some rating scale, but to do more things in their life, together with their relatives and in their accustomed environments.[1]

The risk of mis-focusing on surrogate parameters is high in movement disorder DBS. For example, it was shown in patients with Huntington's disease that pallidal DBS might be effective in reducing chorea, but not in slowing deterioration of gait, bradykinesia, dystonia and neurocognitive functioning (Kang, Heath, Rothlind, & Starr, 2011). As quality of life in Huntington's disease is largely determined by the overall functional capacity and depression (Ho, Gilbert, Mason, Goodman, & Barker, 2009), the improvement of chorea might be of only very little overall benefit to these patients. Similarly, a recent study in PD patients demonstrated that following DBS surgery many patients were not happier with their quality of life, despite—or probably because of—a clear improvement in various outcome variables after DBS implantation (Schupbach et al., 2006). Some of these patients might have tormented periods in their marriages or fail to resume professional activities postoperatively. The factors contributing to these psychosocial misadjustments do not seem to be specific to PD (see Table 15.2), but can be expected after rapid symptom modification in *any* chronic life-determining disease, psychiatric or somatic. For example, a rapid improvement of MD, OCD, or addiction will change previous personal activities of daily living and social relations in a more drastic and rapid way than can be compensated for by the individual or by his or her personal surroundings (Schlaepfer & Fins, 2010). Correspondingly, there is first anecdotal evidence that following DBS surgery an OCD patient became divorced, despite—or maybe because of—a clear improvement in the OCD symptom score.[2] Similarly, a comorbid personality disorder became evident in a MD patient after highly successful DBS treatment, seriously complicating the following treatment and reducing quality-of-life measures to below the pre-DBS state (Schlaepfer & Fins, 2010).

The fact that DBS benefit is determined not just by improvements in a particular target symptom but rather by the impact and consequences of these improvements in a patient's individual context of living has important consequences for DBS research, clinical decision making, and rehabilitation care. Research trials should not only ask whether DBS is effective in terms of demonstrating improvement on movement disorder or psychopathology scores, but whether it indeed allows the very individual patient to live a more satisfying life including psychosocial dimensions. Therefore, quantitative measures have to comprehensively cover also variables like reintegration in social and work environments and psychosocial and global quality of life. Moreover, they have to be complemented with qualitative measures—in particular as changes in disease severity, comorbidities, and psychosocial life situations are often harder to capture in psychiatric disease than in movement disorders.

Table 15.2 EXPERIENCES OF PSYCHOSOCIAL MALADJUSTMENT BY PATIENTS AND CAREGIVERS AFTER SUCCESSFUL DBS IN A CHRONIC DISEASE, EXEMPLARILY DEMONSTRATED FOR PARKINSON PATIENTS WITH STIMULATION OF THE SUBTHALAMIC NUCLEUS (ADAPTED FROM AGID ET AL., 2006)

Experiences of psychosocial maladjustment after successful DBS

- Loss of an aim in life after being less challenged in fighting the chronic disease
- Negative retrospective perspective on the disaster the disease has caused in their lives
- Persistent negative anticipation of future problems
- Marital problems, e.g.,
    - the spouse is forced to switch from the role of a caregiver to the role of an equal partner;
    - the spouse overburdens the patient who is still not able to run his or her own life to the same extent as before the illness;
- Inability to re-integrate into work life, e.g.,
    - by insisting on the sick role
    - by prioritizing leisure and artistic activities
- Incorporation of an alien implanted device into the body image

With respect to clinical treatment decisions, physicians have to make sure that the target symptom of DBS is not selected from the treatment team's perspective of what might be beneficial for the patient, but from the patient's own perspective and goals. To this end, each DBS decision making has to be preceded by a detailed, individualized goal assessment process where the main problems, hopes, and goals of the patient are identified and compared with those effects which are realistically to be expected from DBS. This will help not only to determine possible discrepancies between the patient's and the physicians' goals of DBS treatment, but also to identify unrealistic expectations linked to DBS and to track changes of goals in the course of DBS treatment. Both unrealistic expectations (Bell, Maxwell, McAndrews, Sadikot, & Racine, 2010; Okun & Foote, 2004; Racine, Waldman, Palmour, Risse, & Illes, 2007) and changing goals after DBS insertion (the so-called problem of a moving target (Kubu, 2011)), have been identified as common problems limiting patients' self-perceived benefit of DBS.

If DBS does not end after electrode insertion or with some functional improvements following surgery but requires a complex psychosocial adjustment process, then all DBS patients should be offered a full-blown rehabilitation program, similar, for example, to patients having received orthopedic surgery.

ASSESSING THE RISKS OF DBS

Deep brain stimulation to different targets is associated with severe short-term and long-term risks that need to be assessed on both biological and psychosocial levels (see Table 15.3).

These risks of DBS are mainly derived from the literature on DBS in movement disorders. Extensive analyses of risks in psychiatric disorders are still lacking. For some of the aforementioned risks it still remains unclear whether they result from brain stimulation per se, from a secondary reaction to the stimulation effects, or from the underlying irresistibly progressive disease in case of a underlying neurodegenerative disorder. Moreover, it has to be pointed out that a final risk- (and benefit)-assessment cannot be performed for DBS in general, but needs to be done for each DBS target site separately. Such a site-specific assessment of the efficacy-harm ratio might be of particular importance in psychiatric DBS as often several very different anatomical targets (e.g., nucleus accumbens, habenula, inferior thalamic peduncle, internal capsule, Brodmann area 25) are proposed for the same condition (depression) and as different targets will probably have different benefit-harm ratios. Since it is unlikely that stimulation of one brain area will be sufficient for all types of one psychiatric disease (e.g., one particular single DBS target for all types of depression), psychiatric DBS assessments will require a highly individualized approach that selects the target site depending on each individual's symptom profile (Bewernick et al., 2009b). Thus, deep brain stimulation broadly defined is a very imprecise category for discussing risks and benefits—a fact that has to be kept in mind in societal discussions about psychiatric DBS or in future scientific comparisons of DBS with other different treatment modalities (e.g., DBS vs. ECT or TMS for depression; see Synofzik & Schlaepfer, 2010).

Under the assumption that DBS would be an efficacious treatment, one might do harm to patients not only by performing DBS but also by *not* performing it. The chronic and partly even progressive course of treatment-refractory OCD or MD implies a constant increase in psychological suffering, work disintegration, social withdrawal, and partnership and family relation problems. Thus, also *not performing* DBS in psychiatric patients might one day posit specific, well-reasoned ethical justifications. Moreover, all pharmacological treatments are associated

Table 15.3 ADVERSE EFFECTS OF DBS IN DIFFERENT DOMAINS (ADAPTED FROM SYNOFZIK & SCHLAEPFER, 2011)

Adverse short-term effects related to surgery and implantation

- perioperative complications, e.g., intracerebral hemorrhages, seizures, infections, or hemiparesis (incidence 0,2–2%; Videnovic & Metman, 2008; see Fig. 15.1)

- postoperative complications, e.g., pneumonia, pulmonary embolism, or hepatopathy with a 30-day mortality rate of 0,4 % (Voges et al., 2007)

Adverse short-term effects related to stimulation

- mood elevation/hypomania (Abelson et al., 2005; Funkiewiez et al., 2003; Greenberg et al., 2006b; Mallet et al., 2008)

- sadness, anxiety, panic (Abelson et al., 2005; Shapira et al., 2006)

- aggression (Piasecki & Jefferson, 2004)

- distracting epigastric or gustatory sensations (Greenberg et al., 2006b)

- decreased alertness or cognitive dulling (Greenberg et al., 2006b)

Adverse long-term effects on a neurocognitive level

- worsening of apathy (Drapier et al., 2006)

- depression (Richard, 2007)

- cognitive impairments, e.g., in verbal fluency, color naming, selective attention, and verbal memory (Smeding et al., 2006)

- sudden symptom reoccurrence and aggravation in case of battery depletion, risking, e.g., depressive exacerbations (Greenberg et al., 2006b)
- Battery depletion occurs as a function of programmed stimulation parameters, usually after 5–13 months in the case of higher stimulation current amplitudes such as those required for OCD

- sudden symptom reoccurrence and aggravation in case accidental deactivation

- long-term neuroplastic changes.
Even though DBS effects are generally considered to be reversible (Goodman & Insel, 2009), prolonged stimulation leads to changes in the neural network that endure well beyond the cessation of stimulation, e.g., by exerting long-term effects on synaptic plasticity (Carmichael & Price, 1996; Haber, 2003). Evidence so far indicates that these changes seem to be associated with persistent benefit rather than persistent harm in psychiatric patients (Mayberg et al., 2005), yet this might potentially be conversed in cases where patients experience harmful effects of DBS.

(continued)

Table 15.3 (CONTINUED)

Adverse short- and long-term effects on a psychosocial level

- psychosocial misadjustment even and especially in case of an effective treatment response (M. Schupbach et al., 2006)

- suicidality (Appleby, Duggan, Regenberg, & Rabins, 2007; Kapoor, 2009; Soulas et al., 2008), in particular in patients who have an elevated risk of depression and suicidal behavior, e.g., OCD patients (Torres et al., 2006).

- harmful psychological consequences of nonresponsiveness to DBS, e.g., severe disappointment and renewed desperation (Abelson et al., 2005).
This risk is particularly high in severely disabling, potentially fatal conditions like psychiatric diseases where patients might see the treatment as "the last rope." It might make patients even more susceptible to suicide.

- psychosocial *ir*reversibility of the procedure.
Even though active DBS can be easily discontinued (e.g., in case of failing efficacy, better alternative future treatment options or changed patient preferences), the patients' psychological states might not simply fully reverse to a pre-DBS baseline condition after longer periods of treatment. In case of effective DBS, patients will go through completely new experiences and will change their ways of living, while in case of nonefficacy severe despair will be added to their life. Thus, the assumption that DBS effects are fully reversible—which is commonly taken as one of the strongest ethical arguments in support of psychiatric DBS (Glannon, 2009; Kringelbach & Aziz, 2009)—might be both inaccurate and even potentially dangerous.

---

with significant adverse effects, for example, agitation, sexual side effects, sedation, sleep disturbances, and night sweats in case of depression treatment, often leading to noncompliance (Keller, Hirschfeld, Demyttenaere, & Baldwin, 2002). The same is less recognized but nevertheless the case for psychotherapy (Nutt & Sharpe, 2008). These adverse effects have to be counterbalanced against those of DBS, in particular as none of these adverse events has been reported in DBS depression treatment so far (Bewernick et al., 2009a; Lozano et al., 2008; Malone et al., 2008; Mayberg et al., 2005; Schlaepfer et al., 2008).

MAINTAINING RESPECT FOR AUTONOMY IN DBS DECISION MAKING

While some neuropsychiatric patients are often clearly debilitated in their decision-making capacities (e.g., patients with advanced dementia or acute schizophrenia), others are not more impaired than any other patient group with a chronic, devastating disease (Dunn et al., 2011; Synofzik & Clausen, 2011). Thus, significant constraints to a patient's capacity of decision making might often result not so much from respective underlying neuropsychiatric disease, but from the nature of the patient's particular situation and the uncertainty in the risk-benefit-profile of current DBS interventions (Dunn et al., 2011).

The situation of most patients—neurologic or psychiatric (Synofzik & Clausen, 2011)—is often characterized by a severely disabling underling disease, without the prospects of amelioration by available treatments. This situation evokes a feeling of despair that might lead patients to minimize potential harms of DBS and/or to overestimate the chance of personal benefit from DBS. Such therapeutic misconceptions are particularly problematic in those neuropsychiatric diseases where, at the current point of time, no rigorous data are available on the true long-term benefit from DBS and where individual prognostic predictors for the chance and dimensions of benefit are missing. To ensure an adequate consent process in this situation, physicians have to make sure that patients understand the risks and the uncertainty of benefits of DBS for the given condition, and they must be very careful neither to overemphasize potential benefits nor to deemphasize risks. This requires that physicians step beyond typical consent forms, which are not individualized to specific patients' perceptions and expectations, and enter into a broader in-depth interaction with the patient (and, preferably, also the family members), elucidating and discussing their questions, concerns, motivations, and potential misperceptions (both in terms of risks or benefits) (Dunn et al., 2011).

Within such deliberative decision-making process, the physician should be explicitly asked to scrutinize the patient's preferences and to give recommendations (Emanuel & Emanuel, 1992; Synofzik & Schlaepfer, 2011). This kind of behavior should not be mistaken as paternalistic, but rather as a kind of beneficient guidance that seems justifiable based on the current lack of knowledge about potential benefits in psychiatric DBS and the risk of significant associated harm. It builds on a notion of patient autonomy, which is not understood in terms of absolutes but in terms of a balance between the principles of autonomy and beneficence (Bell et al., 2009; Emanuel & Emanuel, 1992; Synofzik & Schlaepfer, 2011).

FACING THE RAPID EXPANSION OF NEUROPSYCHIATRIC DBS

Following the first promising results from small cohort studies in patients with OCD, MD, or Tourette's syndrome, the applications of DBS in neuropsychiatric patients are rapidly increasing: It is now studied in patients with Alzheimers disease (Laxton et al., 2010), substance abuse (Muller et al., 2009), and schizophrenia (Mikell et al., 2009), as well as in subjects exhibiting obesity or violent behavior (Franzini et al., 2005). This rapid spread evokes strong ethical concerns regarding the possibility of premature expansion to new conditions without appropriate justification and research. Moreover, it might indicate a widespread acceptance of neuropsychiatric DBS treatment before adequate safety and efficacy data are obtained: It seems likely that already in the next few years psychiatric DBS will move from larger specialized academic hospitals to smaller centers and even private practice (Bell et al., 2009).

The quality of most psychiatric DBS data, however, needs to be improved before a more widespread use can be justified: Current findings largely rely on small cohorts or single case studies, variable methodologies, differing outcome measures, and variable methods of assessment within and between academic centers. In particular, the excessive reliance on single-patient case reports makes this nascent DBS domain highly vulnerable to bias. It opens up the possibility that only positive results (with often insufficiently controlled study designs) will be published, while negative data that might also have important implications will not be brought to the public. This well-known problem of selective reporting is already indirectly evidenced in this young field: Several single-case studies have been published only because of interesting secondary effects, whereas the primary outcome effects were not achieved (Schlaepfer & Fins, 2010). This indirectly indicates that there might be many other unpublished single-case DBS interventions for which the primary outcomes were not achieved and no interesting secondary effects were observed. Such a selective reporting will lead to a severe distortion of available evidence in this field, which might harm future neuropsychiatric patients (Dickersin & Rennie, 2003). Moreover, the overreliance on small case studies will favor a premature expansion to new conditions at the expense of rigorous well-controlled trials, which would be needed in a first step to validate the initial preliminary findings.

Due to the uncertainty of benefit and the lack of knowledge on short- and long-term risks, we have proposed that at least at the current stage, psychiatric DBS should only be offered in academic centers with extensive experience in treatment resistant neuropsychiatric disorders (Synofzik & Schlaepfer, 2011). This would allow physicians to enroll patients in standardized reporting protocols and larger controlled clinical trials. Moreover, all DBS interventions in single psychiatric patients who do not meet inclusion criteria of such study protocols should be registered in a comprehensive register of initiated clinical trials. Such a DBS registry, which needs to include in particular all single cases and less rigorous trials, would prevent selective reporting, aggregate data on beneficial and detrimental effects, and facilitate regulatory oversight (Synofzik, Fins, & Schlaepfer, 2011; epub ahead of print. See also Table 15.4).

Table 15.4 ADVANTAGES OF A COMPREHENSIVE TRIAL REGISTRY IN NEUROPSYCHIATRIC DBS THAT PLACES PARTICULAR EMPHASIS OF INCLUDING NOT ONLY LARGE WELL-CONTROLLED AND PROPSPECTIVE TRIALS, BUT ALSO SINGLE CASE STUDIES, RETROSPECTIVE OBSERVATIONS AND ANECDOTAL FINDINGS (ADAPTED FROM SYNOFZIK, FINS & SCHLAEPFER, 2011, EPUB AHEAD OF PRINT)

Advantages of a comprehensive trial registry in neuropsychiatric DBS

- accumulate beneficial effects
- aggregate adverse events
- collect long-term data
- collect data on ineffective DBS/failed trials
- give back the benefit of aggregated information to the patient
- identify side effects that might serve as therapeutic effects
- detect modifications of prespecified primary outcome measures
- coordination of research trials and research groups
- reduce idiosyncrasy in research and in IRB decision making
- provide orientation for IRB review
- compensate for the short-comings of early psychosurgery
- attract industry funders

DBS FOR NEUROENHANCEMENT?

Recent experiences from psychiatric patients have suggested that DBS of the limbic system can induce high levels of context-independent euphoria in a rapid and well titratable fashion by a simple linear increase of DBS voltage (Haq et al., 2010; Synofzik, Schlaepfer, & Fins, 2012). This impressive effect provokes a fundamental ethical question which extends well beyond the boundaries of medical ethics: If happiness can be easily induced by DBS of the brain's reward center, why should we limit its application to treatment of severe, treatment-refractory diseases? For example, in the future, there might be good reasons to use it for enhancing mood states in a patient who does not yet meet the criteria for severe (and even treatment-refractory) disease. Or one might use it to enhance mood states in a patient whose mood states are not affected by the underlying disease—for example, in an OCD patient showing chronic dysphoria unrelated to OCD. In fact, even affectively and cognitively intact persons without any psychiatric disease might want to use DBS for enhancing their mood. DBS of the limbic system might allow such persons to selectively choose stimulation parameters depending on how they want to feel, for example, calm for every day or more revved-up for a party (President's Council on Bioethics, 2004). Thus, the enhancement potential of DBS is no longer only speculative in nature and might in fact exceed the enhancement potential of current psychopharmacological substances (de Jongh, Bolt, Schermer, & Olivier, 2008; Synofzik, 2009). But can the use of DBS for enhancement purposes be ethically justified?

It does not seem that there are any convincing arguments that enhancing one's cognitive or mood states would be *intrinsically* wrong (Synofzik, 2009; Synofzik & Schlaepfer, 2008). Enhancement by neurotechnologies should, at least in principle, be viewed in the same general category as other means which humans have developed to improve themselves, such as education, exercise, information technology, or nutrition (Greely et al., 2008). These means are certainly not intrinsically unethical. They can be discouraged, however, based on *extrinsic* reasons, such as lack of proven efficacy, safety concerns, or compromising of a person's decisional competencies. It is for these reasons why DBS is not yet ripe for enhancement use: At the current stage, there is no evidence about potential benefits of DBS in healthy subjects, there is a high risk of severe harm, and the risk of altering decisional competencies is not yet sufficiently excluded (Synofzik, Schlaepfer, & Fins, 2012). But we should discuss this issue again in the future once there is sufficient evidence that DBS can indeed serve as highly beneficial, safe and cost-effective tool for improving mood and cognition.

CONCLUSIONS

In discussing the ethical challenges of functional neurosurgery, allusions to the history of early psychosurgery and vague fears of modifying personality traits are of only very limited value. They should be replaced by specific evidence-based analyses of beneficial effects, risks of harm, and potential impairments to decision-making capacities. These analyses need to be constantly updated and thoroughly performed not only on the level of a whole patient group, but also on the level of the individual patient. The assessment on the individual level needs to pay particular attention to the patient's personal expectations, his perceived benefit, and the psychosocial system he is living in. But before neuropsychiatric DBS treatment can move from experimental to clinical use, a lot of homework still needs to be done in DBS research. In particular, the favorable efficacy to side effect profile, which is indicated for a significant share of patients in the currently published preliminary studies, needs to be confirmed in larger and better-controlled studies. Standards of outcome reporting have to be established, and a comprehensive central data registry needs to be implemented with particular emphasis on including all single cases. In this way, neuropsychiatric DBS has the potential to become a promising and legitimate treatment approach for some of the most disabling disorders known to humanity.

NOTES

1. I am grateful to Dr. Cynthia S. Kubu for pointing this out to me.
2. Wayne Goodman, personal communication, DBS Expert Workshop Program, Bonn, January 20–21, 2011.

REFERENCES

Abelson, J. L., Curtis, G. C., Sagher, O., Albucher, R. C., Harrigan, M., Taylor, S. F., et al. (2005). Deep brain stimulation for refractory obsessive-compulsive disorder. *Biol Psychiatry, 57* (5), 510–516.

Advanced Neuromodulation Systems (ANS) Inc. (2007). BROADEN Clinical Study. A Study of a Non-Pharmacological Device for Depression. www.broadenstudy.com/sb/index.html.

Agid, Y., Schupbach, M., Gargiulo, M., Mallet, L., Houeto, J. L., Behar, C., et al. (2006). Neurosurgery in Parkinson's disease: The doctor is happy, the patient less so? *J Neural Transm Suppl,* (70), 409–414.

Appleby, B. S., Duggan, P. S., Regenberg, A., & Rabins, P. V. (2007). Psychiatric and neuropsychiatric adverse events associated with deep brain stimulation: A meta-analysis of ten years' experience. *Mov Disord, 22* (12), 1722–1728.

Beauchamp, T., & Childress, J. (2008). *Principles of biomedical ethics,* 6th ed. New York: Oxford University Press.

Bell, E., Mathieu, G., & Racine, E. (2009). Preparing the ethical future of deep brain stimulation. *Surg Neurol.*

Bell, E., Maxwell, B., McAndrews, M. P., Sadikot, A., & Racine, E. (2010). Hope and patients' expectations in deep brain stimulation: healthcare providers' perspectives and approaches. *J Clin Ethics, 21* (2), 112–124.

Bewernick, B. H., Hurlemann, R., Matusch, A., Kayser, S., Grubert, C., Hadrysiewicz, B., et al. (2009a). Nucleus accumbens deep brain stimulation decreases ratings of depression and anxiety in extremely treatment-resistant depression. submitted.

Bewernick, B. H., Hurlemann, R., Matusch, A., Kayser, S., Grubert, C., Hadrysiewicz, B., et al. (2009b). Nucleus accumbens deep brain stimulation decreases ratings of depression and anxiety in treatment-resistant depression. *Biol Psychiatry.*

Carmichael, S. T., & Price, J. L. (1996). Connectional networks within the orbital and medial prefrontal cortex of macaque monkeys. *J Comp Neurol, 371* (2), 179–207.

de Jongh, R., Bolt, I., Schermer, M., & Olivier, B. (2008). Botox for the brain: Enhancement of cognition, mood and pro-social behavior and blunting of unwanted memories. *Neurosci Biobehav Rev, 32* (4), 760–776.

Deuschl, G., Schade-Brittinger, C., Krack, P., Volkmann, J., Schafer, H., Botzel, K., et al. (2006). A randomized trial of deep-brain stimulation for Parkinson's disease. *N Engl J Med, 355* (9), 896–908.

Dickersin, K., & Rennie, D. (2003). Registering clinical trials. *JAMA, 290* (4), 516–523.

Drapier, D., Drapier, S., Sauleau, P., Haegelen, C., Raoul, S., Biseul, I., et al. (2006). Does subthalamic nucleus stimulation induce apathy in Parkinson's disease? *J Neurol, 253* (8), 1083–1091.

Dunn, L. B., Holtzheimer, P. E., Hoop, J. G., Mayberg, H., Weiss Roberts, L., & Appelbaum, P. S. (2011). Ethical issues in deep brain stimulation research for treatment-resistant depression: Focus on risk and consent. *AJOB Neuroscience, 2* (1), 29–36.

Emanuel, E. J., & Emanuel, L. L. (1992). Four models of the physician-patient relationship. *JAMA, 267* (16), 2221–2226.

Fins, J. J. (2003). From psychosurgery to neuromodulation and palliation: history's lessons for the ethical conduct and regulation of neuropsychiatric research. *Neurosurg Clin N Am, 14* (2), 303–319, ix–x.

Ford, P. J., & Henderson, J. M. (2006). Functional neurosurgical intervention: neuroethics in the operating room. In J. Iles (Ed.), *Neuroethics: Definig the issues in theory, practice, and policy* (pp. 213–228). Oxford: Oxford University Press.

Ford, P. J., & Kubu, C. S. (2005). Caution in leaping from functional imaging to functional neurosurgery. *Am J Bioeth, 5* (2), 23–25; discussion W23–W24.

Ford, P. J., & Kubu, C. S. (2006). Stimulating debate: eEhics in a multidisciplinary functional neurosurgery committee. *J Med Ethics, 32* (2), 106–109.

Franzini, A., Marras, C., Ferroli, P., Bugiani, O., & Broggi, G. (2005). Stimulation of the posterior hypothalamus for medically intractable impulsive and violent behavior. *Stereotact Funct Neurosurg, 83* (2–3), 63–66.

Funkiewiez, A., Ardouin, C., Krack, P., Fraix, V., Van Blercom, N., Xie, J., et al. (2003). Acute psychotropic effects of bilateral subthalamic nucleus stimulation and levodopa in Parkinson's disease. *Mov Disord, 18* (5), 524–530.

Glannon, W. (2008). Neurostimulation and the minimally conscious state. *Bioethics, 22* (6), 337–345.

Glannon, W. (2009). Stimulating brains, altering minds. *J Med Ethics, 35* (5), 289–292.

Goodman, W. K., & Insel, T. R. (2009). Deep brain stimulation in psychiatry: concentrating on the road ahead. *Biol Psychiatry, 65* (4), 263–266.

Greely, H., Sahakian, B., Harris, J., Kessler, R. C., Gazzaniga, M., Campbell, P., et al. (2008). Towards responsible use of cognitive-enhancing drugs by the healthy. *Nature, 456* (7223), 702–705.

Greenberg, B. D., Malone, D. A., Friehs, G. M., Rezai, A. R., Kubu, C. S., Malloy, P. F., et al. (2006b). Three-year outcomes in deep brain stimulation for highly resistant obsessive-compulsive disorder. *Neuropsychopharmacology, 31* (11), 2384–2393.

Greenberg, B. D., Malone, D. A., Friehs, G. M., Rezai, A. R., Kubu, C. S., Malloy, P. F., et al. (2006a). Three-year outcomes in deep brain stimulation for highly resistant obsessive-compulsive disorder. *Neuropsychopharmacology, 31* (11), 2394.

Haber, S. N. (2003). The primate basal ganglia: parallel and integrative networks. *J Chem Neuroanat, 26* (4), 317–330.

Haq, I. U., Foote, K. D., Goodman, W. G., Wu, S. S., Sudhyadhom, A., Ricciuti, N., et al. (2010). Smile and laughter induction and intraoperative predictors of response to deep brain stimulation for obsessive-compulsive disorder. *Neuroimage.*

Hariz, M. I., Blomstedt, P., & Zrinzo, L. (2010). Deep brain stimulation between 1947 and 1987: the untold story. *Neurosurg Focus, 29* (2), E1.

Hildt, E. (2006). Electrodes in the brain: Some anthropological and ethical aspects of deep brain stimulation. *International Review of Information Ethics, 5* (9), 33–39.

Ho, A. K., Gilbert, A. S., Mason, S. L., Goodman, A. O., & Barker, R. A. (2009). Health-related quality of life in Huntington's disease: Which factors matter most? *Mov Disord, 24* (4), 574–578.

Isaias, I. U., Alterman, R. L., & Tagliati, M. (2008). Outcome predictors of pallidal stimulation in patients with primary dystonia: the role of disease duration. *Brain, 131* (Pt 7), 1895–1902.

Kang, G. A., Heath, S., Rothlind, J., & Starr, P. A. (2011). Long-term follow-up of pallidal deep brain stimulation in two cases of Huntington's disease. *J Neurol Neurosurg Psychiatry, 82* (3), 272–277.

Kapoor, S. (2009). Subthalamic nucleus stimulation in severe obsessive-compulsive disorder. *N Engl J Med, 360* (9), 931–932; author reply 932.

Keller, M. B., Hirschfeld, R. M., Demyttenaere, K., & Baldwin, D. S. (2002). Optimizing outcomes in depression: focus on antidepressant compliance. *Int Clin Psychopharmacol, 17* (6), 265–271.

Kringelbach, M. L., & Aziz, T. Z. (2009). Deep brain stimulation: avoiding the errors of psychosurgery. *JAMA, 301* (16), 1705–1707.

Kubu, C. S. (2011). *Outcomes: What counts as better and who decides it?* Paper presented at the Presentation at the DBS Expert Workshop, Bonn, January 20–21, 2011.

Kuhn, J., Bauer, R., Pohl, S., Lenartz, D., Huff, W., Kim, E. H., et al. (2009). Observations on unaided smoking cessation after deep brain stimulation of the nucleus accumbens. *Eur Addict Res, 15* (4), 196–201.

Kuhn, J., Lenartz, D., Huff, W., Lee, S., Koulousakis, A., Klosterkoetter, J., et al. (2007). Remission of alcohol dependency following deep brain stimulation of the nucleus accumbens: Valuable therapeutic implications? *J Neurol Neurosurg Psychiatry, 78* (10), 1152–1153.

Kuhn, J., Lenartz, D., Mai, J. K., Huff, W., Klosterkoetter, J., & Sturm, V. (2008). Disappearance of self-aggressive behavior in a brain-injured patient after deep brain stimulation of the hypothalamus: Technical case report. *Neurosurgery, 62* (5), E1182; discussion E1182.

Kupsch, A., Benecke, R., Muller, J., Trottenberg, T., Schneider, G. H., Poewe, W., et al. (2006). Pallidal deep-brain stimulation in primary generalized or segmental dystonia. *N Engl J Med, 355* (19), 1978–1990.

Laxton, A. W., Tang-Wai, D. F., McAndrews, M. P., Zumsteg, D., Wennberg, R., Keren, R., et al. (2010). A phase I trial of deep brain stimulation of memory circuits in Alzheimer's disease. *Ann Neurol, 68* (4), 521–534.

Lozano, A. M., Mayberg, H. S., Giacobbe, P., Hamani, C., Craddock, R. C., & Kennedy, S. H. (2008). Subcallosal cingulate gyrus deep brain stimulation for treatment-resistant depression. *Biol Psychiatry, 64* (6), 461–467.

Mallet, L., Polosan, M., Jaafari, N., Baup, N., Welter, M. L., Fontaine, D., et al. (2008). Subthalamic nucleus stimulation in severe obsessive-compulsive disorder. *N Engl J Med, 359* (20), 2121–2134.

Malone, D. A., Jr., Dougherty, D. D., Rezai, A. R., Carpenter, L. L., Friehs, G. M., Eskandar, E. N., et al. (2008). Deep brain stimulation of the ventral capsule/ventral striatum for treatment-resistant depression. *Biol Psychiatry.*

Mantione, M., van de Brink, W., Schuurman, P. R., & Denys, D. (2010). Smoking cessation and weight loss after chronic deep brain stimulation of the nucleus accumbens: Therapeutic and research implications: case report. *Neurosurgery, 66* (1), E218; discussion E218.

Mayberg, H. S., Lozano, A. M., Voon, V., McNeely, H. E., Seminowicz, D., Hamani, C., et al. (2005). Deep brain stimulation for treatment-resistant depression. *Neuron, 45* (5), 651–660.

Medtronic Inc. (2006). Medtronic to Pursue Major Clinical Trial of Deep Brain Stimulation as Depression Treatment. http://findarticles.com/p/articles/mI_m0EIN/ is_2006_April_25/aI_n26838044?tag=content-inner;col1.

Mesnage, V., Houeto, J. L., Welter, M. L., Agid, Y., Pidoux, B., Dormont, D., et al. (2002). Parkinson's disease: neurosurgery at an earlier stage? *J Neurol Neurosurg Psychiatry, 73* (6), 778–779.

Mikell, C. B., McKhann, G. M., Segal, S., McGovern, R. A., Wallenstein, M. B., & Moore, H. (2009). The hippocampus and nucleus accumbens as potential therapeutic targets for neurosurgical intervention in schizophrenia. *Stereotact Funct Neurosurg, 87* (4), 256–265.

Miller, G. (2009). Neuropsychiatry. Rewiring faulty circuits in the brain. *Science, 323* (5921), 1554–1556.

Müller, S., & Christen, M. (2011). Deep brain stimulation in Parkinsonian patients— Ethical evaluation of cognitive, affective and behavioral sequelae. *AJOB Neuroscience, 2* (1), 3–13.

Muller, U. J., Sturm, V., Voges, J., Heinze, H. J., Galazky, I., Heldmann, M., et al. (2009). Successful treatment of chronic resistant alcoholism by deep brain stimulation of nucleus accumbens: First experience with three cases. *Pharmacopsychiatry, 42* (6), 288–291.

Nutt, D. J., & Sharpe, M. (2008). Uncritical positive regard? Issues in the efficacy and safety of psychotherapy. *J Psychopharmacol, 22* (1), 3–6.

Nuttin, B. J., Gabriels, L. A., Cosyns, P. R., Meyerson, B. A., Andreewitch, S., Sunaert, S. G., et al. (2003). Long-term electrical capsular stimulation in patients with obsessive-compulsive disorder. *Neurosurgery, 52* (6), 1263–1272; discussion 1272–1264.

Okun, M. S., & Foote, K. D. (2004). A mnemonic for Parkinson disease patients considering DBS: a tool to improve perceived outcome of surgery. *Neurologist, 10* (5), 290.

Piasecki, S. D., & Jefferson, J. W. (2004). Psychiatric complications of deep brain stimulation for Parkinson's disease. *J Clin Psychiatry, 65* (6), 845–849.

Porta, M., Brambilla, A., Cavanna, A. E., Servello, D., Sassi, M., Rickards, H., et al. (2009). Thalamic deep brain stimulation for treatment-refractory Tourette syndrome: two-year outcome. *Neurology, 73* (17), 1375–1380.

President's Council on Bioethics (2004). Transcript of session on "Neuroscience, Brain and Behavior V: Deep Brain Stimulation." http://bioethics.georgetown.edu/pcbe/ transcripts/june04/session6.html.

Racine, E., Waldman, S., Palmour, N., Risse, D., & Illes, J. (2007). "Currents of hope": Neurostimulation techniques in U.S. and U.K. print media. *Camb Q Healthc Ethics, 16* (3), 312–316.

Richard, I. H. (2007). Depression and apathy in Parkinson's disease. *Curr Neurol Neurosci Rep, 7* (4), 295–301.

Russo, F. (2007). How to change a personality. *Time,* January 18. www.time.com/time/magazine/article/0,9171,1580389,00.html.

Schiff, N. D., Giacino, J. T., Kalmar, K., Victor, J. D., Baker, K., Gerber, M., et al. (2007). Behavioural improvements with thalamic stimulation after severe traumatic brain injury. *Nature, 448* (7153), 600–603.

Schlaepfer, T. E., Cohen, M. X., Frick, C., Kosel, M., Brodesser, D., Axmacher, N., et al. (2008). Deep brain stimulation to reward circuitry alleviates anhedonia in refractory major depression. *Neuropsychopharmacology, 33* (2), 368–377.

Schlaepfer, T. E., & Fins, J. J. (2010). Deep brain stimulation and the neuroethics of responsible publishing: When one is not enough. *JAMA, 303* (8), 775–776.

Schupbach, M., Gargiulo, M., Welter, M. L., Mallet, L., Behar, C., Houeto, J. L., et al. (2006). Neurosurgery in Parkinson disease: A distressed mind in a repaired body? *Neurology, 66* (12), 1811–1816.

Schupbach, W. M., Maltete, D., Houeto, J. L., du Montcel, S. T., Mallet, L., Welter, M. L., et al. (2007). Neurosurgery at an earlier stage of Parkinson disease: A randomized, controlled trial. *Neurology, 68* (4), 267–271.

Schurr, P. H., & Merrington, W. R. (1978). The Horsley-Clarke stereotaxic apparatus. *Br J Surg, 65* (1), 33–36.

Shapira, N. A., Okun, M. S., Wint, D., Foote, K. D., Byars, J. A., Bowers, D., et al. (2006). Panic and fear induced by deep brain stimulation. *J Neurol Neurosurg Psychiatry, 77* (3), 410–412.

Smeding, H. M., Speelman, J. D., Koning-Haanstra, M., Schuurman, P. R., Nijssen, P., van Laar, T., et al. (2006). Neuropsychological effects of bilateral STN stimulation in Parkinson disease: A controlled study. *Neurology, 66* (12), 1830–1836.

Soulas, T., Gurruchaga, J. M., Palfi, S., Cesaro, P., Nguyen, J. P., & Fenelon, G. (2008). Attempted and completed suicides after subthalamic nucleus stimulation for Parkinson's disease. *J Neurol Neurosurg Psychiatry, 79* (8), 952–954.

Synofzik, M. (2009). Ethically justified, clinically applicable criteria for physician decision-making in psychopharmacological enhancement. *Neuroethics, 2* (2), 89–102.

Synofzik, M., & Clausen, J. (2011). The ethical differences between psychiatric and neurologic DBS: smaller than we think? *AJOB Neuroscience, 2* (1), 37–39.

Synofzik, M., Fins, J. J., & Schlaepfer, T. E. (2011; epub ahead of print). A neuromodulation experience registry for deep brain stimulation studies in psychiatric research Brain Stimulation.

Synofzik, M., & Schlaepfer, T. E. (2008). Stimulating personality: Ethical criteria for deep brain stimulation in psychiatric patients and for enhancement purposes. *Biotechnol J, 3* (12), 1511–1520.

Synofzik, M., & Schlaepfer, T. E. (2010). Neuromodulation—ECT, rTMS, DBS. In H. Helmchen & N. Sartorius (Eds.), *Ethics in Psychiatry. European contributions.* (pp. 299–320). Heidelberg: Springer.

Synofzik, M., & Schlaepfer, T. E. (2011). Electrodes in the brain-Ethical criteria for research and treatment with deep brain stimulation for neuropsychiatric disorders. *Brain Stimul, 4* (1), 7–16.

Synofzik, M., Schlaepfer, T. E., & Fins, J. J. (2012). How happy is too happy? Euphoria, Neuroethics and Deep Brain Stimulation of the Nucleus Accumbens. *AJOB Neuroscience, 3*(1), 30–36.

Torres, A. R., Prince, M. J., Bebbington, P. E., Bhugra, D., Brugha, T. S., Farrell, M., et al. (2006). Obsessive-compulsive disorder: Prevalence, comorbidity, impact, and help-seeking in the British National Psychiatric Morbidity Survey of 2000. *Am J Psychiatry, 163* (11), 1978–1985.

Videnovic, A., & Metman, L. V. (2008). Deep brain stimulation for Parkinson's disease: Prevalence of adverse events and need for standardized reporting. *Mov Disord, 23* (3), 343–349.

Voges, J., Hilker, R., Botzel, K., Kiening, K. L., Kloss, M., Kupsch, A., et al. (2007). Thirty days complication rate following surgery performed for deep-brain-stimulation. *Mov Disord, 22* (10), 1486–1489.

Welter, M. L., Houeto, J. L., Tezenas du Montcel, S., Mesnage, V., Bonnet, A. M., Pillon, B., et al. (2002). Clinical predictive factors of subthalamic stimulation in Parkinson's disease. *Brain, 125* (Pt 3), 575–583.

Witt, K., Kuhn, J., Timmermann, L., Zurowski, M., & Woopen, C. (2011). Deep brain stimulation and the search for identity. *Neuroethics*, in press.

Yamamoto, T., Kobayashi, K., Kasai, M., Oshima, H., Fukaya, C., & Katayama, Y. (2005). DBS therapy for the vegetative state and minimally conscious state. *Acta Neurochir Suppl, 93*, 101–104.

16

Transcranial Magnetic Stimulation

Future Prospects and Ethical Concerns in Treatment and Research

JARED HORVATH, JENNIFER PEREZ, LACHLAN FORROW,
FELIPE FREGNI, AND ALVARO PASCUAL-LEONE

Transcranial Magnetic Stimulation (TMS) is a noninvasive neurostimula-tory and neuromodulatory technique increasingly utilized in clinics and research laboratories around the world. Exploiting the properties of electro-magnetic induction, TMS can transiently or lastingly modulate cortical excit-ability (either increasing or decreasing it) via the application of localized, time-varying magnetic field pulses. Until recently, the ethical considerations guiding the practice and application of TMS were largely concerned with aspects of subject safety and patient population in controlled clinical trials or experimental protocols. While safety remains of paramount importance, the recent approval by the Food and Drug Administration (FDA) in the United States of the Neuronetics NeuroStar TMS device for the treatment of certain patients with medication-resistant depression has engendered a surfeit of unexamined ethical concerns. Some of these are nuanced elaborations of pre-viously addressed issues, but others—such as matters of training and certifi-cation, marketing, and possible off-label use—represent recent and complex concerns. These issues are likely to be relevant to a rapidly growing number of patients, as the possible uses for TMS and other developing forms of brain stimulation (both invasive and noninvasive) expand to include the treatment of a wide range of neuropsychiatric conditions. This chapter provides an over-view of these emerging issues and discusses ways in which some of them might best be addressed.

We begin this chapter with a brief look at the mechanistic and physical parame-ters integral to the TMS device. We continue with a review of the development and progression of brain stimulation from its infancy in the 1700s to the formation of TMS in 1985. From there, we move to an exploration of the ethical considerations relevant during each of the major evolutionary stages of TMS (between 1985 and the present). We conclude with a look at the future prospects of TMS and the concomitant ethical issues.

A TMS PRIMER

TMS is a relatively new neurophysiologic technique that allows for noninvasive stimulation of the human brain (Pascual-Leone et al., 2002; Walsh & Pascual-Leone, 2003). Since its introduction in present form approximately 25 years ago, TMS has proven safe so long as adequate precautions are taken and safety guidelines are adhered to. Often used in conjunction with other neuroscientific methods, TMS can be used to study intracortical, cortico-cortical, and cortico-subcortical interactions. Such studies, in turn, may aid in the determination of causal relationships between focal brain activity and emergent behavior. In addition, repetitive transcranial magnetic stimulation (rTMS) has been shown to modulate brain activity beyond the duration of application and to induce changes in brain function that show early evidence of being efficacious in the treatment of numerous neurological and psychiatric illnesses.

The physical principles that underlie TMS were discovered in 1831 by English physicist and chemist Michael Faraday. Put simply, Faraday observed that a pulse of electric current sent through a wire coil generates a magnetic field. The rate of change of this magnetic field determines the induction of a secondary current in a nearby conductor. In TMS, the stimulating coil is held over a subject's head and, as a brief pulse of current is passed through it, the resulting magnetic field passes through the subject's scalp and skull without much attenuation. This varying magnetic field induces a current in the subject's brain that can depolarize neurons and generate various physiologic and behavioral effects depending on the targeted brain area. Therefore, in TMS, neural elements are not primarily affected by the exposure to a magnetic field, but rather by the current induced in the brain by electrodeless, noninvasive electric stimulation via electro-magnetic induction. In short, by means of magnetic fields, it is possible to induce relatively large electrical currents in a conscious subject without inducing significant discomfort.

The design of magnetic stimulators is relatively straightforward, consisting of a main unit and a stimulating coil. The main unit is composed of a charging system, one or more energy storage capacitors, a discharge switch, and circuitry used to control pulse shape, energy recovery, and other variable functions (Figure 16.1). The factors essential to the effectiveness of a magnetic stimulator are the speed of magnetic field rise time and the maximization of the peak coil energy. Therefore, large energy storage capacitors and efficient energy transfer from the capacitor to the coil are important (typically energy storage capacity is around 2000J with a 500J transfer from the capacitor to the stimulating coil in less than 100 microseconds via a *thyristor*, an electrical device capable of switching large currents in a short time). The peak discharge current needs to be several thousand amperes in order to induce currents in the brain of sufficient magnitude to depolarize neural elements (about 10 mA/cm^2).

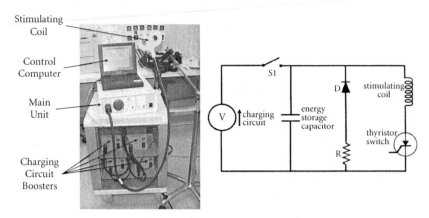

Figure 16.1 (Left) TMS equipment, including the charging circuit boosters, the main stimulating unit, the control computer, and the stimulating coil. (Right) Diagram of the electric circuit used to induce the TMS pulse.

During TMS, only the stimulating apparatus comes into contact with the subject. The stimulating apparatus consists of one or more well-insulated coils of copper wire (frequently housed in a molded plastic cover). These coils can be arrayed in a variety of shapes and sizes. The geometry of the coil determines the focality of brain stimulation. For instance, figure-eight coils are constructed with two sets of windings placed side-by-side, which interact to create a relatively focal means of TMS brain stimulation. Current knowledge, largely based on mathematical modeling, suggests that TMS with a small figure-eight coil (each wing of the coil measuring approx. 4 cm in diameter) can achieve a spatial resolution of about 5 mm^3 of brain volume (Wagner et al., 2004). Stimulation is restricted to rather superficial layers in the convexity of the brain. Deeper penetration is possible using specially shaped coils, for example the H-coil (Brainsway Ltd. Israel). However, selective direct deep brain stimulation is not possible, and superficial areas of the brain closer to the plane of the coil are always exposed to greater induced currents than deeper brain regions. Digitization of the anatomical structure of the subject's head and registration of the TMS stimulation sites onto an MRI of the subject's brain addresses issues of anatomical specificity by identifying the exact location of the intended target in each experimental subject. The use of optical digitization and frameless stereotactic systems represent a further improvement by providing online information about the brain area targeted by a given coil position on the scalp (Figure 16.2).

The precise mechanisms underlying the brain effects of TMS are largely unknown (Pascual-Leone et al., 2002; Robertson et al., 2003; Wagner et al., 2007; Wassermann et al., 2008). Currents induced in the brain by TMS primarily flow parallel to the plane of the stimulation coil (approximately parallel to the brain's cortical surface when the coil is held tangentially to the scalp). Therefore, in contrast with electrical cortical stimulation, TMS preferentially activates neural elements oriented horizontally to the brain surface. Exactly which neural elements

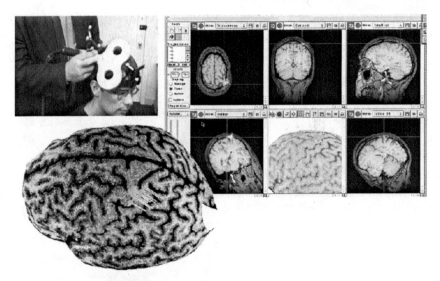

Figure 16.2 (Top Left) TMS being applied by an expirimenter. Both the coil and the subject are affixed with a tracking device allowing head/device location to be monitered in real time. (Top Right) A subjects MRI, linked with the tracking device, displaying the cortical area being targeted. (Bottom Left) Real time overlay between TMS position and cortical area stimulated.

are activated by TMS remains unclear and, in fact, might be variable across different brain regions and different subjects. However, recent animal studies are starting to probe these important mechanistic questions (Allen et al., 2007; Wagner et al., 2007; Valero-Cabre et al., 2008).

THE PRE-TMS HISTORY: AN ETHICAL PERSPECTIVE

The field of electrophysiology was born in 1771 with the discovery of bioelectricity by the Italian physician and physicist Luigi Galvani. Soon thereafter, prompted by the actions of Galvani's nephew Giovanni Aldini, important ethical questions began to emerge. So enamored was Aldini with the potential of his uncle's discovery that in 1802 he began touring the European countryside promoting his belief that bioelectricity could unlock the secrets to dead tissue reanimation. To prove this theory, Aldini staged public demonstrations during which he would force muscular contraction via the application of direct electric current to dead dogs, mules, and, in 1804, the rigor mortis-affected body of the recently executed convict George Forester (Parent, 2004). While Aldini aimed to show that bioelectricity *could* regenerate dead flesh, did he ever consider the ethical aspects of whether his theory *should* be explored? Indeed, many scientific advancements harness great potential—but that scientific potential comes with great ethical responsibility. To what extent may researchers go on to test and try to falsify their theories? How might a system of ethics be maintained, not only to protect research subjects, but also to protect the integrity of, and respect for, science itself?

The history books are rife with examples of emergent technologies that, after initial unethical use, quickly fall out of favor and, eventually, out of mind. However, due to the enormous potential of electrical stimulation, scientific interest in the technique remained strong even as tolerance of Aldini's macabre public spectacles dwindled. Over the ensuing decades, many great physicists and biologists explored the importance of and mechanisms behind the electricity generated inside animal muscle and nerve fibers. By 1874 this scientific expedition reached an apex: electrical stimulation of the human brain. Suffering from a two-inch diameter cancerous ulcer in her skull, Mary Rafferty, a Cincinnati housewife, decided to seek help from American physician Robert Bartholow. Before patching up his patient's wound, Bartholow, inspired by previous stimulatory work in animals done by David Ferrier, opted to apply electrical current to Rafferty's exposed dura. After the successful induction of slight muscular twitches, Bartholow increased the applied current until distress, convulsion, and eventually, coma were reported (Zago et al., 2008; Harris & Almerigi, 2009). Rafferty would die 72 hours later, her death bringing about another consequential dilemma pervasive throughout medical science: In the quest for knowledge, too often experimenters can become blind to the potentially deleterious effects of their actions. Overzealousness can, and often does, cloud typically sound judgment, illustrating the importance of and need for outside, independent risk-to-benefit assessment. Today such assessment is provided by the Institutional Review Boards (IRB). However, the responsibility for the proper conduct and sound ethical grounding of a given experiment ultimately remains in the hands of the investigators.

Another important milestone in the field of brain stimulation came with the development of Electroconvulsive Therapy (ECT) in 1937. This evolution ushered the field into the modern medical device age. Originally developed by Italian physicians Cerletti and Bini to assuage the manic symptoms of schizophrenic inpatients, ECT quickly gained popularity as an effective treatment for many psychological conditions. The cursory utilization of ECT as a psychiatric panacea led to a plethora of adverse (and possibly preventable) psychological and physical side effects, thereby engendering a strong negative public attitude toward this particular therapy.

In 1976, due in no small part to the enormous public backlash against ECT and similar treatments, the FDA opted to assume regulatory control over all emerging medical devices. Interestingly, ECT itself, along with several other devices for brain stimulation, remains today out of the purview of FDA regulation, as it was on the market prior to the establishment of the FDA. It is worth noting that ECT is considered by many to be the most effective treatment for depression (Surgeon General, 1999); however, the therapy remains hampered by the unfavorable public opinion generated by the un-regulated expansion of ECT during its nascence. An interesting point: Lest readers should assume all regulation is beneficial, it is important to consider the ramifications of excessive regulation. For example, the recent trend of *medical tourism* (international travel undertaken to obtain medical treatment not available in one's home country) has been blamed by many on the *over*regulation of the medical field by the FDA. It is thus important to keep in mind that some responses to ethical problems that have been identified can in turn create other, new ethical problems.

In 1972, another incident in the annals of brain-stimulation raised relevant ethical concerns. Building upon Jose Delgado's treatise *Toward a Psychocivilized Society*, which advanced his ethically questionable hypothesis that aggressive tendencies inside the criminal element could be eradicated via deep-brain electrical stimulation, two doctors from Tulane University took it upon themselves to "cure" a patient of his homosexuality. Combining electrical stimulation of patient B-19's septum with forced heterosexual interactions (provided by a female prostitute), Moan and Heath reported a full 10-month eradication of homosexual behavior (Moan & Heath, 1972). Here we see what many modern practitioners consider the darkest hour for brain stimulation, not only because of the experiment itself, but also because of the support and funding it received.

Although his work advanced questionable hypotheses, Delgado made important scientific and technical advances in the field. He was a pioneer in brain-implant technology, and many of his discoveries helped to establish the implantable stimulatory devices used today to treat epilepsy, dystonia, and Parkinson's disease. In addition, he showed that stimulation of the motor cortex resulted in physical reaction (e.g., movement of the limbs) and that stimulation of different regions of the limbic system could induce a host of emotions, ranging from fear to lust, and even euphoria, which raised important therapeutic questions about the potential of stimulation to treat depression.

In perhaps his most famous experiment, Delgado implanted chips in the brains of fighting bulls at a ranch in Cordoba, Spain in 1963. Holding a handheld transmitter that activated the brain chips, Delgado stood in the bullring and was able to halt the charging bulls with a press of a button (Horgan, 2005). This event and others are temporally organized and contextualized in Figure 16.3: More specifically, ethically relevant milestones in both electrical and magnetic brain stimulation leading up to the advent of TMS are itemized in hopes of illustrating that technological advances necessarily generate additional ethical issues and concerns.

While we consider the ethics of the past of brain stimulation, it is important to remember that we reflect on the past with our current understanding of disease of mind. Regarding the work of Delgado and Moan and Heath, few people today would argue that sexual orientation is a disease. That is, our ethical perception is inextricably linked to the way society defines the core concepts of health and disease. It is paramount, therefore, not only to consider the ethics of the past when debating the ethics of TMS future, but also to consider the sociological factors that affect our past and present ethical notions.

THE EVOLUTION OF TMS: FOUNDATIONS FOR AN ETHICAL FRAMEWORK

The first electro-magnetic stimulation device capable of reliable transcranial brain stimulation was developed by Anthony Barker and his colleagues at the United Kingdom's University of Sheffield and was introduced to the world in 1984–1985. Since that time, the device has gone through several major stages of evolution, each presenting unique achievements and considerations (see Table 16.1).

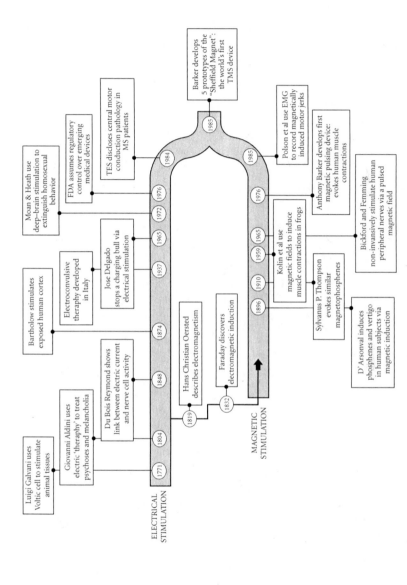

Figure 16.3 A timeline of notable uses and discoveries in electrical and magnetic stimulation leading up to the development of the first reliable magnetic brain stimulator in 1984–1985.

Table 16.1 AN ETHICAL OVERVIEW OF TMS EVOLUTION

	TMS Milestones	**Safety/Ethical Concerns**
STAGE I	• First use of TMS to create virtual lesion: used to test cognitive function/deficit • Figure-8 Coil developed: aids in focal stimulation	• Seizure Induction • Short-Term Neurological Effects
STAGE II	• First use of the repetitive TMS protocol • First use of the paired-pulse protocol	• First reports of TMS linked seizures emerge • Phase I safety studies begin
STAGE III	• First rTMS proof-of-principle trial undertaken with depression • TMS combine with fMRI BOLD imaging • TMS combine with PET imaging	• Long-Term neurological effects • International safety guidelines outlined • TMS/Drug interactions prove negligible
STAGE IV	• First use of Theta-Burst protocol • Neurostar study for depression commences	• Varied marketing and clinical issues • Medium tourism for TMS treatment of depression • Patient care
STAGE V AND VI	• Neurostar okayed by the FDA for treatment of specific forms of depression	• Advertising concerns • Off-lable treatment concerns • Long-term treatment data can/should be sought

Stage I: Establishing Diagnostic Utility

The early years of TMS can perhaps best be described with the phrase *tentative growth proceeded under a cautious temperament.* Beginning in the late 1970s, a handful of clinicians explored the utility of applying high-intensity electrical currents (*transcranial electrical stimulation*, TES) to diagnostically measure and track progressive central motor conduction time in patients suffering from multiple sclerosis. This technique, although efficacious, was limited by the extreme discomfort it elicited in patients and subjects. The introduction of TMS into the medical field provided a propitious solution to this problem. Via single-pulse stimulation, practitioners could perform similar conduction time diagnostic tests in a much safer, far less-painful manner (Green, 1997). Over the proceeding years, many physicians adopted the indicative use of TMS (along the way, several mechanistic innovations were developed, including the figure-of-eight coil), but not many intensively explored its functional use beyond the realm of diagnostics in the setting of very controlled clinical neurophysiology and motor control laboratories.

As can be predicted, the major ethical considerations during this first stage of TMS largely concerned overall device safety. The possibility of seizure induction loomed large, as did the prospect of generating acute side effects. Because TMS originated as a diagnostic and investigational device, not many animal models were used to study the technique. Because of this, and perhaps borrowing a lesson from the stimulation history books, TMS practitioners promoted the field rather cautiously (as compared with ECT), taking great care not only in the stimulatory paradigms utilized but also in the observation and tracking of patient progress and reported side effects. However, it should be noted that this progression was not as stringently regulated as would be expected for an emerging pharmaceutical intervention.

Stage II: Exploring Therapeutic Possibilities

Early generation single-pulse TMS was utilized not only as a diagnostic clinical neurophysiology tool, but also, through carefully designed experiments, as a means of inquiring about the role of different brain areas in specific behaviors. In 1985, Barker et al. demonstrated that TMS could be used to elicit an involuntary finger movement following stimulation of the motor cortex (Barker et al., 1985). Four years later, in 1989, Amassian et al. (1989) showed that single-pulse TMS, applied to the occipital cortex during the presentation of a visual stimulus, could induce errors in the detection of letters.

In the early 1990s, practitioners developed a rapid-pulse stimulatory paradigm known today as repetitive transcranial magnetic stimulation (rTMS). Qualitatively different than single-pulse methodology, rTMS offered investigators the opportunity to effect a more sustained neurological intervention, thereby allowing for deeper and more nuanced studies of higher cognitive functions and complex neural interactions. For instance, in 1991, the journal *Neurology* published an article by Alvaro Pascual-Leone et al. titled *Induction of Speech Arrest and Counting Errors with Rapid-rate Transcranial Magnetic Stimulation*. This paper, one of the first published utilizing a repetitive transcranial magnetic stimulation (rTMS) paradigm, demonstrated the ability of the repetitive paradigm to effectively locate and intervene with complex cognitive processes.

In 1994, Pascual-Leone et al. published an article in the scientific journal *Brain*, entitled *Response to Rapid-Rate Transcranial Magnetic Stimulation of the Human Motor Cortex*. This study measured the effects of rTMS on the latency and amplitude of motor evoked potentials (MEP) in healthy subjects. Interestingly, Pascual-Leone et al. found that the repetitive stimulation paradigm engendered a lasting cortical effect that remained *after* stimulation cessation (lasting between three and four minutes in this study). This finding revealed the potential of rTMS to induce sustained cognitive effects (as opposed to the more transient effects induced by single-pulse and virtual lesion studies) conducive to the field of therapeutic medicine.

This expansion of the field (from single-pulse diagnostic exploration to repetitive-pulse therapeutic intervention) increased the number of TMS ethical concerns nearly exponentially. In addition to lasting matters of safety, practitioners and clinicians were suddenly confronted with innumerous clinical considerations, including which neurological and psychiatric disorders might be responsive to stimulatory treatment, issues of patient versus health control population response patterns, and potential interactions between rTMS and medications.

Typically, when a new device or pharmaceutical shows therapeutic promise, a barrage of increasingly descriptive, slowly compounding safety studies are undertaken. Single-pulse TMS had forgone this phase as it was being used *diagnostically* and had maintained a stellar safety record for nearly five years. Repetitive TMS, on the other hand, is a fundamentally different procedure. Therefore, atop the already established single-pulse safety concerns, a number of additional considerations uniquely relevant to rTMS required attention. Under the pressure of an ever-increasing number of active rTMS studies, practitioners were confronted with the daunting task of playing ethical catch-up, conducting an intense series of phase I studies intended to determine the immediate and short-term consequences of repetitive pulse stimulation (Counter et al., 1990; Dhuna et al., 1991; Pascual-Leone et al., 1993). Whereas most medical phase I studies are enacted only *after* extensive lab testing and animal modeling, TMS had evolved outside of the typical developmental procedure. Consequently, in addition to the standard phase I questions involving human subject safety, researchers were forced to explore the effects of different stimulatory paradigms, the consequences of various stimulatory sites, and the impact of procedural methods on the clinical population.

Stage III: The Clinical Push

One of the first proof-of-principle TMS studies specifically focused on a therapeutic application of repetitive TMS in a neuropsychiatric disease was published by Klobinger et al. (1995). At the University of Bonn in Germany, Klobinger and colleagues studied 15 patients with drug-resistant depression; of the ten patients receiving nonsham stimulation, all showed significant improvement.

This transition to proof-of-principle studies did not eradicate previous ethical concerns; rather, it added new considerations to the ever-relevant previous concerns. As with any controlled study, placebo (*sham stimulation* in the case of TMS) must be covertly administered to some participants. This necessity forced researchers to question the risk-to-benefit standards behind subject blinding. In addition, the inclusion of lengthy and repetitive patient treatment protocols (as opposed to transitory research protocols) raised questions about the long-term cortical effects of TMS. This issue is perhaps best illustrated in a letter titled *Shocking Safety Concerns* published in *Lancet* several months after a successful depression proof-of-principle trial by Pascual-Leone et al. (1996): "Safety studies have largely allayed fears of secondary epilepsy with this technique, and guidelines have been formulated to protect against this complication. However, there is growing epidemiological evidence that low frequency, low intensity electromagnetic fields are associated with cancer...Some studies suggest that magnetic field strengths in the millitesla range may have a promoter or copromoter effect in carcinogenesis, as well as increasing growth in cancer cell lines...The brain, the possibility of delayed malignant disease has not been considered" (Brown, 1996).

The long-term safety of medical devices and treatments, rather unfortunately, has a tendency of falling through the cracks: Many practitioners quietly assume that *someone else* will take care of the admittedly difficult and time consuming task of longitudinal research. This is not to say the task is always swept under the carpet. In fact, there have been many long-term studies undertaken leading to safer conditions for particular patient populations (e.g., Benzoyl Peroxide, a common treatment for acne, has been shown to increase the risk of skin cancer if utilized too often or if exposed to excess amounts of direct sunlight; see Hogan et al., 1991). Unfortunately, too often practitioners push forward and only learn about the long-term effects of particular treatments long after it is too late (e.g., prolonged use of Vioxx, a drug prescribed for the acute treatment of pain, was found to increase the risk of heart disease only after the drug had been widely used; see Mukherjee et al., 2001). To avoid similar cases associated with TMS, it's important that TMS practitioners fully inform patients of the uncertainty of long-term effects, maintain contact with past patients, and report any long-term occurrences, no matter how seemingly isolated the case appears.

Clearly, the issues raised in Brown's letter are valid and reflective of the apprehension maintained by many clinicians and scientists. Unfortunately, at the time this missive appeared—based largely on the false premise that the safety of rTMS had been demonstrated—the majority of clinical studies maintained a focus on efficacy rather than safety. To compound the issue, many of these studies were performed in countries without strict regulatory demands. In order to maintain a circumspect developmental trajectory, it became imperative to outline clear safety and procedural standards.

To this end, a group of the leading TMS authorities convened in 1996 and in 1998 published the first detailed safety and ethical guidelines for TMS in both the clinical and the laboratory settings (Wassermann, 1998). The goals of this assemblage were not only to develop a common language between practitioners (which would make experimental findings interpretable by all), but also to ensure a centralized procedural methodology. Again, borrowing a lesson from the unchecked stimulation methodologies used in the past (e.g., ECT), the investigators attending the conference set written standards to maintain prudence and discretion in a field that could easily be derailed by one or two rogue practitioners. It is interesting to note that a decade later, at the second convening of TMS authorities, these initial standards were found to have stood the test of a decade's worth of work. When followed, these protocols maintained a strong TMS safety record. While safety standards must be regularly reviewed and updated with evolving experience, the safety record of TMS to date confirms that potential future ethical problems can often be largely avoided with proper preventive ethics, foresight, and restraint in the present (Forrow et al., 1993).

Stage IV: Clinical Expansion

With safety guidelines in place and numerous successful proof-of-principle trials affirming potential benefits, the popularity and proliferation of TMS began to grow. Clinics began popping up worldwide, offering off-label therapies for a myriad of neurological and psychological pathologies. In fact, in the clinical population, TMS was shown to hold promise for the treatment of Parkinson's disease (Wu et al., 2008), writer's cramp (Siebner et al., 1999), chronic pain (Jensen et al., 2008), rehabilitation for motor neglect (Oliveri et al., 1999), motor stroke (Mansur et al., 2005) aphasia (Martin et al., 2004; Naeser, Helm-Estabrooks, et al., 2005; Naeser, Kobayashi, et al., 2005), auditory hallucinations (Hoffman et al., 2000), epilepsy (Fregni et al., 2006), visceral pain (Fregni et al., 2005), and depression (Pascual-Leone et al., 1996; Eschweiler et al., 2000; George et al., 2000; Fregni & Pascual-Leone, 2005; Gross et al., 2007; O'Reardon et al., 2007).

As is common with potential new treatments for which many of the risks have not been clearly defined, rTMS was initially examined in patients who had exhausted other, more established forms of treatment. This necessarily biased the rTMS treatment population toward patients with relatively severe forms of neurological or psychiatric disorders refractory to conventional therapies. In fact, for many patients, rTMS became a last hope intercession, which raises an ethical issue relevant to many emerging technologies: When the treatment population consists of desperate individuals willing to endure potentially unsafe procedures in a bid to obtain relief, it becomes doubly important for practitioners to maintain an objective focus on potential signs of diminished autonomy as a result of neuropsychiatric illness, desperation, or both, affecting matters of informed consent (Minogue et al., 1995; Miller & Brody, 2003). Extra efforts and safeguards may need to be put into place to avoid problems, mistakes, or eventual patient backlash when using novel treatment methods on vulnerable populations.

It is important to note that because studies to date have provided increasing evidence for the relatively benign profile of side effects of rTMS if appropriate guidelines are adhered to (Rossi et al., 2009), rTMS today is being offered to broader populations, including patients with less severe conditions and those who have not necessarily proven refractory to all available alternatives.

As rTMS became more established, on-label treatment for depression was approved by the regulatory agencies of numerous countries, including Brazil, Israel, Australia, and Canada. This swift expansion created an explosion of ethical concerns in the world of brain stimulation. Perhaps the largest and most important of these concerns dealt with the economic market. As with any medical endeavor, when monetary profit becomes a reality, the potential for the blurring of once sharply defined parameters of patient viability and therapeutic promise arises. How are practical guidelines to be enforced with the migration of patients over borders to obtain treatment unavailable in their home countries? How is honesty to remain in side effect reporting when said reports directly affect business? These questions remain important today and the need to revisit and continually update TMS safety guidelines and recommendations for global clinical implementation is imperative.

In addition to clinical expansion, TMS added various new stimulation paradigms capable of changing the risk-to-benefit profiles. For example, consider theta-burst stimulation (TBS) (Huang & Rothwell, 2004). Similar to high-frequency stimulation, TBS mimics the powerful oscillations of hippocampal neurons, which can quickly induce long-term potentiation (LTP) and long-term depression (LTD) effects in cortical neurons. In humans, it is possible (via TBS) to induce longer-lasting and more profound modulation of cortical excitability with a much lower number of stimuli and a much shorter duration of stimulation. Other approaches to enhance the physiologic impact of TMS include primed-slow rTMS (Iyer et al., 2003), various asynchronous stimulation paradigms, modification of stimulation pulse shapes, and even development of more powerful stimulation devices (Rossi et al., 2009). Consider also varied coil patterns, several effective cranial stimulation sites, and flexible power levels, and a picture of TMS as a broad, permutated methodology emerges. With this evolution, issues of proper technician and practitioner procedural training and certification requirements became important. Knowing increasing numbers of individuals were seeking (and would continue to seek) TMS therapy, ensuring patients received the best possible treatment by a doctor well versed in not only the appropriate methodology but also in all relevant treatments associated with the device became of paramount concern (and remains so today).

Finally, as with any new treatment, it is important to use caution when considering TMS indications. Usually in medicine, physicians and patients maintain excessive hopes with regards to new treatments. This hope might translate into unsafe utilization of TMS in populations of patients with no clear indication. This ethical consideration should be pondered and investigated by patients and physicians alike.

Stage V: The 2008 Consensus Conference and an Ethical Evaluation of TMS Safety

The guidelines established in 1998 for the application of TMS in research and clinical settings were reviewed in a consensus conference that took place in 2008. The resulting publication summarized the main points of the conference and suggested limits for the combination of rTMS frequencies, intensities, and train durations to reduce risks for inducing seizures, the most concerning (yet very rare) adverse event (Rossi et al., 2009). The article provided the scientific community with an up-to-date evaluation of the safety record of research and clinical applications of TMS. In addition, the article introduced several important questions, such as the use of novel parameters, while underscoring several long-standing questions that remained unanswered, such as the risk of TMS in patients with neuropsychiatric disorders.

In the decade spanning the time between consensus conferences, the use of TMS expanded exponentially. New paradigms of stimulation were developed, research applications expanded, research sessions grew in numbers and duration, and technical advancements allowed TMS to be coupled with electrophysiological and imaging techniques, including electroencephalography (EEG), positron emission tomography (PET), and functional magnetic resonance imaging (fMRI). While the use of TMS grew increasingly more complex, many of the major safety concerns remained. However, thousands of healthy subjects and neuropsychiatric patients underwent TMS in the time between the consensus conferences, allowing for more extensive assessment of TMS-related side effects.

Depending on the specific stimulation technique and parameters, TMS, especially rTMS, may be painful. The most commonly reported side effects of TMS include local pain, headache, and discomfort. The intensity of these effects, however, varies from subject to subject, but all subjects should know going into the procedure that TMS may not be pleasant. A review by Machii et al. reported head-ache or neck pain in approximately 40% of cases when rTMS was delivered to nonmotor areas (2006). Likewise, a meta-analysis of the safety of TMS for the treatment of depression reported headache and discomfort-related side effects, respectively, in about 28 and 39% of patients. Sham-controlled studies reported similar side effects in about 16 and 15% of patients, respectively (Loo et al., 2007). These discomforts often respond to aspirin, acetaminophen (Tylenol), or other common analgesics.

Because the intensity of sound emitted by the TMS procedure is deceptively high, repetitive TMS can also cause transient hearing loss if the subjects do not wear earplugs during the rTMS studies. A few cases have been reported where adults have experienced transient increases in auditory threshold following TMS (Rossi et al., 2009). These hearing safety concerns have been addressed by provid-ing subjects with earplugs to wear throughout the stimulation procedure.

Furthermore, TMS may induce mood and hormone changes, which are rare and usually resolved within hours of cessation of TMS. While many TMS-related side effects are fairly minor, it is essential to the ethical delivery of TMS that appropriate guidelines are followed and necessary precautions are taken (Rossi et al., 2009).

The major risk of TMS, albeit small, is the risk of producing a seizure. Only 17 seizures induced by rTMS have been reported worldwide among the many thou-sands of patients studied, resulting in a projected risk of less than 1:1000 patients and probably less than 1:10,000 TMS sessions. The parameters of TMS that have produced seizures during experimentation are well known and documented (Rossi et al., 2009). Although the risk of seizure is small if safety guidelines are followed, these guidelines are primarily based on experiments in healthy subjects and the risk of seizures may be higher in patients with neuropsychiatric diseases such as stroke and major depression and in patients on certain neuropsychotropic drugs. For example, patients with infarcts or neurological disorders that cause cortical atrophy should be stimulated with great care as the presence of excess cerebrospi-nal fluid (CSF) can alter the electromagnetic field properties and stimulation near CSF can cause adverse effects (Wagner et al., 2006). Finally, TMS has only been studied for approximately 25 years, and the data on potential long-term effects in humans remains insufficient. Although animal studies using TMS have not indi-cated any risks of brain damage or long-term injury, caution remains imperative.

In addition to the concern for seizure, there also exists the theoretical con-cern that rTMS may have lasting effects on seizure threshold. This is likened to the phenomenon of "kindling," where repeated subconvulsive electrical cor-tical stimuli eventually result in seizure activity (Goddard et al., 1969; Loo et al., 2007). Kindling has been clearly demonstrated in animal models but not in humans (reviewed by Loo et al., 2007). In 1994, Pascual-Leone et al., dem-onstrated that rTMS (20 stimuli at high frequency and intensity, 20 Hz, 150% motor threshold) to the motor cortex in healthy subjects resulted in a lower-ing of threshold such that a motor evoked response could then be elicited by a lower level of TMS stimulation to the motor cortex. The effect was transient and noncumulative and thought to be a result of increased activity at the cortical level rather than in peripheral motor pathways. Whether this result and similar reported changes of cortical excitability (Wassermann et al., 1996; Loo et al., 2007) in response to TMS implies a reduction in seizure threshold remains unknown, but remains an important consideration. Of the thousands of healthy subjects and patients who have undergone TMS, there have been no reports of repeated seizure activity following TMS. Studies utilizing combined TMS and EEG protocols have reported no significant changes in EEG following stimulation of the motor cortex (Pascual-Leone et al., 1993; Wassermann et al., 1996; Loo et al., 2001; Loo et al., 2007). Nevertheless, investigators must still continue to question the possibility of kindling and other long-term effects in response to TMS.

Since safety guidelines were generated from information on TMS in adults, relatively little is known about the appropriate safety guidelines for the application of TMS in children (Frye et al., 2008). More generally, the effects of TMS on the developing brain remain unknown. Thus, even though rTMS offers the potential for treating developmental disorders like autism, childhood depression, and obsessive-compulsive disorder, among others, particular caution is needed when carrying out research on children until more is known about safety in younger populations from studies on both humans and animal models. Some authors have suggested that clinical trials on children who have medication refractory focal epilepsy represent a reasonable entry point to the understanding of how the child and adolescent brain responds to TMS (Fregni, Thome-Souza et al., 2005; Thut et al., 2005).

In every facet of TMS use, side effects are a possibility and the safety of the subject or patient should come first. The clinician or investigator must default on safety, on what is best for the patient or subject. The combination of parameters employed in research and TMS treatment modalities must be carefully devised and recommended safety guidelines must be followed. No side effect should be overlooked, and not observing an effect should not necessarily indicate that an effect is not present—all possibilities must be carefully weighed in the minds of TMS investigators and clinicians alike. Moreover, the risk-to-benefit ratio must always remain a priority in evaluating TMS treatment and research. In fact, the 2008 consensus conference guidelines recommend dividing rTMS studies into three classes, introduced by Green et al., (1997), in the order of necessary protection and expected benefits:

Class 1—Direct benefit, with the potential for high risk
Class 2—Indirect benefit, with moderate risk
Class 3—Indirect benefit, with low risk

It is also crucial to continue to collect safety data. Designing safety studies in humans is difficult, however, and animal models provide us with information limited to a number of factors: the differences between animal and human physiology, the issues brought about by restraining the animal during TMS, and experimental coil design and focality, for example (Rossi et al., 2009). While limited, animal models are still relevant for assessing the effects, mechanisms of action, and safety of TMS. But more safety data is necessary. Most of the safety data that we have about TMS are a posteriori, oftentimes collected before and after a study that has benefit to the subject. For example, a recent study reviewed English-language published studies of TMS (single- and paired-pulse and rTMS) in persons younger than 18 (the majority with neuropsychiatric disorders) from 1990 to 2005. The author of this study found 49 studies, which in total applied TMS in 1036 children. No seizures were reported for any of the 1,036 studied children, although the number of rTMS studies (in contrast with safer single- and paired-pulse TMS studies) were low (Quintana, 2005). While this type of data is unquestionably important, the scientific community must keep in mind the long-term effects of TMS. What happens when people are exposed frequently to TMS? Due to limitations with IRBs, it's not always possible to keep track of how often certain research subjects are exposed to varying TMS protocols, although this information would certainly benefit the attempt to understand and better evaluate the safety of TMS.

Stage VI: FDA Approval, On- and Off-Label Treatments, and the Marketing of TMS

In 2007, O'Reardon et al., published the results of an industry-sponsored, multisite, randomized, sham-stimulation controlled clinical trial in which 301 patients with major depression, who had previously failed to respond to *at least one* adequate antidepressant treatment trial, underwent either active or sham TMS over the left dorsolateral prefrontal cortex (DLPFC). The patients, who were medication-free at the time of the study, received TMS five times per week over four to six weeks.

Initial analysis of the results failed to reach statistical significance for the primary outcome measure, the change in severity of the depression according to the Montgomery-Asberg Depression Scale (MADRS). However, a number of secondary outcome measures, including for example severity of depression according to the Hamilton Depression Rating Scale (HAMD) did show statistically significant benefits of active over sham rTMS. Re-examination of the data demonstrated that a *sub-population* of patients, those who were relatively less resistant to medication, having failed not more than two good pharmacologic trials, showed a statistically significant improvement on MADRS, HAMD, and various other outcomes measured. Eventually, supported by the results of this secondary analysis, Neuronetics, the study-sponsoring company obtained approval from the FDA for the clinical treatment of specific forms of medication-refractory depression in October 2008 (FDA approval K061053; O'Reardon et al., 2007).

It is important to note that FDA approval was only granted to the NeuroStar TMS device, manufactured by the Neuronetics Incorporation and for the protocol of stimulation employed, (high-frequency, 10 Hz—rTMS applied daily for four to six weeks at suprathreshold intensity). It is also important to realize that the FDA approval is restricted to a specific subpopulation of patients with medication resistant depression (i.e., adult patients who have failed to achieve satisfactory improvement from no more than one prior adequate antidepressant medication trial).

In such industry-sponsored studies, it is important to consider conflict of interest issues and to question the scientific and statistical rigor of such a sponsored study. Carefully monitoring of scientific integrity is thus critical and the FDA regulatory process provides important assurances toward that end. Independence of scientific research from undue influences is critical, but there are also important considerations in support of industry-sponsorship. For example, in this instance, with Neuronetics gaining approval, the door opened for other companies, devices, and treatments to receive serious consideration; most importantly, the door opened for patients who lacked an efficacious, safe treatment option to be helped with proper reassurance of the risk-to-benefit balance. In the end, company sponsored experimentation presents a double-edged sword, raising questions of legitimacy while simultaneously paving the way for future growth and allowing the alleviation of symptoms in a wider range of patients.

The narrow indication of the FDA-approved on-label indication also poses interesting ethical questions. The criteria of having failed one, but not more than one, adequate antidepressant medical trial may mean that many patients who could benefit from the on-label therapy are left to consider off-label options. Additionally, the FDA has not approved long-term maintenance TMS care because specific, properly designed studies documenting long-term safety and efficacy have not yet been done. However, if a patient responds well to the initial course of rTMS, the benefit can be expected to lapse after about four to six months (Demirtas-Tatlidede et al., 2008). Thus, what is one to do following initial course of TMS without approval of maintenance TMS? This leaves the issue of prolonged symptom alleviation up for ethical evaluation and for physicians' careful consideration of an individual patient's unique medical characteristics when determining to continue treatment. This is a common issue with psychopharmacologic treatment of psychiatric disorders, including depression. Most FDA-approved treatments for depression are based on relatively short-duration trials demonstrating clinical effectiveness. Clinicians nonetheless often continue FDA-approved antidepressant medications long-term, in the absence of rigorous studies (e.g., randomized, placebo-controlled trials) proving their effectiveness with long-term use. The absence of such studies is in part related to cost and to the limited incentives for pharmaceutical companies to undertake such studies if clinicians are using their products anyway.

But should any and all patients who meet the FDA guidelines be treated? Thorough screening protocols to evaluate susceptibility to possible side effects must be implemented by all on-label clinics to ensure, above all, the safety of the patient receiving treatment. It is always important to keep in mind the risk-to-benefit ratio for the *individual* patient. Admittedly, the risk-to-benefit ratio for an individual patient is in part a subjective measure; thus, practitioners are needed with solid experience not only in the treatment of the patient population to be addressed (depression in this case) but also in the basis and application of the treatment (TMS in this case). Ultimately this requires the development of acceptable training guidelines and accreditation requirements for clinicians who use TMS. In other words, practitioner considerations merit ethical evaluation. Who should deliver TMS? What types of guidelines should be implemented to ensure practitioners are using TMS correctly and safely? There are currently no requirements for TMS training, although it is advised that TMS be delivered by users trained in the basics of brain physiology and TMS mechanisms. In addition, it is recommended that TMS practitioners understand the potential risks of TMS and how to appropriately respond to any adverse effects (Rossi et al., 2009). It is apparent, though, that regulations, or at least guidelines, must be established as TMS itself becomes more widely used.

Establishment of regulations or certification guidelines that set limits on approved TMS practitioners, however, may understandably be challenged. Presumably any certification criteria would be established by experts in the field, but those same experts may be attacked for attempting to restrict the use of TMS to experts like themselves for self-interested reasons rather than for the benefit of patients. Procedures for establishing any regulations must be designed entirely to ensure that patients can have well-founded trust in the expertise of TMS practitioners from a patient safety and quality assurance perspective, not to ensure that the field is controlled by those who early in the field's development have become self-appointed guardians of their own dominant positions.

The existing safety guidelines for TMS differentiate between relative and absolute contraindications, thus providing important guidance in selection of TMS suitable TMS candidates. Patients exhibiting relative contraindications may have, for example, a history of epilepsy or lesions of the brain; they may also be taking medication that lowers the seizure threshold. Patients with an absolute contraindication to TMS are those patients who have implanted metallic hardware in close contact to the discharging coil (e.g., cochlear implants). Performing TMS on patients demonstrating an absolute contraindication carries the risk of inducing malfunctioning of such implanted devices (Rossi et al., 2009). Therefore, when it comes to deciding whether a patient is eligible for TMS, the risk-to-benefit ratio is crucial, and the clinician or investigator should carefully weigh the presence of both types of contraindications while assessing whether TMS is appropriate.

FDA approval does not ensure coverage by health insurance companies. Thus, not all patients who meet the FDA guidelines and may benefit from TMS can afford to be treated. Likewise, many patients seeking off-label therapy cannot afford treatment, as neither on- nor off-label TMS is covered by health insurance in the United States. This leaves payment options to the individual clinics. Is there a way for clinics to set pricing so all patients in need, regardless of socioeconomic status, can access treatment? How can such arrangements be made? As TMS continues to advance as a treatment option, such questions will undoubtedly become increasingly common.

At this stage, further questions must be asked concerning treatment guidelines. What criteria should we adopt before recommending TMS as a possible treatment for depression? The patient must of course be fully informed about the treatment. All pertinent information—small effect sizes in TMS trial, safety data, alternative treatments, even potential future effects—must be fully disclosed to the patient. The truth about what TMS can and cannot offer must be explicitly stated, and every effort to balance a patient's hopes and expectations with the unpredictable reality of science must be made. It may be useful to create model written patient information materials about TMS, and/or an easily-accessible website that provides up-to-date information for the lay public, from objective and authoritative sources, about TMS and what is known about its risks and benefits in various contexts.

Another subject worth exploring regards the advertisement of TMS. FDA approval of the Neuronetics NeuroStar TMS Therapy came with privilege of therapeutic advertising. The selling of a product engenders many ethical questions. For example, who exactly should the advertisements target? If the advertised information is too broad, clinicians will spend 90% of their time screening out the myriad patients who don't fit the narrow FDA indications. If the advertised information is too specific, clinicians risk alienating potential customers (as nearly all public sentiment toward electrical stimulation springs from the still stigmatized ECT). Also, how should the matter of side effects be dealt with? Surely TMS treatment for depression can be made appealing based on what it does *not* do (e.g., it does not cause some of the usual side effects associated with some common antidepressant medications), but perhaps it would be more prudent to offer TMS for what the literature indicates it can do. Certainly a proper balance of potential benefits versus potential risks and side effects has to be presented. Establishment of consensus standards regarding the content of promotional advertising for TMS, even though they would almost certainly be nonbinding, could be useful before practitioners or device companies move forward individually on this front on a large scale.

Figure 16.4 temporally organizes many of the above discussed ethical concerns and places each within it proper developmental stage. As in Figure 16.3, hopefully this organization illustrates that advances in technique and application do not eliminate previous ethical concerns, but, instead, adds to their number.

PRESENT AND FUTURE ETHICAL CONSIDERATIONS FOR THE SCIENTIFIC USE OF TMS

The current policy of the FDA regarding off-label applications of TMS is that in any study, regardless of the device, the population being studied, or the stimulation parameters used, it is the responsibility of the IRB to determine whether an investigation poses a significant risk to subjects. Therefore, to ensure full protection of research subjects, it is essential that research motives are candidly communicated by the investigator to the IRB so that the subjects' safety may be accurately assessed.

The issue of subject blinding for TMS research studies presents a unique ethical challenge. While the investigator must maintain transparency with the subject, too much knowledge about the procedure or previous TMS procedures may effectively un-blind all but the most naïve of research subjects until further advancements are made in developing a more authentic sham condition.

In cognitive neuroscientific research settings, TMS can produce desired and undesired changes, most typically mild and transient memory problems and other cognitive deficits. In studies designed to produce such effects, TMS can alter

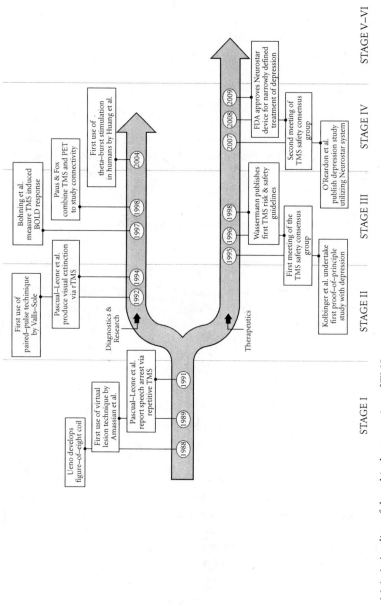

Figure 16.4 A timeline of the ethical progression of TMS.

subjects' performance on specific tasks. Typically the effects produced by such procedures are not long lasting and do not raise particular safety issues (Rossi et al., 2009). Still, these studies challenge the risk-to-benefit ratio. Indeed, the benefit would be increased knowledge about the brain and mechanisms of plasticity, but to what extent is it still ethical to induce desired changes in a subjects' brain?

Studies of rTMS at rates of repetition well within safety limits have been conducted on normal subjects to investigate phenomena varying from self-recognition (Keenan et al., 2000) and sequence learning (Robertson et al., 2001) to moral cognition and decision making (Knoch et al., 2006). However, some studies utilize rTMS at parameters nearing the limit of safety guidelines, and some appear to exceed reasonable ethical criteria. For example, as mentioned above, some researchers have begun to investigate possible mechanisms of depression and potentiation using rTMS protocols in the normal human brain (Iyer et al., 2003; Huang et al., 2005). These studies, while potentially producing valuable scientific knowledge, raise significant questions about the risk-to-benefit ratio for research subjects. The scientific value of such research is that learning how to induce LTD and LTP could provide a new understanding of brain function that might have significant implications for diagnosis and treatment of diseases like depression, epilepsy, Parkinson's disease, and other neurological disorders. But it is not yet known with certainty that all induced adverse brain effects are immediately and completely reversible. In addition, it is known that increasing the length and rate of stimulation increases the risk, albeit low, of seizure. What criteria should be used to determine whether a proposed protocol in normal subjects has acceptably low risk? Exactly what should prospective research subjects be told about these unknown risks? Currently each individual IRB reviewing a protocol makes its own determination, and it may be helpful to convene a group of TMS experts, experts in research ethics, and lay individuals to discuss these issues and attempt to establish guidelines for IRBs.

The therapeutic utility of rTMS, particularly in the treatment of drug resistant patients, might be increased by identifying genetic predictors of efficacy. Several studies are starting to investigate the roles of functional polymorphisms (e.g., BDNF Val 66Met) and their involvement in the rTMS response. This can be important in the context of efficacy, but also for the safety of TMS. For instance, understanding the antidepressant response of rTMS with respect to a person's genotype may one day help to tailor rTMS treatment for mood disorders (Bocchio-Chiavetto et al., 2008). Protocols combining TMS with genetic testing, which ask whether genetic factors affect the role of excitability and plasticity of neuronal circuits, may also be applied to the therapeutic enhancement of function after stroke and chronic neurodegenerative disease. For example, Kleim et al., (2006) used single pulse TMS to demonstrate that BDNF genotype is associated with changes in excitability of primary motor cortex that occur after practicing a motor task. In another study, Cheeran et al., (2008) found that susceptibility to TMS probes of plasticity, including continuous and intermittent theta burst stimulation and homeostatic plasticity in the 1 Hz rTMS model, was significantly different in individuals carrying the Val66Met BDNF polymorphism than susceptibility in individuals carrying the Val66Val polymorphism. The results of the study suggest that inter-individual differences in plasticity may be due in part to the effect of BDNF on synapses to undergo LTP/LTD. In turn, such results may help illuminate the factors that influence the natural response of the brain to injury and disease as well as unfold therapeutically relevant advancements in the understanding of the inter-individual response to TMS.

Combining TMS with genetic testing, though, presents compound ethical challenges: One must not only keep in mind the ethical considerations applicable to TMS research, but also evaluate the complex host of issues pertaining to genetic testing.

While current research does in fact indicate that genetic factors may influence the brain's plasticity and response to TMS, is the current evidence strong enough to merit genetic testing on all TMS subjects? The broad range of genetic factors (not to mention epigenetic effects), coupled with the many possible TMS variations and parameters, leaves scientists with an intricate array of testing possibilities. While we may never know exactly what to test for, do we know enough about the subjects of TMS or genetics on their own to continue exploratory testing of the combination of both subjects? Again, it is necessary to revisit the risk and benefits of the research. Given the postulated role of synaptic plasticity in brain injury and disease, and the epidemiologic evidence concerning the role of BDNF in the synaptic plasticity of the adult brain, there does seem to be a potential therapeutic potential for future interventions. However, that potential comes at a cost. Genetic material must first be obtained, and drawing blood certainly poses a discomfort for research participants. Genetic material must also be stored. In some cases, cell lines are made to allow others to carry on the research. Whatever the protocol, the ethics of genetic testing must be carefully examined. To address some of these issues, Knoppers and Chadwick (2005) highlight the notion of honest and reciprocal exchange between researcher and participant, which they liken to the physician-patient relationship. The autonomy and the contribution of the research participant must be recognized and respected to the fullest in protocols calling for genetic testing. Informed consent and communication with potential participants must be clear, and the objectives transparent. Personal and cultural values of potential participants must be respected. Moreover, participants should fully understand that they have the option to take part in a protocol, whether to have their DNA banked, coded, analyzed, even replicated and potentially commercialized. Researchers, of course, must also offer the highest security for genetic data.

CONCLUSION

The history of TMS helps to highlight the ethical questions that must be asked as TMS continues to gain momentum as a promising medical therapy and research tool. Looking toward the future, we must keep the past in mind, and continually reflect not only on safety guidelines, including the appropriate training of TMS personnel, but also many other ethical issues raised in both clinical and research applications of TMS. Our ethical thinking will inevitably evolve with the science itself and with changes in cultural and societal values, but the issues we have outlined here will provide a basis for better approaching the uncertain. As in any evolving field, the most essential are not those questions we do have the answers to, but rather the questions we're still trying to answer.

Should we fail to ask and carefully answer these questions proactively, there is a risk TMS will at some point be misused with serious adverse consequences for one or more patients or research subjects. If so, the entire field may be halted in its tracks by outcries of public criticism that resonate powerfully with the overwhelmingly negative perceptions that still plague the use of ECT. Given TMS's enormous promise, this would be truly tragic, potentially depriving large numbers of patients of significant therapeutic benefit. Fortunately, because such a tragic situation is so clearly foreseeable, it is also largely preventable if we act without delay (Forrow et al., 1993).

REFERENCES

Allen, E. A., Pasley, B. N., Duong, T., & Freeman, R. D. (2007). Transcranial magnetic stimulation elicits coupled neural and hemodynamic consequences. *Science, 317,* 5846, 1918–1921.

Amassian, V. E., Cracco, R. Q., Maccabee, P. J., Cracco, J. B., Rudell, A., & Eberle, L. (1989). Suppression of visual perception by magnetic coil stimulation of human occipital cortex. *Electroencephalography and Clinical Neurophysiology 74,* 458–462.

Barker, A. T., Jalinous, R., & Freeston, I. L. (1985). Non-invasive magnetic stimulation of human motor cortex. *Lancet, 1,* 1106–1107.

Bocchio-Chiavetto, L., Miniussia, C., Zanardini, R., Gazzolia, A., Bignotti, S., Specchia, C., et al . (2008). 5-HTTLPR and BDNF Val66Met polymorphisms and response to rTMS treatment in drug resistant depression. *Neuroscience Letters, 437,* 130–134.

Brown, P. (1996). Shocking safety concerns. *Lancet, 348,* 9032, 959.

Cheeran, B., Talelli, P., Mori, F., Koch, G., Suppa, A., Edwards, M., et al. (2008). A common polymorphism in the brain-derived neurotrophic factor gene (BDNF) modulates human cortical plasticity and the response to rTMS. *Physiologia, 526,* 23, 5717–5725.

Counter, S. A., Borg, E., Lofquist, L., & Brismar, T. (1990). Hearing loss from the acoustic artifact of the coil used in extracranial magnetic stimulation. *Neurology, 40,* 1159–1162.

Demirtas-Tatlidede, A., Mechanic-Hamilton, D., Press, D. Z., Pearlman, C., Stern, W. M., Thall, M., et al. (2008). An open-label, prospective study of repetitive transcranial magnetic stimulation (rTMS) in the long-term treatment of refractory depression: reproducibility and duration of the antidepressant effect in medication-free patients. *Journal of Clinical Psychiatry, 69,* 6, 930–934.

Dhuna, A. K., Gates, J. R., & Pascual-Leone, A. (1991). Transcranial magnetic stimulation in patients with epilepsy. *Neurology, 41,* 1067–1072.

Eschweiler, G. W., Wegerer, C., & Schlotter, W. (2000). Left prefrontal activation predicts therapeutic effects of repetitive transcranial magnetic stimulation (rTMS) in major depression. *Psychiatry Res, 99,* 161–172.

Farrow, L., Arnold, R. M., & Parker, L. S. (1993). Preventive ethics: Expanding the horizons of clinical ethics. *The Journal of Clinical Ethics, 4,* 4, 287–294.

Freen, R. M., Pascual-Leone, A., & Wassermann, E. M. (1997). Ethical guidelines for rTMS research. *IRB, 2,* 1–7.

Fregni, F., DaSilva, D., Potvin, K., Ramos-Estebanez, C., Cohen, D., Pascual-Leone, A., et al . (2005). Treatment of chronic visceral pain with brain stimulation. *Annals of Neurology, 58,* 6, 971–972.

Fregni, F., Otachi, P.T.M., do Valle, A., Boggio, P. S., Thut, G., Rigonatti, S. P., et al. (2006). A randomized clinical trial of repetitive transcranial magnetic stimulation in patients with refractory epilepsy. *Annals of Neurology, 60,* 4, 447–455.

Fregni, F. & Pascual-Leone, A. (2005). Transcranial magnetic stimulation for the treatment of depression in neurologic disorders. *Current Psychiatry Reports, 7,* 381–390.

Fregni, F., Thome-Souza, S ., Nitsche, M. A., Freedman, S. D., Valente, K. D., & Pascual-Leone, A. (2006). A controlled clinical trial of cathodal DC polarization in patients with refractory epilepsy. *Epilepsia, 47,* 2, 335–342.

Frye, R. E., Rotenberg, A., Ousley, M., & Pascual-Leone, A. (2008). Transcranial magnetic stimulation in child neurology: Current and future directions. *Journal of Child Neurology, 23,* 1, 79–96.

George, M. S., Nahas, Z., & Molloy, M. (2000). A controlled trial of daily left prefrontal cortex TMS for treating depression. *Biolgicalo Psychiatry, 48,* 962–970.

Goddard, G. V., McIntyre, D. C., & Leech, C. K. (1969). A permanent change in brain function resulting from daily electrical stimulation. *Experimental Neurology, 25,* 295–330.

Green, R. M., Pascual-Leone, A., & Wassermann, R. M. (1997). Ethical guidelines for rTMS research. *IRB: A Review of Human Subjects Research, 19,* 2, 1–7.

Gross, M., Nakamura, L., Pascual-Leone, A., & Fregni, F. (2007). Has repetitive transcranial magnetic stimulation (rTMS) treatment for depression improved? A systematic review and meta-analysis comparing the recent vs. the earlier rTMS studies. *Acta Psychiatr Scand, 116,* 3, 165–173.

Harris, L. J., & Almerigi J. B. (2009). Probing the human brain with stimulating electrodes: the story of Roberts Bartholow's experiment on Mary Rafferty. *Brain and Cognition, 70,* 1, 92–115.

Hoffman, R. E., Boutros, N. N., Hu, S., Berman, R. M., Krystal, J. H., & Charney, D. S. (2000). Transcranial magnetic stimulation and auditory hallucinations in schizophrenia. *Lancet, 355,* 9209, 1073–1075.

Hogan, D. J., To, T., Wilson, E. R., Miller, A. B., Robson, D., Holfeld, K., at al . (1991). A study of acne treatments as risk factors for skin cancer of the head and neck. *The British Journal of Dermatology, 125,* 4, 343–348.

Horgan, J. (2005). The forgotten era of brain chips. *Scientific American, 293,* 4, 66–73.

Huang, Y. Z., Edwards, M. J., Rounis, E., Bhatia, K. P., & Rothwell, J. C. (2005). Theta burst stimulation of the human motor cortex. *Neuron, 45,* 2, 201–206.

Huang, Y. Z., & Rothwell, J. C. (2004). The effect of short-duration bursts of high-frequency, low-intensity transcranial magnetic stimulation on the human motor cortex. *Clincial Neurophysiology, 155,* 5, 1069–1075.

Iyer, M. B., Schleper, N., & Wassermann, E. M. (2003). Priming stimulation enhances the depressant effect of low-frequency repetitive transcranial magnetic stimulation. *Journal of Neuroscience, 23,* 34, 10867–10872.

Jensen, M. P., Hakimian, S., Sherlin, L. H., & Fregni, F. (2008). New insights into neuromodulatory approaches for the treatment of pain. *Journal of Pain, 9,* 3, 193–199.

Kleim, J. A., Chan, S., Pringle, E., Schallert, K., Procaccio, V., Jimenez, R., et al . (2006). BDNF Val66Met polymorphism is associated with modified experience-dependent plasticity in human motor cortex. *Nature Neuroscience, 9,* 6, 735–737.

Knoch, D., Gianotti, L.R.R., Pascual-Leone, A., Treyer, V., Regard, M., Hohmann, M., at al . (2006). Disruption of right prefrontal cortex by low-frequency repetitive transcranial magnetic stimulation induces risk-taking behavior. *Journal of Neuroscience, 26,* 24, 6469–6472.

Knoppers, B. M., & Chadwick, R. (2005). Human genetic research: emerging trends in ethics. *Genetics, 6,* 75–79.

Kolbinger, H. M., Hoflich, G., Hufnagel, A., Moller, H. J., & Kaspers, S. (1995). Transcranial magnetic stimulation (TMS) in the treatment of major depression: a pilot study. *Human Psychopharmacology—Clinical and Experimental, 10,* 4, 305–310.

Loo, C. (2007). A review of the safety of repetitive transcranial magnetic stimulation as a clinical treatment for depression. *International Journal of Neuropsychopharmacology, 11,* 1, 131–147.

Loo, C., Sachdev, P., Elsayed, H., McDarmont, B., Mitchell, P., Wilkinson, M., et al. (2001). Effects of a 2- to 4-week course of repetitive transcranial magnetic stimulation (rTMS) on neuropsychologic functioning, electroencephalogram, and auditory threshold in depressed patients. *Biological Psychiatry, 49,* 615–623.

Machii, K., Cohen, D., Ramos-Estebanez, C., Pascual-Leone, A. (2006). Safety of rTMS to non-motor cortical areas in healthy participants and patients. *Clinical Neurophysiology, 117,* 2, 455–471.

Mansur, C. G., Fregni, F., Boggio, P. S., Riberto, M., Gallucci-Neto, J., & Santos, C. M. (2005). A sham stimulation-controlled trial of rTMS of the unaffected hemisphere in stroke patients. *Neurology, 64,* 10, 1802–1804.

Martin, P. I., Naeser, M. A., Theoret, H., Tormos, J. M., Nicholas, M., Kurland, J., et al. (2004). Transcranial magnetic stimulation as a complementary treatment for aphasia. *Semin Speech and Language, 25,* 2, 181–191.

Miller, F. G. and Brody, H. (2003). A critique of clinical equipoise. Therapeutic misconception in the ethics of clinical trials. *Hastings Center Report, 33,* 3, 19–28.

Minogue, B.P., Palmer-Fernandez, G., Udell, L., & Waller, B.N. (1995). Individual autonomy and the double blind controlled experiment: the case of desperate volunteers. *Journal of Medicine and Philosophy, 20,* 43–55

Moan, C. E., & Heath, R. G. (1972) Septal stimulation for the initiation of heterosexual activity in a homosexual male. *Journal of Behavior Therapy and Experimental Psychiatry, 3,* 23–30.

Mukherjee, D., Nissen, S., & Topol, E. J. (2001) Risk of cardiovascular events associated with selective COX-2 inhibitors. *JAMA, 286,* 954–959.

Naeser, M. A., Martin, P. I., Nicholas, M., Baker, E. H., Seekins, H., & Helm- Estabrooks, N. (2005). Improved naming after TMS treatments in a chronic, global aphasia patient—case report. *Neurocase, 11,* 3, 182–193.

Oliveri, M., Rossini, P.M., Traversa, R.Cicinelli, P., Filippe, M.M., Pasqualetti, P., et al. (1999). Left frontal transcranial magnetic stimulation reduces contralesional extinction in patients with unilateral right brain damage. *Brain, 122,* 9, 1731–1739.

O'Reardon, J. P., Solvason, H. B., Janicak, P. G., Sampson, S., Isenberg, K. E., Nahas, Z., et al. (2007). Efficacy and safety of transcranial magnetic stimulation in the acute treatment of major depression: a multisite randomized control trial. *Biological Psychiatry, 62,* 11, 1208–1216.

Parent, A. (2004). Giovanni Aldini: From animal electricity to human brain stimulation. *The Canadian Journal of Neurological Sciences, 31,* 4, 576–584.

Pascual-Leone, A., Gates, J. R., & Dhuna, A. (1991). Induction of speech arrest and counting errors with rapid-rate transcranial magnetic stimulation. *Neurology, 41,* 5, 697–702.

Pascual-Leone, A., Houser, C.M., Reese, K., Shotland, L.I., Grafman, J., Sato, S., et al. (1993). Safety of rapid-rate transcranial magnetic stimulation in normal volunteers. *Electroencephalography and Clinical Neurophysiology, 89,* 2, 120–130.

Pascual-Leone, A., Davey, M., Wassermann, E. M., Rothwell, J., & Puri, B. (Eds.) (2002). *Handbook of transcranial magnetic stimulation*. London: Edward Arnold.

Pascual-Leone, A., Rubio, B., Pallardo, F., & Catala, M.D. (1996). Rapid-rate transcranial magnetic stimulation of left dorsolateral prefrontal cortex in drug-resistant depression. *Lancet, 348,* 9022, 233–237.

Pascual-Leone, A., Valls-Sole, J., Wassermann, E. M., & Hallett, M. (1994). Responses to rapid-rate transcranial magnetic stimulation of the human motor cortex. *Brain, 117,* 847–858.

Quintana, H. (2005). Transcranial magnetic stimulation in persons younger than the age of 18. *Journal of ECT, 21,* 2, 88–95.

Robertson, E. M., Theoret, H., & Pascual-Leone, A. (2003). Studies in cognition: The problems solved and created by transcranial magnetic stimulation. *Journal of Cognitive Neuroscience, 15,* 948–960.

Rossi, S., Hallett, M., Rossini, P. M., Pascual-Leone, A. (2009). Safety, ethical considerations, and application guidelines for the use of transcranial magnetic stimulation in clinical practice and research. *Clinical Neurophysiology, 120,* 12, 2008–2039.

Siebner, H. R., Tormos, J. M., Ceballos-Baumann, A. O., Auer, C., Catala, M. D., Conrad, B. (1999). Low-frequency repetitive transcranial magnetic stimulation of the motor cortex in writer's cramp. *Neurology, 52,* 3, 529–537.

Surgeon General (1999). *Mental health: A report of the surgeon general*. Chapter 4: Adults and mental health. Surgeongeneral.gov.

Thut, G., Ives, J. R., Kampmann, F., Pastor, M. A., & Pascual-Leone, A. (2005). A new device and protocol for combining TMS and online recordings of EEG and evoked potentials. *Journal of Neuroscientific Methods, 141,* 2, 207–217.

Valero-Cabre, A., Pascual-Leone, A., & Rushmore, R.J. (2008). Cumulative sessions of repetitive transcranial magnetic stimulation (rTMS) build up facilitation to subsequent TMS-mediated behavioral disruptions. *European Journal of Neuroscience, 27,* 3, 765–774.

Wagner, T., Fregni, F., Eden, U., Ramos-Estebanez, C., Grodzinsky, A., & Zahn, M. (2006). Transcranial magnetic stimulation and stroke: A computer-based human model study. *Neuroimage, 30,* 3, 857–870.

Wagner, T., Valero-Cabre, A., & Pascual-Leone, A. (2007). Noninvasive human brain stimulation. *Annual Review of Biomedical Engineering, 9,* 527–565.

Wagner, T.A., Zahn, M., Grodzinsky, A.J., & Pascual-Leone, A. (2004). Three-dimensional head model simulation of transcranial magnetic stimulation. *IEEE Transactions on biomedical Engineering, 51,* 1586–1598.

Walsh, V., & Pascual-Leone, A. (2003). *Neurochronometirics of mind: Transcranial magnetic stimulation in cognitive science*. Cambridge, MA: MIT Press.

Wassermann, E. M. (1998). Risk and safety of repetitive transcranial magnetic stimulation: Report and suggested guidelines from the International Workshop on the Safety of Repetitive Transcranial Magnetic Stimulation, June 5–7, 1996. *Electroencephalography and Clinical Neurophysiology, 108,* 1–16.

Wassermann, E. M., Epstein, C., & Ziemann, U. (2008). *Oxford handbook of transcranial magnetic stimulation*. Oxford: Oxford University Press.

Wu, A. D., Fregni, F., Simon, D. K., Deblieck, C., & Pascual-Leone, A. (2008). Noninvasive brain stimulation for Parkinson's disease and dystonia. *Neurotherapeutics, 5,* 2, 345–361.

Zago, S., Ferrucci, R., Fregni, F., & Priori, A. (2008). Bartholow, Sciamanna, Alberti: pioneers in the electrical stimulation of the exposed human cerebral cortex. *Neuroscientist, 14,* 5, 521–528.

17

Implanted Neural Interfaces

Ethics in Treatment and Research

LEIGH HOCHBERG AND THOMAS COCHRANE

Until recently, treatments for neurological and psychiatric impairment have been largely limited to drugs, rehabilitation therapy, or psychotherapy. These treatments are often only modestly successful in helping patients to recover function. Research into stem cell and other biologic therapies are promising but remain in the research phase (see Kimmelman, chapter 18 in this volume). In contrast to the above approaches, the past twenty years have seen remarkable improvements in restoring lost nervous system function with the use of neurotechnologies: devices which interface directly with the nervous system for the monitoring and modulation of neural tissue. While some neurotechnologies do not require surgery (e.g., peripheral nerve stimulators for the treatment of stroke-related foot drop), this chapter focuses on *implanted neural interfaces* (INIs): devices placed in or around the brain and spinal cord for the restoration of mobility, hearing, vision, speech, balance, mood, or cognition.

Ethicists have taken note of these devices because of the potential they have for dramatically altering the treatment of brain disease, and also because of their potential for neurobehavioral or cognitive enhancement in healthy people. In addition, although these technologies will revolutionize treatment of the nervous system, the ethical issues involved in INI research and treatment are substantial. They include issues that will be familiar to anyone involved in modern medicine and biomedical research, and some that are more specific to INIs. In this chapter, we will not attempt a full exposition of the familiar issues, but will survey aspects of INIs that evoke particular ethical attention.

CURRENT AND FUTURE IMPLANTED NEURAL INTERFACES

Generally speaking, the nervous system does three things. It (1) senses the internal and external environment; (2) stores, processes, and manipulates the sensed information; and then (3) produces behavior. Information processing and behavior can be reflexive and simple, or it can be conscious and complex. Nervous system disease and injury affects one or more of these functions, impairing (1) sensation; (2) cognition, mood, or memory; or (3) mobility and communication. In turn, INIs restore function by repairing or replacing sensory organs, cognitive and emotional processes, or motor function or speech. INIs that stimulate brain regions in order to regulate or normalize their function are called stimulators. In chapter 15 in this volume, Synofzik reviews recent progress in the use of DBS for psychiatric indications and explores the ethical issues raised by these applications. INIs that perform the lost or impaired function are called prostheses, and will be the primary focus of this chapter, following a general overview of INIs. We begin with a brief description of INIs in clinical use and active development, which will provide a background for our subsequent discussion of INI-related ethical issues.

IMPLANTED NEURAL INTERFACES IN CLINICAL USE

Auditory Prostheses

The earliest INIs were cochlear implants. Acquired or hereditary diseases of the inner ear can cause severe hearing impairments while leaving the auditory nerve intact. Cochlear implants are electronic devices that replace inner ear function by interfacing directly with the healthy auditory nerve. These devices consist of an external microphone to sense sound, a processor to convert the sound into electrical impulses that can be transmitted by the auditory nerve and decoded by the brain, and a direct electrical interface with the auditory nerve. Cochlear implants can restore hearing after acquired hearing loss, or provide hearing to congenitally deaf children. Research on these devices began the early 1960s, and the first commercial device was approved by the US Food and Drug Administration (FDA) in 1984. These devices initially had a percutaneous connection between the external microphone and the implanted electrodes, but newer devices wirelessly transmit acoustic information to the implanted device. Cochlear implants are the most successful INI to date—over 188,000 had been placed worldwide as of April 2009, and newer devices have considerably improved the fidelity and complexity of the acoustic information perceived in speech and music (National Institute on Deafness and Other Communication Disorders, 2009).

Deep Brain Stimulators

Stimulating electrodes that traverse 50mm or more to reach their targets in the deep nuclei of the brain (e.g., thalamus, subthalamic nucleus) are known as deep brain stimulators (DBS). These devices have been approved by the FDA for the treatment of movement disorder symptoms in Parkinson's disease and essential tremor. DBS is also being studied in patients with chronic pain, depression, obsessive-compulsive disorder, and disorders of consciousness. While the mechanisms by which DBS relieves motor dysfunction or psychological abnormality are not well understood, stimulation can be thought of as restoring more normal information transfer in the brain, possibly by interfering with abnormal (pathologic) neuronal network activity. (Johnson et al., 2008; Shah et al., 2009)

Spinal Cord Stimulators for Pain

Chronic pain affects hundreds of thousands of people, some of whom get inadequate relief from opioid and other pain relievers. Over the past 40 years, the placement of spinal cord stimulators (SCS)—epidural stimulating electrodes with an implanted signal generator—have become a common approach for managing refractory neuropathic pain (North et al., 2007; Cameron, 2004). Linderoth and Meyerson (1995) estimated that more than 100,000 people have been implanted with spinal cord stimulators to treat pain symptoms, with 12,000 people per year receiving SCS. Interestingly, little is known about the mechanisms by which electrical stimulation alters the activity of pain networks or the perception of pain (Mailis-Gagnon et al., 2004).

Functional Electrical Stimulation (FES)

The paralysis of skeletal and smooth muscle following spinal cord injury can lead to immobility and a need for mechanical ventilation. For the reanimation of limbs in people with spinal cord injury, stimulating electrodes can be implanted around peripheral nerves. With simple controllers activated by a still-mobile part of the body (for example, a shoulder shrug), people with tetraplegia can regain functional use of the arm and hand (Kilgore et al., 2006; Hincapie et al., 2008), and people with paraplegia have been able to stand for extended periods of time (Peckham & Knutson, 2005). It is anticipated that better controllers for FES devices will come from integration with other INIs, such as recordings from motor cortex as described below. In patients with diaphragmatic paralysis, recent implanted technologies can stimulate the diaphragm directly—so-called diaphragmatic pacing—and provide for automated breathing, reducing or eliminating the need for external mechanical ventilation (Onders, Elmo, & Ignagni, 2007).

Epilepsy Monitoring and Suppression Systems

Seizure—the pathologic over-activity of cerebral neurons leading to intermittent cortical dysfunction, an alteration in consciousness, and/or involuntary movements—has long been studied with INIs. In the surgical management of medically refractory epilepsy, grids of subdural electrodes or deep brain electrodes are placed over or into the cortical area of interest, hoping to localize an epileptic zone of tissue which can then be resected in order to reduce seizure frequency. One technology now in human trials (Sun, Morrell, & Wharen, 2008) detects seizures using electrocorticographic (subdural) or deep brain electrodes. The detection of a seizure would trigger the device to stimulate the abnormally firing brain region and disrupt the electrical activity before it spreads and causes clinical seizure.

An unrelated but clinically available device is the vagus nerve stimulator (VNS), which is designed to reduce seizures by stimulating the vagus nerve in the neck (Cohen-Gadol et al., 2003).

INVESTIGATIONAL IMPLANTABLE NEURAL INTERFACES

Visual Prostheses

Visual prostheses attempt to capture and transmit visual data to the brain, analogous to the way cochlear implants transmit acoustic data to the brain. Visual data, usually recorded by cameras, are converted and used to directly stimulate of the retina or optic nerve in patients with blindness due to ocular pathology. In addition, technologies are being developed which stimulate the lateral geniculate nucleus (Pezaris & Reid, 2007) or visual cortex (Dobelle, 2000), both of which may eventually help people with optic nerve disease or injury. Although visual prostheses are not yet commercially available, progress is being made in clinical investigations. (Rizzo et al., 2001; Winter, Cogan, & Rizzo, 2007; Horsager et al., 2008).

Vestibular Prostheses

The disruption of balance—dysfunction of the internal sensory organs that recognize orientation and acceleration in space—can be disabling. Preclinical and early clinical studies using electrical interfaces with the vestibular nerve are emerging (Wall et al., 2003). As with the above technologies, vestibular prostheses are designed to replace missing sensory information, or to reduce the influence of malfunctioning vestibular or neural function.

Cognitive Prostheses

There is early research into the possibility of replacing the function of the human hippocampus—a brain region critical for acquisition of new memories—using implantable electronics (Berger et al., 2005). Although this technology is in its infancy, our understanding of the neural underpinnings of learning, memory, and cognition continue to expand. There appears to be no theoretical barrier to the replacement of at least some complex cognitive functions. A so-called cognitive prosthesis would assist information processing by receiving neural impulses from other brain regions, processing and/or storing that information, and relaying it to other neural tissues.

Motor and Communication Prostheses

For patients with spinal cord injury, subcortical or brainstem stroke, neuromuscular disease including amyotrophic lateral sclerosis, limb amputation, and a host of other diseases and injuries, most or all of the disability results from immobility. In the most extreme situations—brainstem stroke or advanced ALS—the locked-in syndrome can develop (Laureys et al., 2005), in which a fully awake and alert person becomes unable to move or speak. For all such patients, the fundamental problem is a limited ability or inability to transmit internal information to the outside world in the form of communication or movement. Brain-computer interfaces (Kubler et al., 2005) are designed to achieve these goals, and INIs have been developed to convert the cortical neuronal activity naturally associated with intended movement into the control of computer cursor (Kubler et al., 2005; Musallam et al., 2004; Taylor et al., 2002, 2003; Serruya et al., 2002) and robot arms (Velliste et al., 2008). Figure 17.1 shows an electrode array of the kind implanted in motor cortex for such purposes. This work has shown great promise in nonhuman primates. Two groups have placed recording electrodes in the motor cortex of humans with tetraplegia, enabling control of computer cursors (Hochberg et al., 2007; Kennedy et al., 2000) and recently have demonstrated intracortical control of a prosthetic hand, a robot arm, and a point-and-click computer interface for typing, browsing the web, and activating other environmental control devices (Hochberg et al., 2007; Kim et al., 2008; Hochberg et al., 2012). Figure 17.2 shows a tetraplegic patient operating a computer with an INI.

ETHICAL ISSUES IN IMPLANTABLE NEURAL INTERFACES

Risks

All surgery carries preoperative, intraoperative, and postoperative risks. These include the risks of general or regional anesthesia, endotracheal intubation, and sedative and paralytic medications. Bleeding and infection can accompany any procedure that disrupts skin integrity. Neurosurgical procedures such as the

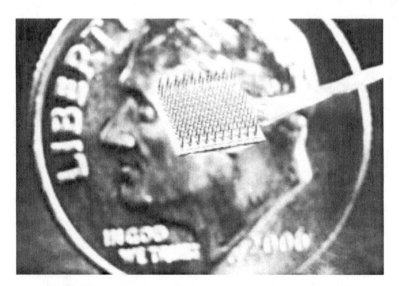

Figure 17.1 The "Utah array", in front of a U.S. dime shown for scale. The array is 4x4mm, and contains 100 electrodes of 1.5mm length. Each electrode can record from one or more inidividual cortical neurons.

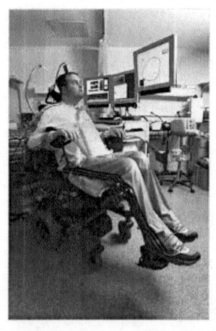

Figure 17.2 The first participant in the BrainGate pilot clinical trials. The trial participant, who was tetraplegic secondary to cervical spinal cord injury, attempted to move his hand in a circle. The array implanted in his dominant motor cortex transmitted neuronal ensemble activity, via the cable seen attached to a pedestal on the top of his head, to computers (seen in the background) which decoded the neural activity into the movement of a computer cursor on a screen (seen in front of the participant). (Courtesy, braingate2.org.)

placement of deep brain stimulators and other INIs (including investigational motor and communication prostheses that record with intracortical electrodes) introduce the possibility of stroke, intracerebral hemorrhage, seizure, loss of neurologic or neuropsychiatric function, and death. Like early cardiac pacemakers, cochlear implants, and deep brain stimulators, some newer investigational INIs have a percutaneous component or connector, which increases the risk of infection—this risk likely is reduced once the device is fully implanted and no longer requires a component that chronically disrupts the skin.

An exhaustive list of risks need not be provided here. Suffice it to say that the manner in which such a list is conceived and quantified, and then presented and discussed with patients or subjects, is critical to providing complete information and to obtaining fully informed consent for elective surgical procedures.

Informed Consent

An essential precondition for any medical or surgical intervention—whether therapy or research—is consent of the patient or research subject. Research and treatment with INIs are no exception. The importance of informed consent is greater when a therapy is experimental or new, when benefits are uncertain, or when the potential harms of the intervention are uncertain, common, or severe (Berg et al., 2001).

In patients or subjects with normal decisional capacity, *informed consent* requires at a minimum (1) the ability to understand the current condition and prognosis, (2) an understanding of the available options, (3) the risks and benefits of each option, and (4) the ability to choose among the alternatives. Many patients who might become involved in INI research and treatment are cognitively normal. Adults with acquired blindness, deafness, or spinal cord injury typically have the capacity to consent to research or treatment with visual, auditory, or motor prostheses. For these patients, the informed consent standard is relatively straightforward in its application. In research applications, institutional review boards (IRBs) ensure that study protocols meet standards for appropriateness, and review the formal consent process that must be used for every subject. In nonexperimental treatment applications (such as cochlear implants and deep brain stimulators) usual standards of surgical informed consent apply.

The ethical framework for proper conduct of early pilot trials of an investigational device, including INIs, deserves special attention. There is often no direct therapeutic benefit to patients in early trials of a device in development. The device may not work at all, early testing may indicate that the device needs to be revised, or the trial may even be performed specifically for the purpose of learning how the device should be altered for more optimal performance. Furthermore, while lists of potential device-related or procedure-related risks can be generated, and the frequency of adverse events can be estimated, by definition the risks and benefits of an investigational device are incompletely known. Participants in pilot trials enroll in part out of a desire to help other people with the same condition, but many may labor under the therapeutic misconception—the misapprehension by research subjects and researchers that trial participation will have direct therapeutic benefit (Appelbaum et al., 1987). It may not be possible to eliminate the therapeutic misconception entirely, but researchers should be honest and direct with subjects and themselves about potential therapeutic benefit—or the lack thereof.

The intangible benefits of participating in research (the satisfaction of contributing directly to medical science, the "concierge" coordination of medical care afforded to trial participants by their grateful clinician-researchers, etc.) are usually not discussed—and should not be raised by investigators—in order to avoid a distorting or coercive influence on decision making. Finally, investigators and IRBs must be confident that the potential benefit of research for future patients (e.g., the development of novel restorative therapies) justifies the risks to current subjects, even if those risks might be readily accepted by potential research subjects.

Decisional capacity, as discussed by Karlawish and Kim (chapters 6 and 7 in this volume) is a crucial issue for research and treatment with INIs, as capacity for informed consent is compromised or absent in some INI candidates. Children, patients with brain injury, and patients with very frequent seizures or medication side effects that impair cognition may have limited ability to consent. Psychiatric conditions such as depression and anxiety, when severe, can interfere with reasoning. Patients with neurodegenerative conditions such as Alzheimer's disease have steadily declining decisional capacity. In these settings, when truly informed consent cannot be obtained, research can only proceed after obtaining surrogate consent (usually a family member). While surrogate consent is commonplace for proven therapies, obtaining surrogate consent for investigational research can be particularly challenging, and the ethical and legal boundaries of such consent are somewhat unclear and contentious (Berg, 1996).

Communication with patients in the classic locked-in state (awake and alert, but only able to move the eyes) or an incomplete locked-in state (tetraplegic and anarthric, but still able to move some body part, such as a finger, voluntarily) is slow and effortful in the best of circumstances (see Fins & Schiff, chapter 13 in this volume). But when patients can communicate reliably (e.g., by signaling yes or no, or by selecting letters of the alphabet using assistive technology), it is possible to assess cognitive status and decisional capacity. However, the discussion of a potential therapy or research protocol may take weeks, with the physician or researcher providing information about the INI, including its risks and benefits, allowing the patient time to consider the intervention and to develop questions, which can then be answered and new questions asked. The extended time required for obtaining informed consent from patients with classic or incomplete locked-in syndrome requires additional effort from the investigator, but this should be viewed as an additional opportunity to obtain truly informed consent (which is a process that extends well beyond the review and signing of an IRB-approved form). Family or friends can sometimes aid in discussions with a locked-in patient because they can often anticipate questions that the patient would have asked—allowing the patient to observe a back-and-forth conversation about the proposed intervention. The investigator must take care to avoid the shortcut of relying entirely upon a surrogate's translation of a locked-in patient's communication method, and should make every effort to understand the patient herself.

Rarely, patients can become completely locked in, lacking the ability to move any body part, including the eyes. In this most severe communication impairment, only surrogate consent is possible—although some are investigating measurement of salivary pH to allow these patients yes/no communication (Wilhelm, Jordan, & Birbaumer, 2006). The problem of surrogate consent is no different for INIs than it would be for any other medical procedure, but it is intriguing to consider that implantation of an INI that restored communication would eliminate the subsequent need for surrogate consent in such a patient.

While the INIs discussed above are placed after long consideration and in the setting of chronic medical conditions, in the future such devices will become available and appropriate for use during acute disease or injury. The ability of patients and surrogates to learn and process new information and to make well-reasoned, thoughtful decisions can be impaired by the disorientation and emotional upheaval of sudden illness and disability. When researchers eventually begin planning for use of INIs in the acute setting, careful attention to issues of consent will be needed.

Effects of INI Research on Clinical Decision Making

It is possible that the mere existence of research protocols and potential new therapies for some conditions could dramatically affect the ways in which patients make decisions about established life-sustaining therapies (LSTs) such as artificial ventilation, hydration, and nutrition. In the absence of such therapies, many patients with spinal cord injury, brainstem stroke, or neuromuscular disease such as amyotrophic lateral sclerosis (ALS) decide that their condition is so burdensome that they prefer not to initiate or continue LST. Although this decision leads directly to their death, we allow patients to make this choice because we generally allow competent patients to refuse unwanted treatments. Because many healthy persons can appreciate the rationale behind such a patient's decision to withdraw LST, many physicians responsible for the care of such patients consider this an acceptable choice.

However, it is now conceivable that within a matter of years, today's patients with locked-in syndrome could regain effective communication and independent movement using INIs or other brain-computer interfaces. Indeed, INIs and other BCIs (Sellers & Donchin, 2006) are approaching clinical utility, although the availability and support for such devices may remain limited for some time. In turn, this development may begin to influence decisions about LST—persuading some to continue LST when they otherwise would have refused it. At least one editorial has expressed worry that the promise of such (as yet unavailable) communication-enhancing technology could cause some patients with ALS to continue LSTs until the point that they become totally locked in (Phillips, 2006).

We agree that widespread change in the guidance physicians provide regarding LST for locked-in patients would be premature in 2012, but do not agree that efforts to restore communication for these patients are misguided. It is true that much is unknown about quality of life in the locked-in state, but evidence suggests that it is better than healthy individuals (including physicians) predict (Laureys et al., 2005). It should also be recognized that an individual decision to continue LST is never irreversible. Patients who prolong LST in order to pursue the possibility of communication using INI or other technologies would be able to refuse the LST at a later time (either themselves, or by advanced directive or surrogate) if their quality of life is not sufficiently improved by the technology (Aita & Kai, 2006; Borasio, Gelinas, & Yanagisawa, 1998).[1]

Access to Research and Therapies

INI researchers face some dilemmas with respect to access to research protocols and posttrial access to interventions. These technologies are expensive and some require the regular attention of skilled researchers or technicians. Even if research shows them to be beneficial (as with cochlear implants), decisions must then be made about who will have access to INIs, who will pay for them, and who will make those decisions. Our tools for quantifying benefit are often crude, and there are disagreements about payment for some obviously beneficial interventions—as an example, some insurers will not pay for motorized wheelchairs if patients have access to caregivers who can push a manual wheelchair. With newer INIs, the benefits are not yet certain, and those who reimburse the cost of care must confront whether high-cost therapies should be provided—a problem that has received recent attention with respect to expensive cancer therapies (Brock, 2006).

Some candidates for INIs have catastrophic and untreatable diseases, making even the high risks of an experimental INI seem inconsequential from their perspective. What should a researcher say to a locked-in patient who might benefit from an INI, but who doesn't qualify for a research protocol due to a predetermined, but semiarbitrary exclusion criterion such as age? We believe researchers should carefully consider the risks and benefits of research participation to the individual patient. When appropriate, researchers should request an exception to the exclusion criteria for this single patient. The desire to help individuals in this way must be weighed against the need for a scientifically valid study, since the pursuit of unbiased knowledge about the INI and its potential safety and feasibility is what justifies the risks taken by research participants.

Current INI motor prostheses rely on frequent software adjustments, expensive external hardware, and technician assistance. If the investigational device is helpful (e.g., if the device provides a better communication method than the patient's usual assistive technology), what happens when the research protocol comes to an end and the technician is no longer available to the patient? It may or may not be advisable to remove an INI—decisions to do so must be made by research participants themselves, in consultation with the physician-investigator. End-of-trial transitions are more challenging in research involving implanted devices. INI researchers must consider the transition explicitly during the study design phase, and explicit and transparent discussion about this transition is an indispensable component of the informed consent process.

Enhancement of Neural Function

Given the pace of progress in INI research, some have wondered about the potential for such devices to enhance the function of an otherwise healthy individual. Here we will ignore the question of what makes an intervention an enhancement, as distinguished from therapy, and simply consider some key concerns about using INIs for enhancement. Some have imagined a future in which INIs enhance normal brain function—perhaps visual prostheses could permit sight in the infrared spectrum (enabling one to judge the warmth of an object from a distance), auditory prostheses might enable super-sensitive hearing, including the high-pitched sounds heard by a poodle or a porpoise. Even more imaginatively, bidirectional speech prostheses could permit silent, brain-based telepathy between individuals. In reviewing the subject of neural enhancement, Chatterjee discusses four potential categories of concern in his chapter: safety, character, inequity, and coercion.

The *safety* of INIs is primarily a function of the risks of surgical implantation and the chronic risk of an implanted foreign body. As the technology develops these risks will decline, and potential benefits will increase. At some point reasonable healthy individuals will be willing to take on those risks in pursuit of the benefits. What will clinicians and researchers say then? There are some combinations of risk and benefit that no reasonable, able-bodied person would assume (e.g., a neural implant that gives only hands-free access to the local movie timetable) and there are some risk-benefit ratios that most persons happily assume (e.g., ca hot cup of coffee to increase alertness while driving.) In between, there is a broad range of risk-benefit ratios about which reasonable persons can disagree. At one extreme of risk-benefit ratio, we should allow individuals to make their own choices about which risks to assume in pursuit of which benefits; at the other extreme, the device should not be available. Almost inevitably, there will come a time when INIs will be safe enough that we should allow individuals to make their own choice to use them for enhancement. It is interesting to note the Dalai Lama's comments during his lecture at the Society for Neuroscience Annual Meeting in November 2005: "If it was possible to become free of negative emotions by a riskless implementation of an electrode—without impairing intelligence and the critical mind—I would be the first patient." Surprisingly similar discussions occur regarding the ethics of elective surgery for cosmetic enhancement. We allow people to take substantial risks in pursuit of enhancements that seem to pale in comparison with the potential enhancement benefits of INIs.

Some involved in debates about enhancement suggest that there are burdens that should not be eliminated even if it becomes possible to do so, because it would adversely affect our human *character* (President's Council on Bioethics, 2003). Like Chatterjee, we find this argument against enhancement unpersuasive as a guide to policy. On what basis would policy makers decide which burdens are essential, and therefore mandatory? As a rule, we generally allow individuals to decide for themselves whether a burden should be accepted or avoided, and which aspects of their character to develop or neglect.

Concerns about *inequity*, on the other hand, are potentially persuasive as a guide to policy. Allowing only some individuals access to enhancement could lead to worsening societal inequities. Almost any INI-based enhancement is likely to be expensive at first, and only those with significant financial resources will be able to access it. For example, an INI that improved memory or calculation, provided better access to information or superior physical capabilities, would confer even greater ability to amass wealth or power. It is not difficult to envision a widening societal and economic gap between those who have such INIs and those who do not. It is not immediately clear how this problem should be managed. Inequality of *resources* is not necessarily cause for concern, but the inequality of *opportunity* that could arise in the setting of significant cognitive or physical enhancement certainly should be. This is an issue that will require attention in the future.

Coercion to enhance performance is also a concern that will require attention. Pharmacological enhancements already promise the ability to work longer hours and with greater mental focus, raising concern about workplace coercion—subtle and otherwise. INIs could provide even more impressive enhancements. For example, will the last few surgeons without an INI prosthesis to enhance their surgical abilities be able to resist the pressure to get one? How should physicians and society respond if chiefs of surgery begin to insist that all surgeons undergo such enhancement? For now, INI-based enhancements are more theoretical than real, but given the pace of progress in the field, it's not too early to have thoughtful public discussion about the appropriate use of INI for enhancement.

Military Applications and Research Support

Current INI research is firmly focused on the restoration of function to people with limb loss or neurologic disease/injury. This explains the interest of military organizations, which ensure and provide care and rehabilitation for wounded soldiers. Funding from the Department of Veterans Affairs, and the Department of Defense through the Defense Advanced Research Projects Agency (DARPA), has been instrumental in the development of INIs and superb prosthetic limbs. A goal of recent funding is to enable cortical control of an advanced prosthetic arm and hand for patients with upper extremity amputation (Ling, 2009).

The restoration of function for wounded warfighters should be uncontroversial, but if that same technology were later reconfigured toward the development of a neurally enhanced solider, what ethical dilemmas arise in this secondary use of the technology? We observe here that there is a perennial problem for scientists involved in *any* research that could be used to improve military operations, not just INIs. The ethical issues raised by such secondary uses of neuroscientific technology are covered in Chapter 3 by Russo and colleagues.

Animal research

INI research and the revolutionary therapies it promises would simply not be possible without animal research, including research with nonhuman primates. Indeed, the development of cortically controlled prosthetic devices, as just one example, results from more than 40 years of publicly funded basic research (funded largely by the NIH and the Department of Veterans Affairs). The translation of fundamental neuroscience and bioengineering knowledge into improved devices will require both human clinical testing and animal preclinical testing. Like all researchers who are engaged in animal research to alleviate human suffering (or other animal suffering, in the case of veterinary research), the responsibility to minimize any animal discomfort caused by the research is central to properly and effectively conducted research.

Most modern observers agree that we have a duty to prevent unnecessary discomfort in all its forms, including the discomfort of animals. Researchers must recognize the potential for discomfort created by their work, and must take steps to (1) avoid it whenever possible, (2) minimize it when it cannot be avoided, and (3) ensure that the results of the work justify any discomfort that is created. Research that entails even brief periods of discomfort to the animal should only be performed when it is justified by the promise of important results—a requirement that should preclude trivial, poorly designed, or unnecessarily duplicative research. Whenever INI research requires the use of animals, care must be taken to use appropriate analgesia or sedation, to ensure appropriate living conditions for the animals, and to observe institutional guidelines for animal research. Institutional approval of animal research is necessary, but by itself does not absolve researchers of the responsibility to attend to animal discomfort and to avoid or ameliorate it whenever possible.

Elimination of Disabilities and Concerns about Diversity and Value

Some believe that there are disabilities that should not be treated. Indeed, there is some disagreement about whether some conditions should be considered disabilities at all. For example, some in the deaf community object strenuously to the use of cochlear implants to treat congenital or acquired deafness, especially in children (Hyde & Power, 2006). Historically, deaf persons—the vast majority of whom are cognitively normal—have been stigmatized, segregated, and deprived of educational opportunity. As a result, many in the deaf community share strong cultural connections with deaf friends and family members, and choose not to view their deafness as a disability. A minority also perceive attempts to treat hearing loss as attempts to eliminate deaf culture, and to diminish the personal worth of deaf persons. This view seems strange to physicians and family members who wish to provide hearing to those who lack it, but given the historical and social context, it should not be disregarded or dismissed in a casual way. Researchers and clinicians working with INIs and other technologies to restore lost or absent function should be aware of this perspective among some in the diverse community of people with disabilities, and carefully avoid making quick judgments about the autonomy and quality of life of persons with disabilities.

IMPLICATIONS FOR POLICY AND PRACTICE

Do INIs pose truly new ethical concerns, or do they simply represent novel versions of old problems in ethics? All of the issues discussed in this chapter— informed consent, conflicts of interest, privacy, coercion, military applications of scientific research, end-of-life decision making, and enhancement—have long been familiar to physicians, researchers, and ethicists. At this point, the ethical issues specific to INIs are all old ones, which leads us to conclude that INI technologies are not so ethically distinct from other medical interventions that separate bodies of ethical analysis or oversight are required. Current governmental policies and regulations designed to ensure safety and research integrity are applicable to current and foreseeable INIs.

Although the ethical issues raised by INIs are not new, the technologies themselves are, and the full medical and societal implications of INIs are not yet fully understood. With this uncertainty comes risk, to current and future patients and subjects. Even the *perception* of unethical conduct could halt or slow beneficial research (a lesson learned the hard way in the field of gene therapy), which in turn could harm the interests of future patients who would have benefited from the research. The best way to avoid harms to these future patients is to avoid harms to current subjects and patients, to pay scrupulous attention to the ethical conduct of research, and to allow as much public scrutiny and transparency as possible (Fins & Schiff, 2010). Federally funded INI research undergoes additional layers of ethical review, but individual researchers and physicians must still pay special attention to the ethical conduct of INI research and therapies. Researchers must not assume that IRB approval and consent forms alone are sufficient protections.

There will be an increasing need for societal consensus regarding the appropriate use of restorative and enhancing INI technologies. Researchers, physicians, patients, research subjects, and funding agencies are in the best position to assess the medical, societal, and ethical risks and benefits of INI technology. The INI research and treatment community can only benefit from careful attention to surgical risks and conflicts of interest, and the early anticipation and discussion of dilemmas that may arise from future INI developments.

AUTHOR NOTE

The views expressed in this chapter are solely those of the authors. They do not necessarily represent the views of the educational or federal institutions with which the authors are affiliated. The authors thank Joseph J. Pancrazio and Geoffrey Ling for their insightful comments and suggestions regarding early drafts of this chapter.

NOTE

1. Our claim that the decision to begin LST is not irreversible may not apply to all cultures. As an example, complete LIS may be more common in Japan, where there are different societal and legal preconditions for withdrawal of LST (Aita & Kai, 2006; Borasio, Gelinas, & Yanagisawa, 1998) than in the United States.

REFERENCES

Aita, K., & Kai, I. (2006). Withdrawal of care in Japan. *Lancet, 368* (9529), 12–14.

Appelbaum, P., Roth, L., Lidz, C., Benson, P., & Winslade, W. (1987). False hopes and best data: Consent to research and the therapeutic misconception. *Hastings Center Report, 17* (2), 20–24.

Berg, J. (1996). Legal and ethical complexities of consent with cognitively impaired research subjects: Proposed guidelines. *Journal of Law, Medicine and Ethics, 24* (1), 18–35.

Berg, J., Appelbaum, P., Lidz, C., & Parker, L. (2001). *Informed consent: Legal theory and clinical practice, 2nd ed.* New York: Oxford University Press.

Berger, T., Brinton, R., Marmarelis, V., Sheu, B., & Tanguay, A. (2005). Brain-implantable biomimetic electronics as a neural prosthesis for hippocampal memory function. In T. Berger & D. Glanzman (Eds.), *Toward replacement parts for the brain* (pp. 241–276). Cambridge, MA: The MIT Press.

Borasio, G., Gelinas, D., & Yanagisawa, N. (1998). Mechanical ventilation in amyotrophic lateral sclerosis: a cross-cultural perspective. *J Neurol, 245* (Suppl 2), S7–S12.

Brock, D. (2006). How much is more life worth? *Hastings Center Rep, 36* (3), 17–19.

Cameron, T. (2004). Safety and efficacy of spinal cord stimulation for the treatment of chronic pain: A 20-year literature review. *J Neurosurg, Spine 3* (100), 254–267.

Cohen-Gadol, A., Britton, J., Wetjen, N., Marsh, W., Meyer, F., & Raffel, C. (2003). Neurostimulation therapy for epilepsy: Current modalities and future directions. *Mayo Clin Proc, 78* (2), 238–248.

Dobelle, W. (2000). Artificial vision for the blind by connecting a television camera to the visual cortex. *ASAIO J, 46* (1), 3–9.

Fins, J., Schiff, N. (2010) Conflicts of interest in deep brain stimulation research and the ethics of transparency. *J Clin Ethics, 21* (2), 125–132.

Hincapie, J., Blana, D., Chadwick, E., & Kirsch, R. (2008). Musculoskeletal model-guided, customizable selection of shoulder and elbow muscles for a C5 SCI neuroprosthesis. *IEEE Trans Neural Syst Rehabil Eng, 16* (3), 255–263.

Hochberg, L., Serruya, M., Friehs, G., Mukland, J., Saleh, M., Caplan, A., et al. (2006). Neuronal ensemble control of prosthetic devices by a human with tetraplegia. *Nature, 442* (7099), 164–171.

Hochberg, L.R., Bacher, D., Jarosiewicz, B., Masse, N.Y., Simeral, J.D., Vogel, Haddadin, S., Liu, J., Cash S.S., van de Smagt, P., and Donoghue, J.P. (2012). Reach and grasp by people with tetraplegia using a neurally controlled robotic arm. *Nature, 485,* 372–375.

Horsager, A., Greenwald, S., Weiland, J., Humayun, M., Greenberg, R., McMahon, M., et al. (2009). Predicting visual sensitivity in retinal prosthesis patients. *Invest Ophthalmol Vis Sci, 50* (4), 1483–1491.

Hyde, M., & Power, D. (2006). Some ethical dimensions of cochlear implantation for deaf children and their families. *J Deaf Stud Deaf Educ, 11* (1), 102–111.

Johnson, M., Miocinovic, S., McIntyre, C., & Vitek, J. (2008). Mechanisms and targets of deep brain stimulation in movement disorders. *Neurotherapeutics, 5* (2), 294–308.

Kennedy, P., Bakay, R., Moore, M., Adams, K., & Goldwaithe, J. (2000). Direct control of a computer from the human central nervous system. *IEEE Trans Rehab Eng, 8* (2), 198–202.

Kilgore, K., Hart, R., Montague, F., Bryden, A., Keith, M., Hoyen, H., et al. (2006). An implanted myoelectrically-controlled neuroprosthesis for upper extremity function in spinal cord injury. *Conf Proc IEEE Eng Med Biol Soc, 1,* 1630–1633.

Kim, S., Simeral, J., Hochberg, L., Donoghue, J., & Black, M. (2008). Neural control of computer cursor velocity by decoding motor cortical spiking activity in humans with tetraplegia. *J Neural Eng, 5* (4), 455–476.

Kübler, A., Nijboer, F., Mellinger, J., Vaughan, T., Pawelzik, H., Schalk, G., et al. (2005). Patients with ALS can use sensorimotor rhythms to operate a brain-computer interface. *Neurology, 64* (10), 1775–1777.

Laureys, S., Pellas, F., Van Eeckhout, P., Ghorbel, S., Schnakers, C., Perrin, F., et al. (2005). The locked-in syndrome: What is it like to be conscious but paralyzed and voiceless? *Progress in Brain Research, 150,* 495–511.

Linderoth, B., & Meyerson, B. (1995). Dorsal column stimulation: modulation of somatosensory and autonomic function. *Seminars in Neuroscience, 7* (4), 263–277.

Ling, G. Revolutionizing prosthetics. http://www.darpa.mil/our_work/dso/programs/revolutionizing_prosthetics.aspx

Mailis-Gagnon, A., Furlan, A., Sandoval, J., & Taylor, R. (2004). Spinal cord stimulation for chronic pain. *Cochrane Database of Systematic Reviews* (3), CD003783.

Musallam, S., Corneil, B., Greger, B., Scherberger, H., & Andersen, R. (2004). Cognitive control signals for neural prosthetics. *Science, 305* (5681), 258–262.

National Institute on Deafness and Other Communication Disorders. (2009). *NIH Publication No. 09–4798*. Bethesda: National Institute of Health.

North, R., Shipley, J., Prager, J., Barolat, G., Barulich, M., Bedder, M., et al. (2007). Practice parameters for the use of spinal cord stimulation in the treatment of chronic neuropathic pain. *Pain Medicine, 8* (Suppl 4), S200–S275.

Onders, R., Elmo, M., & Ignagni, A. (2007). Diaphragm pacing stimulation system for tetraplegia in individuals injured during childhood or adolescence. *J Spinal Cord Med, 30* (Suppl 1), S25–S29.

Peckham, P., & Knutson, J. (2005). Functional electrical stimulation for neuromuscular applications. *Annu Rev Biomed Eng, 7*, 327–360).

Pezaris, J., & Reid, R. (2007). Demonstration of artificial visual percepts generated through thalamic microstimulation. *Proc Natl Acad Sci U S A, 104* (18), 7670–7675.

Phillips II, L . (2006). Communicating with the "locked-in" patient: Because you can do it, should you? *Neurology, 67* (3), 380–381.

President's Council on Bioethics (2003). Beyond therapy: Biotechnology and the pursuit of happiness. Washington, DC.

Rizzo III, J ., Wyatt, J., Humayun, M., & et al. (2001). Retinal prosthesis: An encouraging first decade with major challenges ahead. *Ophthalmology, 108* (1), 13–14.

Santhanam, G., Ryu, S., Yu, B., Afshar, A., & Shenoy, K. (2006). A high-performance brain-computer interface. *Nature, 442*, 195–198.

Schiff, N., Giacino, J., & Kalmar, K., et al. (2007). Behavioural improvements with thalamic stimulation after severe traumatic brain injury. *Nature, 448* (7153), 600–603.

Sellers, E., & Donchin, E. (2006). A P300-based brain-computer interface: Initial tests by ALS patients. *Clinical Neurophysiology, 117* (3), 538–548.

Serruya, M., Hatsopoulos, N., Paninski, L., Fellows, M., & Donoghue, J. (2002). Instant neural control of a movement signal. *Nature, 416*, 141–142.

Shachtman, N. (2007). Pentagon preps mind fields for smarter war stations. *Wired Online*. www.wired.com/science/discoveries/news/2007/03/72996?currentPage=all.

Shah, S., Baker, J., Ryou, J., Purpura, K., Schiff, N. (2009). Modulation of arousal regulation with central thalamic deep brain stimulation. *Conf Proc IEEE Eng Med Biol Soc. 2009*, 3314–3317.

Sun, F., Morrell, M., & Wharen, R. (2008). Responsive cortical stimulation for the treatment of epilepsy. *Neurotherapeutics, 5* (1), 68–74.

Taylor, D., Tillery, S., & Schwartz, A. (2002). Direct cortical control of 3D neuroprosthetic devices. *Science, 296*, 1829.

Taylor, D., Tillery, S., & Schwartz, A. (2003). Information conveyed through brain-control: Cursor versus robot. *IEEE Trans Neural Syst Rehabil Eng, 11* (2), 195–199.

Velliste, M., Perel, S., Spalding, M., Whitford, A., & Schwartz, A. (2008). Cortical control of a prosthetic arm for self-feeding. *Nature, 453* (7198), 1098–1101.

Wall III, C ., Merfeld, D., Rauch, S., & Black, F. (2002–2003). Vestibular prostheses: The engineering and biomedical issues. *J Vestib Res, 12* (2–3), 95–113.

Wilhelm, B., Jordan, M., & Birbaumer, N. (2006). Communication in locked-in syndrome: Effects of imagery on salivary pH. *Neurology, 67* (3), 534–535.

Winter, J., Cogan, S., & Rizzo III, J . (2007). Retinal prostheses: Current challenges and future outlook. *J Biomater Sci Polym Ed, 18* (8), 1031–1055.

18

Biologics and the Human Brain

JONATHAN KIMMELMAN

According to a recent World Health Organization report, neurological disorders account for 6.3% of the global burden of disease—more than malignant neoplasms or HIV/AIDS (WHO, 2006). Neurological diseases also exact an enormous toll on the economy; one recent study found that brain disorders cost European countries €386B each year (Andlin-Sobocki, Jonsson, Wittchen, & Olesen, 2005).

Biologic therapies offer a promising set of strategies for modifying the otherwise inexorable and costly course of neurological disorders. These approaches include gene transfer (application of genetic materials encoding gene products like neurotransmitter metabolizing enzymes, receptors, or growth factors), cell transplantation (delivery of tissues like neural stem cells, fetal tissues, or bone marrow-derived cells), and neurotrophic factors (administration of diffusible peptides that promote survival or regeneration of neural tissues).

Yet their development and application raise a host of challenges for various decision makers in health care and research. In what follows, I offer a bracingly condensed overview of biologic-based approaches to brain disorders and associated ethical, social, and policy issues. A disclaimer: Though ocular diseases, neuropathies, and spinal cord injury are no less neurological than diseases like Parkinson's, the present chapter will focus on neurological disorders of the brain.

OVERVIEW

Many neurological disorders are caused by loss or degeneration of tissue in the brain. Therapies for halting or reversing such processes are severely constrained by several aspects of brain physiology. For instance, neurological tissue has very limited regenerative capacity. Once damaged through injury or disease, options for restoring function are limited. Another difficulty arises from the fact that brain function is dictated by complex and organized neural circuitry. Strategies that deliver therapeutic agents to the entire brain risk deranging delicately balanced functions in off-target brain structures. A third important challenge is the blood–brain barrier, which prevents many drugs from reaching their target.

Biologic based approaches offer opportunities for modifying disease course within these constraints. In theory, delivered tissues can protect brain structures from degeneration, or even restore function. Neural growth factors can stimulate tissue regeneration within specific brain structures, and gene transfer approaches might correct metabolic processes that disrupt normal function in particular brain regions. Because gene vectors or cells, once instilled in the brain, can provide continuous pharmacological activity, they have the potential to overcome the problem of the blood–brain barrier.

Unfortunately, however, clinical translation of biologics for brain disorders has been beset by numerous technical obstacles. Attempts to transplant cells to the brain date as far back as 1890, though experimental tissue transplantation only gathered momentum in the late 1980s (Redmond et al., 1993). After a series of open label studies suggesting significant clinical benefit for patients with Parkinson's disease, various tissue grafts were put to randomized trials in the late 1990s. These agents did not demonstrate significant clinical benefit against a sham comparator; in studies involving fetal tissue grafting, some patients developed disabling dyskinesias (Freed et al., 2001; Olanow et al., 2003). As of this writing, no major fetal tissue graft studies have been initiated since, though new strategies are being contemplated. Tissue grafting has also been tested in Amyotrophic Lateral Sclerosis (ALS) (Mazzini et al., 2010), Huntington's disease (Bachoud-Levi et al., 2006; Hauser et al., 2002; Rosser et al., 2002), stroke (Kondziolka et al., 2005), epilepsy (Edge, 2000; Loscher, Gernert, & Heinemann, 2008), and lysosomal storage diseases (Mossman, 2006). In addition, intravenously delivered cells have been transplanted to patients with stroke and lysosomal storage diseases (Boncoraglio, Bersano, Candelise, Reynolds, & Parati, 2010; Lonnqvist et al., 2001). As of this writing, none of these approaches has demonstrated efficacy in randomized trials.

The earliest trials involving gene transfer for brain based disorders involved delivery of gene transfer vectors to the brains of patients with malignant glioma. Gene transfer trials for nonmaligancies have been conducted in ALS (Aebischer et al., 1996), Canavan's disease (McPhee et al., 2006), Alzheimer's disease (Tuszynski et al., 2005), Parkinson's disease (Christine et al., 2009), Huntington's disease (Bloch et al., 2004), X-linked adrenoleukodystrophy (Cartier et al., 2009), and neuronal ceroid lipofuscinosis (Worgall et al., 2008). With two exceptions, those approaches advanced into randomized trials have not met their endpoints. The first exception involved a malignant glioma treatment. This strategy reached phase 3, but was rejected for regulatory licensure in 2010 (Mitchell, 2010). The second exception involved a sham controlled phase 2 trial of one of several gene transfer strategies for Parkinson's disease (LeWitt et al., 2011). However, the advantages of this approach over established effective care—deep brain stimulation—have yet to be demonstrated (Hutchinson, 2011).

Many neurotrophic factor trials have used gene transfer vectors to deliver agents. However, several teams have conducted trials involving direct, catheter delivery of neurotrophic factors in patients with Parkinson's disease (Gill et al., 2003; Lang et al., 2006), Alzheimer's disease (Crook, Ferris, Alvarez, Laredo, & Moessler, 2005), and ALS (Beck et al., 2005; Kalra, Genge, & Arnold, 2003; Nagano et al., 2005; Ochs et al., 2000). Some of the former studies raised safety concerns (Anon, 2005; Nutt et al., 2003); none have led to well-designed randomized trials demonstrating efficacy.

Trials of biologic based strategies are not an undifferentiated landscape of failure. Studies that did not show clinical activity have yielded insights into disease processes and therapeutic strategies (for specific examples, see Bartus et al., 2011; Kordower, Chu, Hauser, Freeman, & Olanow, 2008). And some studies have produced encouraging results. One example is the regression of disease in a study using intravenous delivery of hemaotopoetic stem cells (with and without gene transfer) for X-linked adrenoleukodystrophy (Cartier et al., 2009). Nevertheless, biologic-based strategies for treatment of brain disorders have yet to deliver on their promise.

Therefore, the primary challenges for neurologists and other health care stakeholders presented by biologics are primarily in the arena of research and innovative care. This chapter will survey distinctive ethical and policy issues surrounding the experimental application of biologics for brain disorders.

CLINICAL EXPERIMENTATION

Clinical research, in general, presents two broad ethical/policy problems. First, the raw materials of trials—human beings—have moral status. Second, error, oversight, or bias in clinical research can adversely affect downstream consumers of evidence. Testing biologics in the human brain raises several distinctive problems for addressing each.

Consent and Capacity

The brain is the very organ of decisional capacity. Studies that intervene on brain processes often involve patients with compromised decisional capacity. According to prevailing ethical frameworks (and various policies), patients lacking decisional capacity ought to be accorded additional protection from risk, since they are unable to protect their own interests through consent (see chapters 6 and 7 by Karlawish and Kim for discussions of capacity and consent). For example, the Declaration of Helsinki—one of the most authoritative international statements on human research ethics—states that nontherapeutic experimental procedures applied in persons lacking decisional capacity ought never to exceed minimal risk (WMA, 2011, paragraph 27). As we will see below, biologic based strategies involve unusually high levels of risk for clinical research. How, then, can brain-directed biologics be advanced within these stringent standards of research risk?

The prevailing answer would be to ensure that trials testing biologic strategies do not apply research procedures that exceed minimal risk, and that all other procedures satisfy conditions of clinical equipoise—briefly, uncertainty within the expert community about the balance of risks and benefits as compared with standard of care in the same patient population. Thus, when first-in-human trials of novel biologics are conducted in decisionally incompetent populations, they should never substitute for a validated treatment. Nor should they apply study procedures, like invasive sham surgeries, that involve greater than minimal risk. As we will see in the next section, the requirement of clinical equipoise may, however, present an impasse for the development of certain brain-based biologics.

Acceptable Risk

Such trials involve extraordinarily high degrees of risk and uncertainty. First, because the brain is relatively inaccessible, trials usually involve invasive surgical procedures (exceptions are studies involving cell infusion, as used in the X-linked adrenoleukodystrophy studies). Surgery adds a relatively fixed level of risk. According to one recent estimate, delivery of cell or genetic material to deep nuclei along four needle trajectories would confer a risk of surgery-related mortality of .5% (Kimmelman et al., 2011).

Second, whereas drugs can be administered and then withdrawn, instillation of biologic agents can be irreversible (or, to put it more accurately, reversibility entails substantial risk because it may involve tissue ablation). Gene transfer vectors or tissue transplants continue to exert pharmacological effects long after delivery. Transplanted tissues or vectors may give rise to malignancies over the long term, or subtle derangements in brain function due to misdifferentiation of tissue. Once toxicities emerge, options for halting them may be limited.

Third, risks of trials are very difficult to anticipate, and disease response difficult to model in animals. Brain processes like cognition, language, or affect may be distinctively human, and thus not well represented in preclinical studies (Mathews et al., 2008). Many human brain diseases have no analog in animals. Compounding this uncertainty, biologic-based strategies are potentially immunogenic, and immunotoxicities are difficult to model in nonhuman animals (Descotes, 2005). A common challenge in development of brain-directed biologics, then, is the question of when preclinical data are sufficiently complete and compelling to justify initiating first-in-human trials.

In the previous section, I mentioned that clinical equipoise establishes a moral boundary of acceptable risk for research procedures performed on decisionally incompetent persons. According to many commentators, it also establishes a boundary of acceptable risk for all therapeutically directed research procedures, regardless of decisional capacity (Weijer & Miller, 2004). In late phase trials, evidence gathered in early phase trials of a novel agent may enable emergence of clinical equipoise. But given the profound uncertainty concerning benefit and the relatively fixed levels of risk for surgical delivery, it may be difficult to sustain a claim that delivery of novel biologics to the brains of patients in early phase trials can be reconciled with clinical equipoise.

Standards of acceptable risk also present recurrent challenges for subject selection. Many brain disorders are degenerative. The rationale for many biologic based strategies is to intervene in this degenerative process. This biological feature of brain disorders presents another ethical conundrum. If investigators conduct studies in patients with advanced disease, there may be too little healthy tissue to protect and investigators may be unable to detect efficacy. Moreover, such patients may be unable to provide valid informed consent. The alternative is to conduct trials in patients with relatively recent disease onset. While addressing the consent and efficacy signal problems, enrolling recent onset patients introduces new risk problems. Such patients may have a reasonable life expectancy, with various treatment options available. Measured against a baseline of expected quality of life, trials have much less favorable risk-direct benefit balance for medically stable volunteers. There are no simple ways out of this dilemma, though the present commentator favors an approach in which agents are introduced into patients with recent disease onset only once interventions have been optimized and tested for safety in refractory patients.

Design Challenges

Biologics directed toward brain disorders present various design challenges for trialists. First, biologic interventions are complex. Research protocols may involve new surgical procedures, novel delivery systems, as well as a previously untested investigational agent. The agent itself can involve many components (a gene transfer agent, for example, consists of a vector plus a transgene), and can require cointerventions like immunosuppressive drugs. Each component adds layers of uncertainty about risk and study design.

For example, complexity prompts questions about which comparator interventions should be used in randomized trials. Parkinson's disease trials involving surgical delivery often produce large and persistent placebo effects (de la Fuente-Fernandez, Schulzer, & Stoessl, 2002). Maintaining allocation concealment in such studies may require sham surgical procedures (e.g., a partial burr hole or, in one case (Lang et al., 2006), actual delivery of saline to brain structures) and a course of immunosuppressive drugs. Is it ethical to expose patients with serious unmet health needs to such risks absent a therapeutic justification? Or, is it ethical to enroll patients in a trial that, in forgoing a sham comparator arm, leaves a significant and well-established threat to internal validity unaddressed? Though highly controversial when first applied in Parkinson's disease trials (Macklin, 1999), the use of sham comparators—provided good biostatistical justification, consent, and risk minimization—is for the most part accepted.

Another design challenge is safety monitoring. As noted, biologics may give rise to late-onset complications that are undetectable in trials. These complications may be particularly difficult to detect, in part because they affect more subtle aspects of brain function (Duggan et al., 2009). To the extent that biologic-based strategies require one-time administration and are effective over an extended period, patients receiving them may return to their doctors less frequently for follow-up (Sugarman & Sipp, 2011). Monitoring late-onset complications will likely present challenges for scientists and regulators; several commentators have urged trialists to conduct neuropsychiatric assessments at baseline and at intervals following treatment (Duggan et al., 2009).

Integrity

Often, patients with brain disorders face bleak prognoses. They—or their caregivers—may be willing to undergo extraordinary levels of risk for the remotest prospect of benefit. This often gives rise to a conflict between the canons of respect for persons (which favors respecting a well-informed patient's preferences for risk) and beneficence (which tends toward protecting patients from undue risk). A good illustration of this tension is provided by an episode in which parents sought to enter their child in a gene transfer trial for Canavan disease. The parents were educated and exceptionally well-informed, investigators were ready and willing to administer vector, and the outlook for the child was otherwise grim. And yet ethics committees initially refused to approve the protocol (Winerip, 1998).

Nevertheless, the aggregate effects of individuals pursuing their personal interests can have adverse social consequences. Clinical research is a collective undertaking; it draws on scarce and common resources. For instance, trials recruit from a limited population of eligible trial volunteers, and rely on the same pool of talent and funding for their execution. They also have externalities. Trial successes and failures can affect many third parties by improving or lowering the reputation of a research line, or by depleting a reserve of resources. Many gene transfer researchers found their work stymied by heavier oversight and diminished private and public sponsorship after a volunteer died in a gene transfer trial in 1999 (Kimmelman, 2010).

This may be in a particularly salient concern for a research area as politically charged as stem cell research or brain interventions. In numerous instances, controversies surrounding the testing of brain-directed biologics have breached the containment of medical journals and landed in newspapers (Edelman, 2003; Kolata, 2001; Pollack, 2005; Winerip, 1998). Accordingly, there may be nonpaternalistic grounds for regulating private transactions between willing patients and investigators on the basis of risk, scientific merit, or resource consumption (London, Kimmelman, & Emborg, 2010).

Moral Status

Human protections begin with the premise that human beings possess a sui generis moral status. This moral status, by some accounts, derives from mental capacities that are uniquely human. Basic as well as translational studies occasionally involve grafting human tissues into the brains animals. For example, in 2001, scientists reported grafting neural tissues into the brains of fetal bonnet monkeys to study brain development (Ourednik et al., 2001). Such studies raise the possibility that experiments may cause emergence of human-like brain functions- and hence human moral status- in chimeric animals.

Though some commentators recoil at the notion of humanizing animals or crossing species boundaries, the most compelling ethical issues surrounding cross-species neural tissue grafting is the problem of moral status. First, is it ethical to confer traits that carry moral status in their train (Greene et al., 2005; Karpowicz, Cohen, & van der Kooy, 2005; Streiffer, 2005)? Second, what kind of moral status should be accorded when chimeras possess some morally relevant human traits but not others (Greene et al., 2005)? Various commentators and policy-making initiatives advocate caution and independent ethical review of studies involving human neural grafting to animals (Greene et al., 2005; Hyun et al., 2007; National Research Council, 2005). Curiously, however, none provide a moral framework for the evaluation of such studies. Perhaps this reflects the great difficulty and political contentiousness of defining precisely what pattern of mental features accounts for human moral status (DeGrazia, 1996) and Farah, chapter 14, this volume.

CLINICAL APPLICATION

Because they have yet to be validated, current ethical issues surrounding brain-directed biologics revolve around clinical trials. Increasingly, however, patients are seeking access to unproven treatments outside of research. One particularly controversial venue is medical tourism. Many overseas medical clinics make claims in online promotional materials that extend well beyond published, systematic investigation (Caulfield, 2004; Stafford, 2009). Medical tourism for brain disorders raises a host of ethical and policy problems.

Among them is the question of whether administration of such risky, nonvalidated treatments outside a formalized study is ethical. Recognizing the fuzziness of boundaries separating innovative care and clinical research, both the Declaration of Helsinki and the Belmont Report sanction innovative care. The former does, however, require informed consent, and both policies encourage formalized study of an innovative approach (National Commission, 1979; WMA, 2011, paragraph 35). In general, protocols for innovative care involving novel (and risky and expensive) approaches should receive prospective ethical and safety review, and some commentators argue that scientists should lobby their local authorities for regulation of such clinics (Kiatpongsan & Sipp, 2009). In June 2010, the International Society of Stem Cell Research (ISSCR) created a publicly accessible, web-based resource reporting on whether clinics were supervised by a regulatory body like the FDA, and whether they maintained ethics committees to protect the interests of patients. The initiative was reportedly suspended after overseas clinics threatened litigation (Ledford, 2011).

Another important question is how clinicians ought to respond to patients (or caregivers) contemplating pursuit of nonvalidated treatments—either within the research setting or outside it through medical tourism. In both instances, patients should be carefully informed about risks and uncertainties. A physician can also furnish patients or caregivers with base rate information—that, for example, that only a small minority of CNS disorders entering clinical development are ever proven safe and effective. In the case of research, patients should understand that there may be burdensome study procedures that are not designed to cause disease response (e.g., use of allocation concealment, a sham comparator, extra blood draws). For patients considering medical tourism, one approach recently favored by various commentators is to help patients become effective information consumers (Kukla, 2007). For example, the ISSCR has published a patient handbook (ISSCR, 2008) that provides a list of questions patients should put to clinics or clinicians offering stem cell treatments.

CONCLUSION AND FUTURE DIRECTIONS

This chapter has emphasized ethical issues faced today. Application of validated brain-directed biologics will, of course, eventually present important challenges for policy and ethics. In 1999, researchers published a study showing they could make smarter, less absent-minded mice by genetically modifying forebrains with vectors carrying the NMDA receptor 2B. In 2004, a different team of researches induced monogamous behavior in a promiscuous rodent species, the meadow vole, by delivering the vasopressin V1a receptor gene to the ventral forebrain (Lim et al., 2004). That same year, researchers converted shiftless monkeys into workaholics by genetically modifying neurotransmitter receptors in the rhinal cortex (Liu et al., 2004).

Elsewhere, this book addresses ethical issues arising from the application of new brain interventions for military ends or performance enhancement. Two other major issues are visible at the horizon.

The first is cost. At first glance, biologic approaches would seem to offer economic advantages over conventional, molecule-based drugs. Whereas the latter may require chronic administration, stem cells are delivered once and might be curative. More likely, however, biologic strategies will be extremely costly. Biologics are associated with several drivers of cost, including elaborate and expensive manufacturing, the requirement of surgical delivery, resistance of payers for covering one-off treatments (where savings may primarily benefit parties other than the insurer), and lengthy, research-intensive development periods (Danzon & Towse, 2002; Fulton, Felton, Pareja, Potischman, & Scheffler, 2009). In addition, a clear pathway for defining biosimilarity for biologics has yet to be established (Paradise, 2011; Reingold et al., 2009). If biologics are ever successfully translated for conditions that afflict large populations, like Alzheimer's disease, they are likely to cause significant strain on health care systems.

A second issue is expansion of morbidity. Many brain-directed biologics target symptoms that originate in specific brain structures. And yet the disease process may affect other brain structures as well. Consider Parkinson's disease. Motor symptoms are caused by degeneration of dopaminergic neurons in the substantia nigra. Gene transfer and cell transplantation approaches are therefore directed toward this structure. However, Parkinson's disease affects other brain regions as well, and is associated with nonmotor symptoms like dementia. By reducing mortality associated motor-based symptoms, biologic strategies may inadvertently prolong the burden of highly morbid nonmotor symptoms.

Getting to these downstream issues, however, will require effective negotiation of intersecting scientific, ethical, and policy challenges. Though none of the challenges catalogued in this chapter are unique to brain-based biologics, they are more recurrent, complex, and hotly debated than in many research arenas.

REFERENCES

Aebischer, P., Schluep, M., Deglon, N., Joseph, J. M., Hirt, L., Heyd, B., et al. (1996). Intrathecal delivery of CNTF using encapsulated genetically modified xenogeneic cells in amyotrophic lateral sclerosis patients. *Nat Med, 2* (6), 696–699.

Andlin-Sobocki, P., Jonsson, B., Wittchen, H. U., & Olesen, J. (2005). Cost of disorders of the brain in Europe. *Eur J Neurol, 12* (Suppl 1), 1–27.

Anon. (2005). *Lancet Neurol, 4* (12), 787.

Bachoud-Levi, A. C., Gaura, V., Brugieres, P., Lefaucheur, J. P., Boisse, M. F., Maison, P., et al. (2006). Effect of fetal neural transplants in patients with Huntington's disease 6 years after surgery: A long-term follow-up study. *Lancet Neurol, 5* (4), 303–309.

Bartus, R. T., Herzog, C. D., Chu, Y., Wilson, A., Brown, L., Siffert, J., et al. (2011). Bioactivity of AAV2-neurturin gene therapy (CERE-120): Differences between Parkinson's disease and nonhuman primate brains. *Mov Disord, 26* (1), 27–36.

Beck, M., Flachenecker, P., Magnus, T., Giess, R., Reiners, K., Toyka, K. V., et al. (2005). Autonomic dysfunction in ALS: A preliminary study on the effects of intrathecal BDNF. *Amyotroph Lateral Scler Other Motor Neuron Disord, 6* (2), 100–103.

Bloch, J., Bachoud-Levi, A. C., Deglon, N., Lefaucheur, J. P., Winkel, L., Palfi, S., et al. (2004). Neuroprotective gene therapy for Huntington's disease, using polymer-encapsulated cells engineered to secrete human ciliary neurotrophic factor: Results of a phase I study. *Hum Gene Ther, 15* (10), 968–975.

Boncoraglio, G. B., Bersano, A., Candelise, L., Reynolds, B. A., & Parati, E. A. (2010). Stem cell transplantation for ischemic stroke. *Cochrane Database of Systematic Reviews, Issue 9. Art. No.:* CD007231.

Cartier, N., Hacein-Bey-Abina, S., Bartholomae, C. C., Veres, G., Schmidt, M., Kutschera, I., et al. (2009). Hematopoietic stem cell gene therapy with a lentiviral vector in X-linked adrenoleukodystrophy. *Science, 326* (5954), 818–823.

Caulfield, T. (2004). Biotechnology and the popular press: Hype and the selling of science. *Trends Biotechnol, 22* (7), 337–339.

Christine, C. W., Starr, P. A., Larson, P. S., Eberling, J. L., Jagust, W. J., Hawkins, R. A., et al. (2009). Safety and tolerability of putaminal AADC gene therapy for Parkinson disease. *Neurology, 73* (20), 1662–1669.

Crook, T. H., Ferris, S. H., Alvarez, X. A., Laredo, M., & Moessler, H. (2005). Effects of N-PEP-12 on memory among older adults. *Int Clin Psychopharmacol, 20* (2), 97–100.

Danzon, P., & Towse, A. (2002). The economics of gene therapy and of pharmacogenetics. *Value Health, 5* (1), 5–13.

de la Fuente-Fernandez, R., Schulzer, M., & Stoessl, A. J. (2002). The placebo effect in neurological disorders. *Lancet Neurol, 1* (2), 85–91.

DeGrazia, D. (1996). *Taking animals seriously: Mental life and moral status.* New York: Cambridge University Press.

Descotes, J. (2005). Immunotoxicology: Role in the safety assessment of drugs. *Drug Saf,* *28* (2), 127–136.

Duggan, P. S., Siegel, A. W., Blass, D. M., Bok, H., Coyle, J. T., Faden, R., et al. (2009). Unintended changes in cognition, mood, and behavior arising from cell-based interventions for neurological conditions: Ethical challenges. *Am J Bioeth, 9* (5), 31–36.

Edelman, S. (2003). Brain-gene op may flop; Parkinson's docs rip "risky" surgery. *New York Post.* September 21.

Edge, A. S. (2000). Current applications of cellular xenografts. *Transplant Proc, 32* (5), 1169–1171.

Freed, C. R., Greene, P. E., Breeze, R. E., Tsai, W. Y., DuMouchel, W., Kao, R., et al. (2001). Transplantation of embryonic dopamine neurons for severe Parkinson's disease. *N Engl J Med, 344* (10), 710–719.

Fulton, B. D., Felton, M. C., Pareja, C., Potischman, A., & Scheffler, R. M. (2009). *Coverage, cost-control mechanisms, and financial risk-sharing alternatives of high-cost health care technologies.* California Institute for Regenerative Medicine.

Gill, S. S., Patel, N. K., Hotton, G. R., O'Sullivan, K., McCarter, R., Bunnage, M., et al. (2003). Direct brain infusion of glial cell line-derived neurotrophic factor in Parkinson disease. *Nat Med, 9* (5), 589–595.

Greene, M., Schill, K., Takahashi, S., Bateman-House, A., Beauchamp, T., Bok, H., et al. (2005). Ethics: Moral issues of human-non-human primate neural grafting. *Science, 309* (5733), 385–386.

Hauser, R. A., Furtado, S., Cimino, C. R., Delgado, H., Eichler, S., Schwartz, S., et al. (2002). Bilateral human fetal striatal transplantation in Huntington's disease. *Neurology, 58* (5), 687–695.

Hutchinson, M. (2011). At last, a gene therapy for Parkinson's disease? *Lancet Neurol, 10* (4), 290–291.

Hyun, I., Taylor, P., Testa, G., Dickens, B., Jung, K. W., McNab, A., et al. (2007). Ethical standards for human-to-animal chimera experiments in stem cell research. *Cell Stem Cell, 1* (2), 159–163.

International Society for Stem Cell Research (ISSCR). (2008). Patient handbook on stem cell therapies. Appendix I of the guidelines for the clinical translation of stem cells. www.isscr.org/clinical_trans/pdfs/ISSCRPatientHandbook.pdf.

Kalra, S., Genge, A., & Arnold, D. L. (2003). A prospective, randomized, placebo-controlled evaluation of corticoneuronal response to intrathecal BDNF therapy in ALS using magnetic resonance spectroscopy: Feasibility and results. *Amyotroph Lateral Scler Other Motor Neuron Disord, 4* (1), 22–26.

Karpowicz, P., Cohen, C. B., & van der Kooy, D. (2005). Developing human-nonhuman chimeras in human stem cell research: Ethical issues and boundaries. *Kennedy Inst Ethics J, 15* (2), 107–134.

Kiatpongsan, S., & Sipp, D. (2009). Medicine: Monitoring and regulating offshore stem cell clinics. *Science, 323* (5921), 1564–1565.

Kimmelman, J. (2010). *Gene transfer and the ethics of first-in-human research: Lost in Translation.* Cambridge: Cambridge University Press.

Kimmelman, J., Duckworth, K., Ramsay, T., Voss, T., Ravina, B., & Elena Emborg, M. (2011). Risk of surgical delivery to deep nuclei: A meta-analysis. *Mov Disord.*

Kolata, G. (2001). Parkinson's research is set back by failure of fetal cell implants. *The New York Times.* March 8. www.nytimes.com/2001/03/08/us/parkinson-s-research-is-set-back-by-failure-of-fetal-cell-implants.html?scp=4&sq=fetal%20AND%20 parkinson%27s&st=cse.

Kondziolka, D., Steinberg, G. K., Wechsler, L., Meltzer, C. C., Elder, E., Gebel, J., et al. (2005). Neurotransplantation for patients with subcortical motor stroke: A phase 2 randomized trial. *J Neurosurg, 103* (1), 38–45.

Kordower, J. H., Chu, Y., Hauser, R. A., Freeman, T. B., & Olanow, C. W. (2008). Lewy body-like pathology in long-term embryonic nigral transplants in Parkinson's disease. *Nat Med, 14* (5), 504–506.

Kukla, R. (2007). How do patients know? *Hastings Cent Rep, 37* (5), 27–35.

Lang, A. E., Gill, S., Patel, N. K., Lozano, A., Nutt, J. G., Penn, R., et al. (2006). Randomized controlled trial of intraputamenal glial cell line-derived neurotrophic factor infusion in Parkinson disease. *Ann Neurol, 59* (3), 459–466.

Ledford, H. (2011). Stem-cell scientists grapple with clinics. *Nature, 474* (7353), 550.

LeWitt, P. A., Rezai, A. R., Leehey, M. A., Ojemann, S. G., Flaherty, A. W., Eskandar, E. N., et al. (2011). AAV2-GAD gene therapy for advanced Parkinson's disease: A double-blind, sham-surgery controlled, randomised trial. *Lancet Neurol, 10* (4), 309–319.

Lim, M. M., Wang, Z., Olazabal, D. E., Ren, X., Terwilliger, E. F., & Young, L. J. (2004). Enhanced partner preference in a promiscuous species by manipulating the expression of a single gene. *Nature, 429* (6993), 754–757.

Liu, Z., Richmond, B. J., Murray, E. A., Saunders, R. C., Steenrod, S., Stubblefield, B. K., et al. (2004). DNA targeting of rhinal cortex D2 receptor protein reversibly blocks learning of cues that predict reward. *Proc Natl Acad Sci U S A, 101* (33), 12336–12341.

London, A. J., Kimmelman, J., & Emborg, M. E. (2010). Research ethics: Beyond access vs. protection in trials of innovative therapies. *Science, 328* (5980), 829–830.

Lonnqvist, T., Vanhanen, S. L., Vettenranta, K., Autti, T., Rapola, J., Santavuori, P., et al. (2001). Hematopoietic stem cell transplantation in infantile neuronal ceroid lipofuscinosis. *Neurology, 57* (8), 1411–1416.

Loscher, W., Gernert, M., & Heinemann, U. (2008). Cell and gene therapies in epilepsy—promising avenues or blind alleys? *Trends Neurosci, 31* (2), 62–73.

Macklin, R. (1999). The ethical problems with sham surgery in clinical research. *N Engl J Med, 341* (13), 992–996.

Mathews, D. J., Sugarman, J., Bok, H., Blass, D. M., Coyle, J. T., Duggan, P., et al. (2008). Cell-based interventions for neurologic conditions: ethical challenges for early human trials. *Neurology, 71* (4), 288–293.

Mazzini, L., Ferrero, I., Luparello, V., Rustichelli, D., Gunetti, M., Mareschi, K., et al. (2010). Mesenchymal stem cell transplantation in amyotrophic lateral sclerosis: A Phase I clinical trial. *Exp Neurol, 223* (1), 229–237.

McPhee, S. W., Janson, C. G., Li, C., Samulski, R. J., Camp, A. S., Francis, J., et al. (2006). Immune responses to AAV in a phase I study for Canavan disease. *J Gene Med, 8* (5), 577–588.

Mitchell, P. (2010). Ark's gene therapy stumbles at the finish line. *Nat Biotechnol, 28* (3), 183–184.

Mossman, K. (2006). The world's first neural stem cell transplant. *Scientific American* www.scientificamerican.com/article.cfm?id=the-worlds-first-neural-s.

Nagano, I., Shiote, M., Murakami, T., Kamada, H., Hamakawa, Y., Matsubara, E., et al. (2005). Beneficial effects of intrathecal IGF-1 administration in patients with amyotrophic lateral sclerosis. *Neurol Res, 27* (7), 768–772.

National Commission for the Protection of Human Subjects of Biomedical and Behavioural Research. (1979). *The Belmont report: Ethical principles and guidelines for the protection of human subjects of research.*

National Research Council (US). (2005). Committee on Guidelines for Human Embryonic Stem Cell Research., & Institute of Medicine. Board on Health Sciences Policy. *Guidelines for human embryonic stem cell research.* Washington, DC: The National Academies Press.

Nutt, J., Burchiel, K. J., Comella, C. L., Jankovic, J., Lang, A. E., Laws, E. R., et al. (2003). Randomized, double-blind trial of glial cell line-derived neurotrophic factor (GDNF) in PD. *Neurology, 60* (1), 69–73.

Ochs, G., Penn, R. D., York, M., Giess, R., Beck, M., Tonn, J., et al. (2000). A phase I/II trial of recombinant methionyl human brain derived neurotrophic factor administered by intrathecal infusion to patients with amyotrophic lateral sclerosis. *Amyotroph Lateral Scler Other Motor Neuron Disord, 1* (3), 201–206.

Olanow, C. W., Goetz, C. G., Kordower, J. H., Stoessl, A. J., Sossi, V., Brin, M. F., et al. (2003). A double-blind controlled trial of bilateral fetal nigral transplantation in Parkinson's disease. *Ann Neurol, 54* (3), 403–414.

Ourednik, V., Ourednik, J., Flax, J. D., Zawada, W. M., Hutt, C., Yang, C., et al. (2001). Segregation of human neural stem cells in the developing primate forebrain. *Science, 293* (5536), 1820–1824.

Paradise, J. (2011). Foreword: Follow-on biologics: Implementation challenges and opportunities. *Seton Hall Law Rev, 41* (2), 501–510.

Pollack, A. (2005). Judge rejects patients' suit to get test drug. *The New York Times.* June 8. www.nytimes.com/2005/06/08/business/08amgen.html?scp=7&sq=gdnf&st=cse.

Redmond, D. E., Jr., Roth, R. H., Spencer, D. D., Naftolin, F., Leranth, C., Robbins, R. J., et al. (1993). Neural transplantation for neurodegenerative diseases: Past, present, and future. *Ann N Y Acad Sci, 695,* 258–266.

Reingold, S. C., Steiner, J. P., Polman, C. H., Cohen, J. A., Freedman, M. S., Kappos, L., et al. (2009). The challenge of follow-on biologics for treatment of multiple sclerosis. *Neurology, 73* (7), 552–559.

Rosser, A. E., Barker, R. A., Harrower, T., Watts, C., Farrington, M., Ho, A. K., et al. (2002). Unilateral transplantation of human primary fetal tissue in four patients with Huntington's disease: NEST-UK safety report ISRCTN no 36485475. *J Neurol Neurosurg Psychiatry, 73* (6), 678–685.

Stafford, N. (2009). Germany tightens law on stem cell treatments. *BMJ, 339,* b2967.

Streiffer, R. (2005). At the edge of humanity: human stem cells, chimeras, and moral status. *Kennedy Inst Ethics J, 15* (4), 347–370.

Sugarman, J., & Sipp, D. (2011). Ethical aspects of stem cell-based clinical translation: Research, innovation, and delivering unproven interventions. In K. Hug & G. Hermerén (Eds.), *Translational Stem Cell Research* (pp. 125–135). Totowa, NJ: Springer.

Tuszynski, M. H., Thal, L., Pay, M., Salmon, D. P., U, H. S., Bakay, R., et al. (2005). A phase 1 clinical trial of nerve growth factor gene therapy for Alzheimer disease. *Nat Med, 11* (5), 551–555.

Weijer, C., & Miller, P. B. (2004). When are research risks reasonable in relation to anticipated benefits? *Nat Med, 10* (6), 570–573.

Winerip, M. (1998). Fighting for Jacob. *The New York Times.* December 6. www.nytimes.com/1998/12/06/magazine/fighting-for-jacob.html.

Worgall, S., Sondhi, D., Hackett, N. R., Kosofsky, B., Kekatpure, M. V., Neyzi, N., et al. (2008). Treatment of late infantile neuronal ceroid lipofuscinosis by CNS administration of a serotype 2 adeno-associated virus expressing CLN2 cDNA. *Hum Gene Ther, 19* (5), 463–474.

World Health Organization (WHO). (2006). *Neurological disorders: Public health challenges.* Geneva: World Health Organization.

World Medical Association (WMA). (2011). WMA Declaration of Helsinki—ethical principles for medical research involving human subjects. www.wma.net/en/30publications/10policies/b3/.

INDEX

Note: Page numbers followed by *t, f* and *n* refer to tables, figures and notes.